SILICON CARBIDE BIOTECHNOLOGY

A Biocompatible Semiconductor for Advanced Biomedical Devices and Applications

SILICON CARBIDE BIOTECHNOLOGY

A Biocompatible Semiconductor for Advanced Biomedical Devices and Applications

Second Edition

Edited by

STEPHEN E. SADDOW

Department of Electrical Engineering,
University of South Florida, Tampa, FL,
United States of America

ELSEVIER

Amsterdam • Boston • Heidelberg • London • New York • Oxford
Paris • San Diego • San Francisco • Singapore • Sydney • Tokyo

Elsevier
Radarweg 29, PO Box 211, 1000 AE Amsterdam, Netherlands
The Boulevard, Langford Lane, Kidlington, Oxford OX5 1GB, UK
50 Hampshire Street, 5th Floor, Cambridge, MA 02139, USA

Notices
Knowledge and best practice in this field are constantly changing. As new research and experience broaden our understanding, changes in research methods, professional practices, or medical treatment may become necessary.

Practitioners and researchers must always rely on their own experience and knowledge in evaluating and using any information, methods, compounds, or experiments described herein. In using such information or methods they should be mindful of their own safety and the safety of others, including parties for whom they have a professional responsibility.

To the fullest extent of the law, neither the Publisher nor the authors, contributors, or editors, assume any liability for any injury and/or damage to persons or property as a matter of products liability, negligence or otherwise, or from any use or operation of any methods, products, instructions, or ideas contained in the material herein.

British Library Cataloguing-in-Publication Data
A catalogue record for this book is available from the British Library

Library of Congress Cataloging-in-Publication Data
A catalog record for this book is available from the Library of Congress

ISBN: 978-0-12-802993-0

For information on all Elsevier publications
visit our website at http://elsevier.com

Typeset by Thomson Digital

IN MEMORIAM

To my first and greatest teacher, Edward M. Saddow Jr, who raised and taught me to always make a difference, be compassionate and respect others, and to be "the one who makes it happen." Thank you for your life and dedication to us all—you will be forever missed.

DEDICATION

This book is dedicated to la mia principessa, Sophia Angelo Saddow, who shows me every day what love truly is. As you grow into the amazing woman I know you will be, always be kind to and respect those around you, question conventional wisdom, seek to make this world a better place, and, above all, live a happy, healthy, long, and productive life under the eyes of God.

CONTENTS

LIST OF CONTRIBUTORS

Shamima Afroz
Department of Electrical Engineering, University of South Florida, Tampa, FL, United States of America

R. Alinovi
Department of Clinical and Experimental Medicine, Parma University, Parma, Italy

Edwige Bano
IMEP-LAHC Minatec, Grenoble, France

E. Bedogni
Chemistry Department – Parma University, Parma, Italy

F. Bigi
Chemistry Department – Parma University, Parma, Italy

A. Cacchioli
Department of Veterinary Science, Unit of Normal Veterinary Anatomy – Parma University, Parma, Italy

Fabiola Araujo Cespedes
Department of Electrical Engineering, University of South Florida, Tampa, FL, United States of America

Hamid Charkhkar
Electrical and Computer Engineering Department, Volgenau School of Engineering, George Mason University, Fairfax, VA, United States of America

Ji-Hoon Choi
IMEP-LAHC; LTM-CNRS Minatec, Grenoble, France

Stuart F. Cogan
Department of Bioengineering, Erik Jonsson School of Engineering & Computer Science, University of Texas at Dallas, Richardson, TX, United States of America

C. Coletti
Center for Nanotechnology Innovation@NEST, Pisa, Italy

L. Cristofolini
Physics and Earth Science Department – Parma University, Parma, Italy

Pasquale D'angelo
IMEM-CNR Institute of Materials for Electronics and Magnetism – Italian National Research Council, Parma, Italy

F. Fabbri
IMEM-CNR Institute of Materials for Electronics and Magnetism – Italian National Research Council, Parma University, Parma, Italy

Louis Fradetal
IMEP-LAHC and LMGP Minatec, Grenoble, France

Christopher L. Frewin
The Department of Biomedical Engineering, Erik Johnsson School of Engineering and Computer Science, University of Texas, Dallas, TX, United States of America

C. Galli
Department of Biomedical, Biotechnological, and Translational Sciences – Parma University, Parma, Italy

Owen J. Guy
Centre for NanoHealth, College of Engineering, Swansea University, Swansea, United Kingdom

Salvatore Iannotta
IMEM-CNR Institute of Materials for Electronics and Magnetism – Italian National Research Council, Parma, Italy

Gretchen L. Knaack
U.S. Food and Drug Administration, Center for Devices and Radiological Health, Office of Science and Engineering Laboratories, Division of Biophysics, Silver Spring, MD, United States of America

P. Lagonegro
IMEM-CNR Institute of Materials for Electronics and Magnetism – Italian National Research Council, Parma University, Parma, Italy

G.M. Macaluso
Department of Biomedical, Biotechnological, and Translational Sciences – Parma University, Parma, Italy

M. Negri
IMEM-CNR Institute of Materials for Electronics and Magnetism – Italian National Research Council, Parma University, Parma, Italy

Maysam Nezafati
Department of Electrical Engineering, University of South Florida, Tampa, FL, United States of America

Katie Noble
Department of Electrical Engineering, University of South Florida, Tampa, FL, United States of America

Maelig Ollivier
IMEP-LAHC; LTM-CNRS Minatec, Grenoble, France

Joseph J. Pancrazio
Department of Bioengineering, Erik Jonsson School of Engineering & Computer Science, University of Texas at Dallas, Richardson, TX, United States of America

S. Pinelli
Department of Clinical and Experimental Medicine, Parma University, Parma, Italy

F. Ravanetti
Department of Veterinary Science, Unit of Normal Veterinary Anatomy – Parma University, Parma, Italy

T. Rimoldi
Physics and Earth Science Department – Parma University, Parma, Italy

Agostino Romeo
IMEM-CNR Institute of Materials for Electronics and Magnetism – Italian National Research Council, Parma, Italy

F. Rossi
IMEM-CNR Institute of Materials for Electronics and Magnetism – Italian National Research Council, Parma University, Parma, Italy

Stephen E. Saddow
Department of Electrical Engineering, University of South Florida, Tampa, FL, United States of America

G. Salviati
IMEM-CNR Institute of Materials for Electronics and Magnetism – Italian National Research Council, Parma University, Parma, Italy

A. Smerieri
Department of Biomedical, Biotechnological, and Translational Sciences – Parma University, Parma, Italy

Valérie Stambouli
LMGP Minatec, Grenoble, France

Giuseppe Tarabella
IMEM-CNR Institute of Materials for Electronics and Magnetism – Italian National Research Council, Parma, Italy

Sylvia Thomas
Department of Electrical Engineering, University of South Florida, Tampa, FL, United States of America

Kelly-Ann D. Walker
Centre for NanoHealth, College of Engineering, Swansea University, Swansea, United Kingdom

PREFACE TO SECOND EDITION

The 21st century has long been hailed as the century of biology, as the extremely complex nature of the biological system is only now beginning to be more fully understood through the use of technology and scientific tools, which allow mankind to observe biological processes at the nanoscale. This has resulted in the ever-pressing need to develop advanced "smart materials" for biomedical applications, and this important field of research has taken center stage worldwide as a new frontier of advanced scientific research. While biocompatible materials have been in use for decades, the quest to merge smart materials such as those used in modern computer integrated circuits with the biological system has proven to be much more difficult than originally imagined. If we take as one example the brain–machine interface, or BMI, which was first reported in the early 1970s and was based on silicon probes, we learn that mankind is not much closer to the prospect of long-term (multiple years or even decades) implantation of these devices. Why? Simply put, most "smart materials" such as most semiconductors are typically toxic to the biological environment, and silicon is no exception. What is needed to realize true nanoscale functionality therefore is a material with the "smart properties" of silicon and the "biocompatible" properties of polymers. In this book such a candidate material, silicon carbide, is presented and the evidence that has been amassed, mostly over the last decade, indicates that this material system may prove to be the dream material for biomedical devices that must do more than simply provide structural support.

Silicon carbide, abbreviated SiC, has a long history as a robust and hard material, which was first used as an abrasive and cutting material in the 19th century and later as a high-temperature semiconductor for advanced applications in the 20th century. Indeed, the first solid-state blue light emission was observed from SiC in 1907, and this material has been under study ever since. One can consult the various available literature to fully understand all of the many aspects of SiC, from how it is formed, to its myriad crystal properties, and finally to the large number and types of applications it is being used in. Fortunately, it is sufficient here to provide a brief overview of SiC so that the reader can understand why this material is so compelling for biomedical applications and may become one of the most used biomaterials in the future. Indeed, the purpose of this second book is to lay the foundation for exactly this – to introduce silicon carbide to biomedical engineers, medical professionals, and scientists, thus bringing together technologists from across many disciplines to help enable the ultimate use of SiC as one of the next generation of smart biomaterials to realize advanced biomedical devices.

The first edition of *Silicon Carbide Biotechnology* focused mainly on *in vitro* material studies in order to introduce the reader to the possibility that SiC can be an important player in the biomedical device community. Indeed, biocompatibility and hemocompatibility data were presented, which either cleared up prior misconceptions of SiC in this arena, or reinforced the known excellent performance of certain SiC material types, called polytypes, to the biological world. While prototype devices were introduced and discussed, the present edition focuses mainly on the significant progress that has been made in the development of SiC biomedical devices, ranging from implants to sensors to *in vitro* DNA assay systems. The organization of this second edition starts with an updated introduction of SiC, with a particular emphasis on biomedical applications, and is then divided into four distinct sections. First, a fully comprehensive treatment of SiC *in vitro* performance is presented with a rigorous study of 3C–SiC biocompatibility, conducted in full compliance with ISO 10993, in chapter: Cytotoxicity of 3C–SiC Investigated Through Strict Adherence to ISO 10993. Next, amorphous SiC (*a*-SiC) is studied using the same *in vitro* methods; additionally, a study of hemocompatibility for both 3C–SiC and *a*-SiC is documented in this chapter: Study of the Hemocompatibility of 3C–SiC and *a*-SiC Films Using ISO 10993-4 where we see both materials have great potential suitability for use in the cardiovascular system.

Having completed the *in vitro* materials studies required by the FDA to allow for subsequent clinical trials, the second section focuses on the important topic of SiC biosensors. Extremely comprehensive studies on the use of graphene on SiC, as well as surface functionalization of SiC, both for biosensor applications, are presented in the next two chapters: Graphene Functionalization for Biosensor Applications and: SiC Biosensing and Electrochemical Sensing: State of the Art and Perspectives, respectively. We see that the use of epitaxial graphene on SiC is not only a powerful method to realize advanced sensors, it ensures that the device substrate is compatible with the biological system, thus improving device performance and reliability. We see that SiC, due to the fact that it is a compound semiconductor, provides more flexibility in terms of biosensor design simply by taking into account differences in how one can functionalize the Si and C surfaces of the material. This extra degree in design freedom is not possible with the mainstay advanced biosensor material, silicon, which has proved to have issues with bio- and hemocompatibility. After these two chapters, very important work to realize a continuous glucose-monitoring (CGM) device *in vivo* is presented, whereby semiinsulating 4H–SiC is used as an implantable RF sensor to detect, in real time, glucose levels *in vitro*. Given the significant number of humans suffering from diabetes worldwide, this work is extremely important and significant progress is being made in the remote sensing of glucose via a "radio frequency identification (RFID)-type" sensor mechanism.

Having completed the sensor portion of this edition, we then move on to the nervous system where developments to realize more robust, long-term performing SiC-based neural implants is discussed. Building on the work presented in the first edition that indicated that 3C–SiC appeared to be an ideal choice for this application, extremely compelling data documenting the *in vivo* performance of 3C–SiC intracortical implants in "wild-type" mice is presented in chapter: *In Vivo* Exploration of Robust Implantable Devices Constructed From Biocompatible 3C–SiC, whereby no evidence of neurological immune system response was observed. To further motivate the development of 3C–SiC for this application, results of a comprehensive study, also conducted under the auspices of ISO 10993, shed light on the reason why 3C–SiC performs so well *in vitro*: the material is highly stable and does not degrade in either normal physiological environments or physiological environments that are enhanced to enable accelerated aging studies to be properly conducted. For any fully functional neural implant using 3C–SiC a suitable electrical insulator is required and this material must have similar biological performance to the support material, 3C–SiC. Amorphous SiC (*a*-SiC) is just such a material and, being part of the "SiC family of biocompatible materials," is an ideal companion material for this (and other) applications. The *in vivo* compatibility of commercial implantable neural interfaces (INI) coated with *a*-SiC is documented in rats and, once again, proves the incredible potential of SiC for this highly challenging application. This work, including functional *in vitro* recording data made using microelectrode arrays coated with *a*-SiC, is presented in chapter: Amorphous Silicon Carbide for Neural Interface Applications. With the neural implant potential of SiC well documented, we then move on to the final section of this edition – the use of SiC nanotechnology for advanced applications such as label-free DNA detection (chapter: SiC Nanowire Based Transistors for Electrical DNA Detection), cancer treatment, and tissue/bone scaffolds (chapter: Silicon Carbide Based Nanowires for Biomedical Applications).

The authors listed at the beginning of each chapter provided the content, including figures and text that comprises the majority of this book. The Editor both solicited these chapters and then edited the content in an effort to achieve a uniform and consistent text. This was done so that the reader could both focus on the very interesting subject matter contained within, and to also see the connection that many of the chapters have with each other. It is the editor's hope that this endeavor was successful and that you, the reader of this book, will find this work both informative and enjoyable to read. Indeed, it was an extremely rewarding experience putting together the second edition of *Silicon Carbide Biotechnology* and the editor hopes that this will be similarly felt by you, the reader.

At Academic Press/Elsevier Science, the project was coordinated by Christina Gifford and the editor is grateful to her for all of her support and kind assistance throughout the project. In addition, Sruthi Satheesh very capably coordinated the production of the book and is gratefully acknowledged for all of her hard work in making sure galley proof corrections were made, copyright approvals completed, and for her overall help in pulling this book together. The Editor of *Silicon Carbide Biotechnology* hopes that users of this book, and especially the students who will become the next generation of biomedical practitioners and technologists, will find the subject matter of the second edition of *Silicon Carbide Biotechnology* as interesting and exciting as the editor does.

Prof. Stephen E. Saddow, Editor
University of South Florida, Tampa, FL, United States of America

ACKNOWLEDGMENTS

This project was made possible by so many people that it is difficult to thank everyone sufficiently and completely, and if I forget to include anyone in this acknowledgment it is truly inadvertent. I am first grateful to my family who raised me to have real values – those of honest work and to always respect and appreciate others. To them I remain forever grateful. None of this would have been possible had it not been for all of the teachers and professors in my life, most notably my PhD advisor, Prof. Chi Lee at the University of Maryland, who taught me that no matter how successful one is, or how famous one becomes, the true measure of a colleague is kindness and respect. Naturally I am indebted to the dedicated students that I have had the good fortune to work with and advise over the past 20 years. It is upon the shoulders of many of these fine students that this work rests. A few deserve a special note of recognition since they were the SiC biomaterials pioneers who laid the foundation for this book – Dr C. Coletti for her pioneering work on the *in vitro* biocompatibility of SiC and upon whose work Dr N. Schettini (hemacompatibility) and Dr C. Frewin (neuronal biocompatibility) directly followed. In addition, Dr A. Oliveros continued this tradition when she developed SiC biosensors along with exploring the biocompatibility of graphene and carbon-based conductors. This core cadre of students made this work possible and I am eternally indebted to them. I wish to also thank Dr C. Locke and Dr Meralys Reyes for their nonwavering support of all of the research in my group, mostly notably by growing and processing the needed SiC films. In addition, I wish to thank all of the present and former students of the USF SiC Group, as well as my close colleagues Dr A. Hoff and Dr S. Thomas of the Department of Electrical Engineering at USF. Dr Hoff was my inspiration to enter the biomedical engineering field and introduced me to Dr M. Jaroszeski of the Chemical and Biomedical Engineering Department, who opened his laboratory to my students, thus making the work of the Bio-SiC group possible. Dr Thomas provides consistent encouragement, and together we advised Dr S. Afroz in her pioneering on SiC *in vivo* glucose sensors, which my current PhD student, Ms Fabiola Araujo Cespedes, is continuing towards the goal of continuous glucose monitoring (CGM). I wish to thank the USF NREC staff, and in particular Richard Everly, for their support and kind assistance to my students as well as my department chairman, Dr T. Weller, who has always been extremely supportive of my research activity and group. I especially want to thank Dr G. De Erausquin, Dr K. Muffly, and Dr L. Hoyte, all of the USF Morsani Medical School, for our very fruitful collaboration in the area of biomedical engineering and devices. I am especially indebted to my good friends and colleagues from IMEM in Parma, Dr S. Iannotta, Dr G. Salviati, and Dr F. Rossi, and their colleagues, for our continuing scientific collaboration in the area of SiC biosensors and nanotechnology. I again thank Dr C. Frewin for always being there to support and share my vision of SiC biotechnology and for his ever-faithful friendship and dedication without which I doubt this book would have been possible. Finally, and most dear to my heart, is the ever present love and support of my wife, Vaine Angelo Saddow, with whom we have two amazing children, Luca Edward and Sophia Angelo, who provide all of the incentive that anyone needs to continue to strive for scientific and technological excellence – in some small way I hope that all of this work will allow their quality of life to be the best that man on earth can provide.

Prof. Stephen E. Saddow, Editor
Tampa, FL

CHAPTER 1

Silicon Carbide Materials for Biomedical Applications

Stephen E. Saddow
Department of Electrical Engineering, University of South Florida, Tampa, FL, United States of America

Contents

1.1 PREAMBLE

Considerable progress has been made since the publication of the first edition of *Silicon Carbide Biotechnology: A Biocompatible Semiconductor for Advanced Biomedical Devices and Applications* in 2011, mostly in the area of SiC nanoparticle and nanowire incorporation in various biomedical devices. In our research group we have made significant progress to translate research performed in the area of SiC materials development for biomedical application into prototype SiC biomedical devices. Examples include 3C–SiC neural implants, functionalized biosensors, and the demonstration of semi-insulating 4H–SiC as a suitable platform for an *in vivo* glucose sensor based on an RF antenna transducer. In this edition, we build upon the materials-based discussion from the first edition placing the focus on SiC biotechnology at the device level. We begin by adding additional information to the *in vitro* biological performance story of 3C–SiC (see chapter: Cytotoxicity of 3C–SiC Investigated Through Strict Adherence to ISO 10993) and add amorphous silicon carbide (*a*-SiC) to the story in chapter: Study of the Hemocompatibility of 3C–SiC and *a*-SiC Films Using ISO 10993-4. Next, the focus is on biosensors, with a review of graphene on SiC substrates for biosensing applications (see chapter: Graphene Functionalization for Biosensor Applications), followed by a comprehensive overview of SiC bio- and electrochemical sensing (chapter: SiC Biosensing and Electrochemical Sensing: State of the Art and Perspectives). We then move to the demonstration of 4H–SiC as a sensor transducer for *in vivo* continuous glucose monitoring and WiFi applications in chapter: SiC RF Antennas for *In Vivo* Glucose Monitoring and WiFi Applications. The story then shifts to the important area of neural probes where two chapters cover the use of SiC for implantable neural interfaces (INI). Chapter: *In Vivo* Exploration of Robust Implantable Devices Constructed From Biocompatible 3C–SiC expands the information provided from Chapters 6 and 7 of the first edition, demonstrating the biological compatibility of 3C–SiC as a neural probe material *in vivo* as well as detailing the integration of 3C–SiC neural probes with a wireless on-board recording chip (chapter: *In Vivo* Exploration of Robust Implantable Devices Constructed From Biocompatible 3C–SiC). The bio- and hemocompatible coating, *a*-SiC, is then investigated as a durable coating for this important neural application (chapter: Amorphous Silicon Carbide for Neural Interface Applications). The book then finishes with two very interesting chapters on the use of SiC nanotechnology for biomedical applications: SiC nanowire-based transistors for electrical DNA detection (chapter: SiC Nanowire Based Transistors for Electrical DNA Detection) and the use of SiC-based nanowires for biomedical applications in chapter: Silicon Carbide Based Nanowires for Biomedical Applications.

The book is organized in such a way that the references are listed at the end of each chapter rather than comprehensively in a bibliography at the end of the book. This was done intentionally so that the reader can search the relevant literature in a "topic-based" fashion. I sincerely hope that you will find this to be one of the more useful aspects of the book as there is quite a body of literature pertaining to SiC biotechnology. The topics of this book cover all the major aspects of SiC biotechnology and it is my belief that you will find this both fascinating as well as compelling. I hope that you, the reader, might decide to get either involved in this very exciting area of biotechnology or, if you are in a position to make technological decisions, consider adopting SiC as a "material of choice" for whatever application you are involved in. In the end, it is all about improving the lives of millions of people worldwide and this will only happen through thinking in new ways and offering new therapeutic pathways to mitigate the impact of disease and injury for us all. Please enjoy the book and I would love to hear from you if you have any feedback, as I plan to put together another edition in about 5 years from now.

1.2 INTRODUCTION TO THE SECOND EDITION

The use of SiC as a suitable interface to important biological systems, such as skin, bone, neurons, blood platelets, etc., was thoroughly discussed in *Silicon Carbide Biotechnology: A Biocompatible Semiconductor for Advanced Biomedical Devices and Applications* [1]. In this edition we build on the solid materials foundation that has already been established and discuss some of the more advanced biomedical device concepts that have been developed using this important material system. In the USF SiC Group at the University of South Florida, which the author leads, several doctoral students have completed their research since publication of the first edition. Dr Chris Locke completed his important work in the area of 3C–SiC microelectromechanical systems (MEMS) devices where he showed that high-quality membranes, cantilevers, and bridges could be realized using 3C–SiC grown on both bulk Si and polycrystalline Si films supported by an oxidized Si substrate [2]. This was followed by Dr Luca Abbati who completed his research on the development of a bidirectional electronics platform to allow for both stimulation and recording of neural signals *in vitro* and *in vivo* [3]. A short while later, Dr Shamima Afroz demonstrated a breakthrough in continuous glucose monitoring whereby she demonstrated that an RF antenna sensor, made on semi-insulating 4H–SiC, could be used to measure changes in blood glucose level in both synthetic and pig blood [4]. This work forms the first part of chapter: SiC RF Antennas for *In Vivo* Glucose Monitoring and WiFi Applications of the present book, where her work is being translated into a working *in vivo* glucose sensor in the industrial, scientific, and medical radio band of the RF spectrum by a current PhD student, Ms Fabiola Auraujo Cespedes. Working with Dr Lennox Hoyte and Dr Karl Muffly of USF Health, Dr Afroz earlier demonstrated the *in vivo* compatibility of 4H-, 3C-, and *a*-SiC via subcutaneous implantation in female farm pigs. No inflammatory response was observed for any of the SiC materials studied in contrast to Au, SiO_2, and Si, which showed indications of progressively higher immune system responses [5]. This portion of her work is included in chapter: *In Vivo* Exploration of Robust

Implantable Devices Constructed From Biocompatible 3C–SiC, where a null inflammatory response to 3C–SiC implanted within a murine model is demonstrated. The next student to complete her research in this area was Dr Alexandra Oliveros, who performed the first *in vitro* studies on the biocompatibility of graphene and the use of self-assembled monolayers (SAMs) to demonstrate that SiC could be biofunctionalized in a similar way to Si (see Chapters 3 and 12 in the first edition, respectively). She extended this work to demonstrate that myoglobin, a blood protein present at the onset of a heart attack, could be detected using an electrochemical sensor based on SAMs and antimyoglobin antibody functionalization [6]. Following this excellent work in the SiC biosensor arena, Dr Joseph Register developed 3C–SiC as a suitable platform for neural implants that could stimulate optically sensitive neurons with light [7]. These SiC optrodes, which possess greater chemical and physical resilience, may replace silicon optrodes utilized in long-term *in vivo* neuroscience applications. Finally, Dr Maysam Nezafati completed the critical study of SiC bio- and hemocompatibility by using in-house methods developed under a DARPA program to evaluate materials for use in neural interfaces by strict adherence to ISO-10993. What is important about this work is that, for the first time, 3C–SiC was tested fully in accordance with the standard, which calls for chemical stability evaluation, biocompatibility assessment via cytotoxicity assays, and finally hemocompatibility studies using platelet-rich plasma in a dynamic Chandler's loop setup [8]. Clearly, since the publication of the first edition a significant amount of progress has been made in the author's research group, spanning the range from materials studies to novel device fabrication and testing.

New studies that are currently under way involve the use of SiC nanowires, in this case 3C–SiC, for numerous biomedical device applications. Dr Paola Lagonegro from the University of Parma and the CNR-IMEM, also in Parma, visited the group in the spring of 2014 with her work focusing on the bio- and hemocompatibility of 3C–SiC nanowires [9]. This work is ongoing with the qualitative outcome being that this material system is compatible with L929 fibroblasts, as per ISO 10993 evaluation, and platelets exposed to the nanowires from platelet-rich plasma. A new study on the interaction of osteoblasts with 3C–SiC has been initiated with Dr Carlos Galli of the University of Parma Hospital and preliminary results indicate that, as expected, planar 3C–SiC performs well *in vitro* with MC3T3 osteoblast precursor bone cells (Galli C. Private communication, University of Parma Medical Center). This is very encouraging as early work presented in the first edition showed that porous SiC was a suitable surface for bone cell growth and could serve as a bone tissue scaffold [1], while also being used for dental implants [10]. The final piece of new research ongoing in the group is to translate the excellent *in vivo* tissue histology and MRI experimental data on lightly doped n-type 3C–SiC neural probes (to be presented in chapter: *In Vivo* Exploration of Robust Implantable Devices Constructed From Biocompatible 3C–SiC) into functional *in vitro* (planar microelectrode arrays, or MEAs) and *in vivo* (implantable MEAs) using 3C–SiC as electrodes and supports [11], and *a*-SiC as a bio- and hemocompatible insulating film [12]. It is anticipated that this work will be completed at the end of 2015 thus demonstrating the continuing growth in SiC biotechnology at the University of South Florida.

Equally important, progress is being made around the world in SiC biotechnology, and numerous examples of published works since publication of the first edition are available in the literature. Examples include the use of SiC nanomaterials such as nanoparticles [13,14] for use in biosensors such as for cancer detection, and SiC nanowires (NWs) [15], also for use in biosensors and artificial retina applications. Indeed, chapter: SiC Nanowire Based Transistors for Electrical DNA Detection in this edition will discuss the impressive work in the area of SiC NW sensors for DNA detection [16], and chapter: Silicon Carbide Based Nanowires for Biomedical Applications will review equally impressive work in the area of SiC NWs for the artificial retina and biosensors [17].

Much progress has been made in the field of epitaxial graphene on SiC surfaces, and the research group of Dr Owen Guy in Swansea has done a significant amount of work in the area of graphene on SiC for biomedical applications. Pioneering work in the area of the ultrasensitive detection of the cancer risk (oxidative stress) biomarker 8-hydroxydeoxyguanosine (8-OHdG) has been reported [18] and will also be discussed as part of chapter: Graphene Functionalization for Biosensor Applications. One of the most prevalent methods to detect a biological species, such as a biomarker, protein, etc., is to functionalize the SiC surface with antibodies that allow for the linkage of the target species with the antibody. Indeed, we have used this method to detect myoglobin as discussed previously. Interesting theoretical work performed by Catellani and Calzolari [19] provides information on how one can functionalize SiC(110) surfaces via porphyrin adsorption, which has been utilized to functionalize SiC NWs for various bio applications by the group of Dr Salvatore Iannotta [20]. Indeed, in the same laboratory the group of Dr Giancarlo Salviati has been doing impressive work in the area of SiC NW functionalization with porphyrin to allow for the treatment of cancerous tumors via X-ray excited photodynamic therapy [21]. The same group has done some nice work to quantify the biological response of SiC NWs to alveolar basal epithelial and human dermal fibroblast cells whereby they demonstrate cytocompatibility of SiC NWs, thus further motivating the use of SiC NWs for biosensor applications [22]. Finally, the functionalization via the specific and selective immobilization of the streptavidin protein on SiC NWs has been demonstrated [23], which is, again, another example of the plethora of biosensor applications enabled by SiC NWs. These are just a few of the highlighted works in the area of SiC nanobiotechnology and more information will be provided in the appropriate chapters later in this work.

Naturally, progress has been made in the area of SiC biomedical devices and systems. One interesting work is in the area of 3C–SiC membranes for applications such as blood pressure sensing [24]. While the paper does not discuss biological testing of the membranes after fabrication, it elaborates on surface science studies demonstrating morphological changes to the 3C–SiC surface after reactive ion etching. It is therefore important to confirm that the excellent biocompatible response of 3C–SiC [1] is maintained after device processing. Another interesting paper discusses the use of 3C–SiC for use in a dog bone resonator [25], which is another example of the use of 3C–SiC in BioMEMS applications. By replacing Si with 3C–SiC, a 57% increase in resonant frequency is predicted in addition to the well-known benefits of using 3C–SiC as a robust material for harsh conditions not suitable for silicon. The application of this work is in the detection of bioparticles, such as DNA, bacteria, viruses, etc., although, this is a simulation discussion and details of the use of a physical device to detect bioparticles was not presented.

One area of significant interest for robust SiC materials deployment is in the field of neural engineering, and in particular implantable neural interfaces (INI). The overwhelming device architecture used for construction of INI devices utilizes MEAs, which are intended to enable bidirectional electrical communication with neurons. The INI has been proposed for use as a key component in brain–machine interface (BMI) systems, which are designed to provide a therapeutic pathway to assist in the ability of humans to overcome many neurological disorders stemming from disease or trauma, or to facilitate the replacement of lost limbs and sensory organs [26]. Additionally, the same INI core technologies are also used by neuroscientists around the world to investigate and construct the machinations of the nervous system. For instance, planar MEA systems are used to study neurons and neural tissue *in vitro* [27]. Indeed, the MEA is an excellent means to screen novel material compatibility with the brain, as testing cells or tissue slices *in vitro* is much more convenient, humane, and economical than implanting MEAs within an animal model, such as a mouse or rat. Research performed by the group of Dr J. Pancrazio has shown that functional recordings of neural activities using an MEA are an excellent means to study the biocompatibility of a material. Their group has indicated that excellent neural recordings were obtained through an MEA using *a*-SiC as an insulator, which was in intimate contact with the functional neural network [28]. Work done by the same group reviewed and combined multiple ISO 10993 methods into one protocol, which is the method we now use in our group to assess a material's biocompatibility (see chapter: Cytotoxicity of 3C–SiC Investigated Through Strict Adherence to ISO 10993) and functional assays [29]. In our group, we have realized several generations of SiC-based MEAs, all utilizing 3C–SiC as the base material. Early versions of this technology were used to stimulate cortical neurons from wild-type mice [30], and more recent versions have incorporated *a*-SiC insulators to realize more robust electrically insulating films [31]. In closing the introduction to the second edition of *Silicon Carbide Biotechnology: A Biocompatible Semiconductor for Advanced Biomedical Devices and Applications*, I would like to mention the collaboration with Dr Mario Gazziro of the Universidade Federal do ABC, Santo André (Sao Paulo), Brazil. This collaboration, sponsored by the Brazilian government's Science without Borders program, seeks to develop a wireless neural implant, based on the 3C–SiC technology developed by my team at the University of South Florida. To date, an application-specific integrated circuit has been designed, fabricated, and tested for its electrical performance [32]. The chip consists of four differential recording channels with 46-dB gain up to 10 kHz with very low total power consumption of only 144 μW. During this project we also undertook to study the compatibility of 3C–SiC neural probes with MRI fields and discovered that low-doped n-type 3C–SiC probes produced no image artifacts during scanning using a 2T magnetic field, which is markedly different than metal electrodes. This work, in collaboration with Dr Luciene Covolan of the Universidade Federal de São Paulo (UNIFESP) and Jackie Malherios of the Universidade de São Paulo (USP), São Carlos, Brazil, will be further discussed in chapter: *In Vivo* Exploration of Robust Implantable Devices Constructed From Biocompatible 3C–SiC.

1.3 SUMMARY TO THE SECOND EDITION

In summary, this edition both provides an update of important developments since the publication of *Silicon Carbide Biotechnology: A Biocompatible Semiconductor for Advanced Biomedical Devices and Applications*, published in late 2011, while initiating discussions on new applications of the biomedical material, SiC. The bulk of the biomedical device-related research in the SiC biotechnology area has utilized 3C–SiC and *a*-SiC, particularly in the field of neuroengineering and neuroscience although bulk hexagonal 4H- and 6H–SiC have also been examined. Continuous glucose monitoring using a semi-insulating 4H–SiC RF sensor has shown excellent *in vitro* performance, and is currently being refined for use *in vivo* at the University of South Florida. Numerous new works in the area of SiC nanobioMEMS have been reported covering a wide range of applications, from cancer detection and treatment to artificial retinas, to DNA assays and functionalized biosensors.

None of this would be possible without the continuing study of SiC as a suitable biomaterial, and a complete study comparing 3C–SiC and *a*-SiC with traditional biomedical device materials has been completed by our group and is reported here in chapter: Cytotoxicity of 3C–SiC Investigated Through Strict Adherence to ISO 10993. In chapter: Study of the Hemocompatibility of 3C–SiC and *a*-SiC Films Using ISO 10993-4 a study of the hemocompatibility of these same materials is reported using a dynamic Chandler loop, with the results being that these materials are indeed hemocompatible insofar as blood platelet attachment and activation are concerned. In chapter: Graphene Functionalization for Biosensor Applications new results pertaining to graphene biosensors on SiC will be reported, followed by a review of some interesting work on SiC electrochemical sensors for biomedical applications (chapter: SiC Biosensing and Electrochemical Sensing: State of the Art and Perspectives), and semi-insulating 4H–SiC for *in vivo* continuous glucose monitoring (chapter: SiC RF Antennas for *In Vivo* Glucose Monitoring and WiFi Applications). Progress in the development of SiC neural interfaces is discussed in chapter: *In Vivo* Exploration of Robust Implantable Devices Constructed From Biocompatible 3C–SiC, where work currently under way to develop a wireless neural interface is presented along with image data from 3C–SiC neural implants exposed to 2T MRI fields in a rat model. Extremely promising work in the area of robust, biocompatible *a*-SiC coatings for neurological applications is then presented (chapter: Amorphous Silicon Carbide for Neural Interface Applications) to close out the neuroengineering portion of this edition. The final two chapters present some very novel research in the area of DNA assays using SiC nanowires (chapter: SiC Nanowire Based Transistors for Electrical DNA Detection) and the use of the same material system for various biomedical device applications closes this edition in chapter: Silicon Carbide Based Nanowires for Biomedical Applications.

It is hoped that you, the reader, will find this "application-driven" approach to the organization of this book easy to follow as well as logical and useful in your work. In an effort to provide continuity with the first edition of *Silicon Carbide Biotechnology: A Biocompatible Semiconductor for Advanced Biomedical Devices and Applications* we have repeated the introductory chapter here so that important general information on SiC and its properties is close by for your edification. If you are already familiar with this chapter you are encouraged to move on to the next chapter now, but I hope that you will find a quick review of SiC for biotechnological applications worthwhile and that you will agree with this assessment after a quick review of the rest of this chapter.

1.4 INTRODUCTION TO THE FIRST EDITION

Silicon carbide has a long history as a robust and hard material, first used as a cutting material in the 19th century and later as a high-temperature semiconductor for advanced applications in the 20th century. The history of silicon carbide (SiC) is quite interesting and the reader is referred to the first chapter in a book dedicated to this subject [33]. It is best to consult the literature to fully understand all of the many aspects of SiC, from how it is formed, to its myriad crystal properties, and finally to the large number and types of applications it is being used in. Fortunately, it is sufficient here to provide a brief overview of SiC so that the reader can understand why this material is so compelling for biomedical applications and may become one of the most used biomaterials in the 21st century. Indeed, the purpose of this book is to lay the foundation for exactly this – to introduce silicon carbide to biomedical engineers, medical professionals, and scientists, thus bringing together technologists from across many disciplines to help realize the ultimate use of SiC as one of the next generation of smart biomaterials to realize advanced biomedical devices.

1.5 SILICON CARBIDE – MATERIALS OVERVIEW

SiC is first and foremost a material that consists of the covalent bonding of Si and C atoms, typically in biatomic layers as shown in Fig. 1.1. These form tetrahedrally oriented molecules of SiC, with a very short bond length

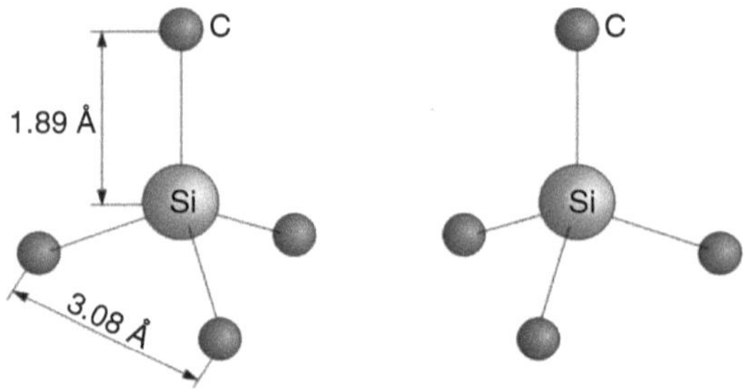

Figure 1.1 ***All SiC crystals are formed via bilayers of C and Si, covalently bonded to form a tetrahedron.*** Four carbon atoms are covalently bonded with a silicon atom in the center. Two types exist – one is rotated 180 degrees around the c-axis with respect to the other, as shown. This tetrahedrally bonded molecule then forms the basic building block of all SiC materials [1].

and, hence, a very high bond strength. This is the origin of the extremely high chemical and mechanical stability of SiC [34,35]. SiC can be formed in amorphous, polycrystalline, and monocrystalline solid forms, and due to the high bond strength and high-temperature operating capabilities of SiC, synthesis of SiC material normally requires high temperatures (greater than 1000°C). The material can be grown in both bulk (boule) crystal form, currently with diameters up to 150 mm (6 in.), and can be heteroepitaxially grown on Si substrates (details of how this is accomplished are provided in the next chapter).

One of the important characteristics of SiC is that the bilayers of Si and C (Fig. 1.1) can be stacked one upon the other in different crystal orientations: cubic, hexagonal, and rhombohedral. With more than 200 known polytypes reported in the literature, the three technologically relevant forms are one purely cubic (β-SiC) and two hexagonal forms (α-SiC), which actually have some cubic symmetry. These three polytypes are shown in Fig. 1.2. The cubic form has the designation 3C–SiC, where the 3 delineates that 3 bilayers of SiC are needed to form the basic structure and C indicates that the crystal form is cubic. The hexagonal forms are 4H–SiC and 6H–SiC, where the 4 and 6 delineate that 4 and 6 bilayers are needed while the H indicates that the crystal form is hexagonal. While interesting in their own right, these various forms of SiC actually have varying applications, where the dominant power electronic device crystal of choice is 4H–SiC due to its having the highest bandgap (3.2 eV). 6H–SiC is ideally suited for solid-state lighting (i.e., LEDs) as its lattice constant is close to the GaN family of alloys used in advanced LEDs that have enabled DVD and blue ray technology, not to mention the solid-state lighting revolution that

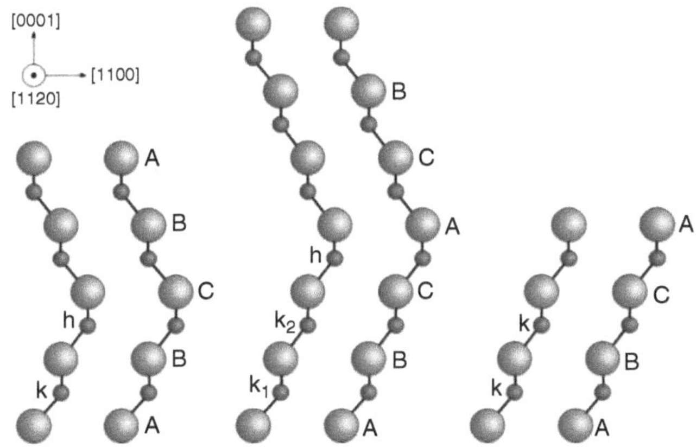

Figure 1.2 ***Atomic stacking sequence of the three technologically relevant SiC polytypes viewed in the [9–10] plane.*** From left to right, 4H–SiC (ABCB), 6H–SiC (ABCACB), and 3C–SiC (ABC). k and h denote crystal symmetry points that are cubic and hexagonal, respectively [1]. Note that only 3C–SiC consists of purely cubic crystal symmetry.

is currently leading to dramatic reductions in power consumption worldwide [36]. A comparison of the properties of SiC relative to Si is shown in Table 1.1 for reference.

1.6 SILICON CARBIDE MATERIAL GROWTH AND PROCESSING

There is a long history of how to grow, process, and characterize SiC materials – the reader is referred to several of the excellent references if further details are desired [37,38]. For the purpose of this book it is sufficient to provide an overview of the technology as it pertains to future biomedical devices, hence a simple discussion follows with sufficient references to fill in the details. We will first review how bulk hexagonal crystals of SiC are grown and, more importantly, what the state of the art is and the key characteristics of these commercially available substrates. Next, the growth and synthesis of thin films, starting with homoepitaxy on hexagonal substrates, followed by heteroepitaxy on Si substrates, and finally polycrystalline growth on various surfaces, will be presented. One of the great benefits of SiC is that it can be deposited in *a*-SiC form at lower temperatures than for single-crystal film growth. This technology will be reviewed in this chapter and the final section will be an introduction to how SiC can be micromachined, and how this points to the potential of SiC as an excellent bioMEMS material.

Table 1.1 Basic properties of SiC compared with Si and diamond [39,40]

Property	4H–SiC	6H–SiC	3C–SiC	Si	Diamond
Energy bandgap at 300 K (eV)	3.20	3.00	2.29	1.12	5.45
Intrinsic carrier concentration at 300 K (cm^{-3})	5×10^{-9}	1.6×10^{-6}	1.5×10^{-1}	1×10^{10}	$\sim10^{-27}$
Critical breakdown electric field (MV/cm)	2.2	2.5	2.12	0.25	1–10
Saturated electron drift velocity ($\times 10^7$ cm/s)	2.0	2.0	2.5	1.0	1.5
Electron mobility (cm^2/V s)	1000	600	800	1450	480
Hole mobility (cm^2/V s)	115	100	40	470	1600
Thermal conductivity at 300 K (W/cm K)	3.7	3.6	3.6	1.49	6–20
Coefficient of thermal expansion at 300 K (10^{-6} K^{-1})	4.3 $\perp$c, 4.7 $\parallel$c	4.3 $\perp$c, 4.7 $\parallel$c	3.2	3.0	1.0
Lattice constant (*a*, *c* in Å)	*a* = 3.0730, *c* = 10.053	*a* = 3.0806, *c* = 15.1173	*a* = 4.3596	*a* = 5.430	*a* = 3.5668
Elastic coefficient[a] (GPa)	$C_{44} = 600$	$C_{11} = 500$, $C_{12} = 92$, $C_{44} = 168$	$C_{11} = 352$, $C_{12} = 120$, $C_{44} = 233$	$C_{11} = 167$, $C_{12} = 65$, $C_{44} = 80$	$C_{11} = 1079$, $C_{12} = 124$, $C_{44} = 578$

[a]Calculated.

1.6.1 Bulk Growth

Single-crystal substrates of SiC have been in commercial production for more than two decades, starting with Cree, Inc.'s first commercial 6H–SiC wafers, which were 25 mm in diameter, in 1990 [41]. Much progress has been made worldwide since then, with both the quality and diameter of the wafers improving over the years; the current commercially available wafers are 100 mm (4 in.) and are available in both n-, p-, and intrinsic (very high purity "semi-insulating") form. Several manufacturers around the world sell SiC wafers and the crystal specifications are readily available [42]. Recently, larger 150-mm (6-in.) diameter wafers of quite high quality have been produced [43], which bodes well for the economic viability of SiC as an electronics technology (Fig. 1.3).

All SiC bulk wafers are grown in a high-temperature (>2000°C) furnace that is either gas fed (silicon and carbon containing precursors

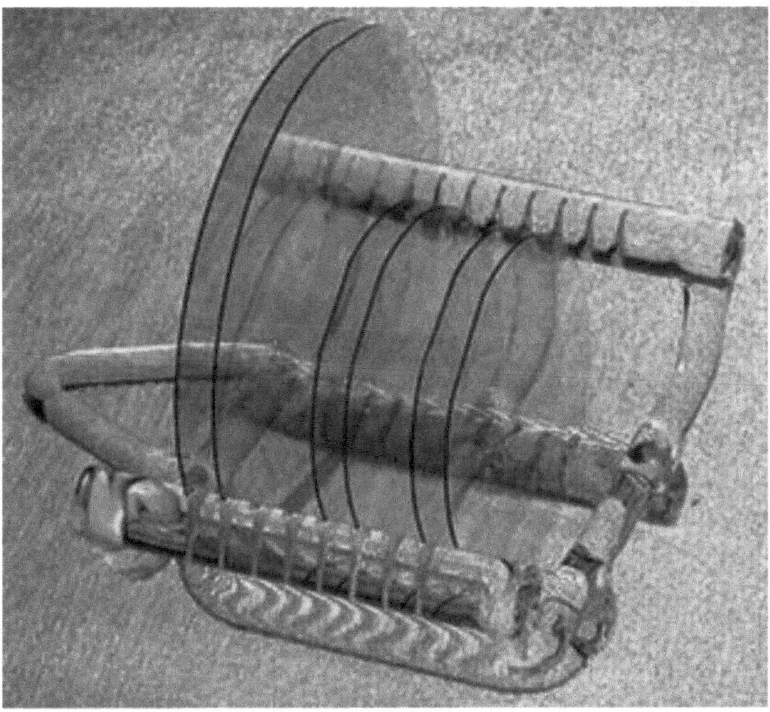

Figure 1.3 ***Commercially available SiC wafers can be found in off- and on-axis orientations, also with a Si or C terminated face.*** They can be either heavily doped or semi-insulating, depending on the application of use. Current wafers are 4 in. (100 mm) in diameter with 6-in. (150-mm) wafers demonstrated and ready for production in 2012. *(Photo courtesy of II-VI, Inc. [44].)*

transported in a carrier gas) or solid fed with SiC powder that is sublimated (evaporated) and condenses on a single-crystal SiC seed [45,46]. Growth rates are typically 0.1–0.3 mm/h and a typical boule thickness is roughly 25–30 mm (O. Kordina Private communication). Resistivity ranges from 1×10^5 to 1×10^7 Ω-cm for semi-insulating crystals, and with high-temperature chemical vapor deposition (HTCVD) bulk growth, resistivity as high as 1×10^{11} Ω-cm has been reported [47]. The wafers are either cut parallel to the basal plane (ie, perpendicular to the growth direction) or cut at a slight miscut angle, typically 3.5 degrees and 4.0 degrees for 6H–SiC and 4H–SiC, respectively. This is done to ensure that the epitaxy replicates the proper polytype.

1.6.2 Thin-Film Growth

Single crystalline SiC films can be grown epitaxially using a number of methods. The preferred method is CVD [48], but liquid phase epitaxy [49] and molecular beam epitaxy [50] have also been used for some time. The application of HTCVD, which is also used to grow bulk crystals, has demonstrated high growth rates [14]. The current state-of-the-art CVD reactors employ what is referred to as a hot-wall design whereas the growth substrate is surrounded by an actively heated graphite susceptor that allows for very high growth rates (as high at 100 μm/h with good crystal quality typically) [51]. The epitaxy of SiC is an active field of research, with numerous improvements in growth rate, crystal quality, doping control, etc., being made, which is critical for advanced device development. In addition, SiC crystals can be hydrogen polished via a hydrogen etching process that is somewhat dependent on both polytype and surface orientation (on-axis and off-axis) [52]. In addition hydrogen surface passivation using the same approach has been demonstrated, which has an important role to play in the control of the SiC crystal surface chemical state, particularly when performing surface functionalization [53].

The growth of SiC thin films via homoepitaxy is a relatively straightforward process that involves the use of a silicon- and a carbon-containing precursor molecule transported to the growth surface via carrier gas, typically hydrogen. The silicon and carbon precursors have traditionally been silane and propane [54], and more recently chlorine-based silane precursors such as trichlorosilane have been used since the addition of chlorine allows for higher growth rates with low Si-cluster defect levels [55]. When growing 3C–SiC, the only cubic polytype, on Si, this heteroepitaxy process typically involves three steps: first, hydrogen surface etching is performed to remove any native oxide that may be present; then, a buffer layer is formed with the so-called carbonization step, which seeks to bond C to Si dangling bonds to create an SiC layer; and finally, the growth step, which involves Si and C atoms delivered in the same was as described earlier for homoepitaxy [56]. Recently, this process was adapted by several groups to allow for the use of the hot-wall CVD reactor and it was discovered that the removal of the hydrogen etching step and the inclusion of a small amount of silane between the carbonization and growth steps resulted in very high-quality films [57] (Fig. 1.4). Use of this method has resulted in growth rates on the order of 10–30 μm/h, and the addition of chlorine has allowed for nearly double the growth rate in a large diameter reactor [58].

A new process that has been demonstrated uses a polycrystalline silicon film as a seed layer for poly-SiC growth [59]. While other groups have

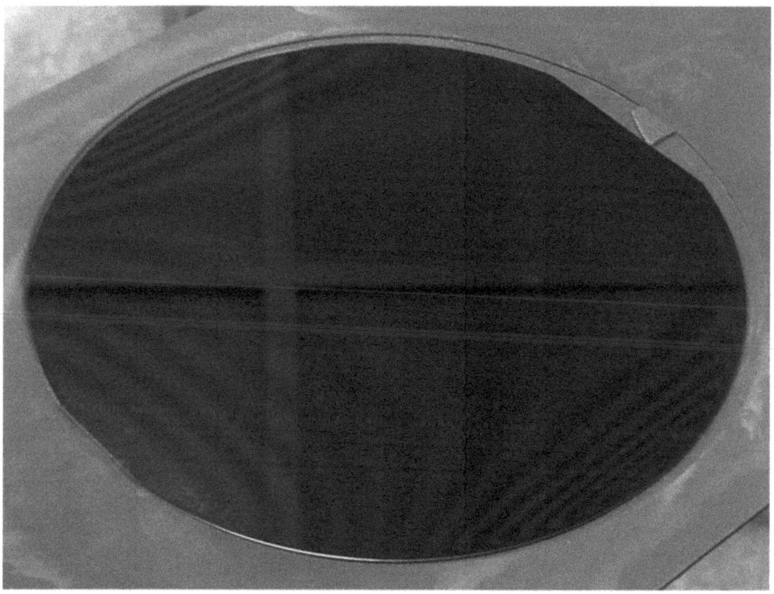

Figure 1.4 ***Monocrystalline and polycrystalline epiwafers of 3C–SiC are routinely grown using the CVD process.*** Typical epiwafer diameters are 6 in. (150 mm) and make use of the low cost and high quality of Si substrates, thus reducing the materials cost. In addition 3C–SiC on Si allows for micromachining to be performed in a similar fashion to Si MEMS technology. *(Photo courtesy of the USF SiC Group and shows a 100-mm [4-in.] 3C–SiC epitaxial layer grown on (100)Si. Epi thickness ~5 μm. Note the specular morphology of this monocrystalline film and Newton fringes due to the slight thickness variation across the epiwafer [reflection of ceiling tiles visible in image].)*

Table 1.2 Measured SiC Mechanical Properties Versus Si [28] [a]

Material type	Hardness (GPa)	Elastic modulus (GPa)
(100)Si	12.46 ± 0.78	172.13 ± 7.76
Lely platelet 15R–SiC	42.76 ± 1.19	442 ± 16.34
3C–SiC on (100)Si	31.198 ± 3.7	433 ± 50
Poly-3C–SiC on (100)Si	33.54 ± 3.3	457 ± 50

[a]Measured via nanoindentation using a Hysitron TriboIndenter with a Berkovich indenter tip.

reported poly-SiC film growth, this work involves the use of an oxide layer whereby the eventual release of poly-SiC MEMS structures can be easily facilitated [60]; the details will be discussed later in this chapter. A study of the mechanical properties of both monocrystalline and polycrystalline 3C–SiC films grown on Si substrates was performed by Volinsky et al. [61]. The results demonstrate the mechanical durability of SiC, as expected, but also point out a very important property – 3C–SiC films have a surprising resistance to fracture and possess a Young's modulus of nearly three times that of Si.

Using a Berkovich indenter, the mechanical properties of thin (~1 μm-thick) 3C–SiC films on (100)Si substrates were studied in our laboratory. Table 1.2 reports the average values of the measured hardness and modulus of elasticity of single-crystal 3C–SiC, polycrystalline 3C–SiC, a 15R–SiC Lely platelet, and (100)Si substrate. At least five indents were performed at each maximum load. A 15R–SiC Lely platelet was used as a comparison since this material is known to have a minimum amount of dislocations (typically dislocation free in the center of the platelet). There is a relatively large absolute variation in measured elastic modulus (±50 GPa), not typically observed in the indentation of softer materials. One must consider the relative variation, which is typically on the order of 10%. The reported mechanical properties are for a maximum indentation depth range between 60 and 250 nm, where quartz calibration is reasonable.

Wear tests on the single-crystal 3C–SiC film were also performed in a 3 × 3 μm area using the low load transducer at 2 μN normal load and 1 Hz frequency. The objective of this procedure was to determine the wear resistance of the sample by repeated scanning of the poly-3C–SiC surface. After a certain number of wear cycles the scan area was zoomed out to 5 × 5 μm to determine the material wear. The film topography of the mono- and polycrystalline 3C–SiC film surface was measured before and after the tip-induced wear test, which was performed for 1045 scans. During this scanning there was very little or negligible wear because only 1–2 nm of material depth was removed. This result confirms the high wear resistance of 3C–SiC films necessary for MEMS applications [28].

1.6.3 Amorphous Silicon Carbide Coatings

Amorphous SiC has long been studied and applied as a biomaterial. The amorphous form of SiC can be deposited using different techniques, most of which are familiar to the materials community, such as sputtering, pulsed laser deposition, evaporation, among others [62]. The advantage of *a*-SiC is that it can be deposited at low temperatures, which allows for the coating of plastics and other low-temperature materials such as polymers [63]. Another important advantage of *a*-SiC is that it can be electrically insulating, a property that many seek to exploit for numerous bioelectric devices [64]. One of the additional benefits, which is at the same time a challenge, is that line-of-sight coating of surfaces is usually the deposition mode, meaning that coating both sides of a material requires a rotation of the material during *a*-SiC coating. Issues that impact *a*-SiC functionality are the morphology of the film (smooth, rough, etc.) and the presence of any pinholes or cracks. In fact, one of the recent uses of *a*-SiC was in the coating of polyimide-based BMI devices to overcome the problem of polymer swelling inside the brain [29]. Perhaps one of the best-known biomedical uses of *a*-SiC coatings was the very successful coating of coronary heart stents to reduce the serious problem of clotting in the cardiovascular system [29]. In this instance, Rzany and Schladach at the University of Erlangen studied *a*-SiC:H (ie, hydrogenated *a*-SiC [65]) as a candidate coating for stainless steel stents and reported that SiC appears to be an ideal biomaterial due to its electronic properties in a physiological environment. They compared the performance of *a*-SiC:H coated and bare stainless steel *in vitro* and saw significantly improved performance for the SiC-coated steel [66] (Fig. 1.5).

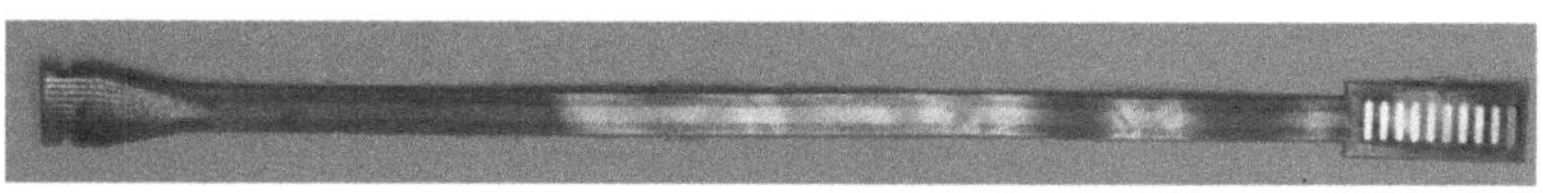

Figure 1.5 ***Amorphous SiC (a-SiC) coatings are readily formed using various processes such as RF sputtering, PLD, plasma spraying, etc.*** These coatings can be realized at low temperatures thus allowing for the coating of soft, low-temperature materials such as polymers and plastics. Photo shows an *a*-SiC-coated polyimide brain–machine interface (BMI) device. *(Photo courtesy of A. Hess et al. [67].)*

1.6.4 SiC Micromachining

The realization of mechanical devices based on microelectronic circuit processing techniques is known as microelectromechanical systems or MEMS. While a full treatment of MEMS is not practical here, a brief summary of the salient features of this technology will allow the reader to have an adequate understanding of the technology. The thermal, mechanical, and chemical strength of the SiC bond manifests itself in a very high material hardness and Young's modulus. These values, as measured by Volinsky et al. [61], were ~31 and 433 GPa, respectively, for 1-μm thick monocrystalline 3C–SiC films grown on (100)Si substrates. Similar measurements made on polycrystalline 3C–SiC (ie, poly-SiC) films of the same thickness and grown on the same substrate were comparable with ~33 and 457 GPa, respectively. For comparison purposes the corresponding values for (100) Si in bulk substrate form (700 μm thick) were only ~12 and ~172 GPa, respectively, roughly two-thirds of the value for the 3C–SiC films studied.

In chapter: Silicon Carbide Based Nanowires for Biomedical Applications a more comprehensive discussion of bioMEMS technology based on SiC will be presented. Basically, SiC MEMS are fabricated using the same micromachining technology as Si MEMS, so there is no need to reinvent the processing required to make highly robust SiC MEMS. The basic process flow for SiC MEMS depends on whether one is using bulk SiC or 3C–SiC on Si as the MEMS material. One advantage of using the first option (all-SiC materials) is that there is no problem with differing thermal expansion coefficients during processing, but the cost and difficulty in forming mechanical structures in this highly robust material is an issue [68]. Amorphous, poly-, and monocrystalline 3C–SiC on Si MEMS can be fabricated using standard Si MEMS techniques, as shown in Fig. 1.6, and allows for a simpler MEMS process technology, in

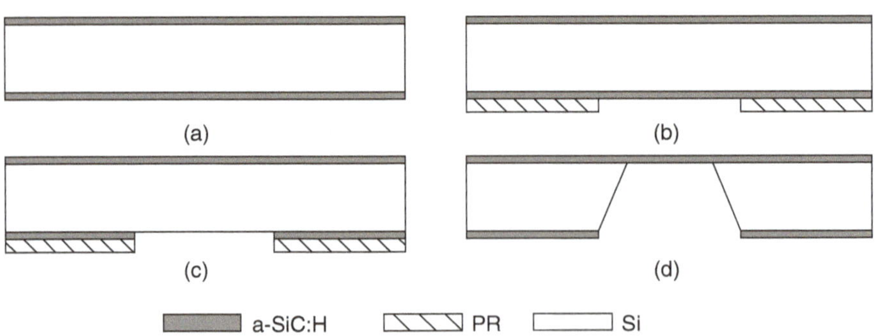

Figure 1.6 ***Schematic cross-sectional diagrams of a typical bulk micromachining process to fabricate SiC membranes.*** (a) Deposit SiC film, (b) fabricate backside etch mask, (c) etch backside SiC film to form Si etch mask, and (d) etch Si in anisotropic wet etchant [69].

particular when a poly-SiC on oxide release layers is utilized as described in the last section of this chapter. In this instance, the resulting MEMS structures are very straight and display high durability during processing [70]. Subsequent 3C–SiC on Si MEMS devices, however, display wafer bow issues resulting from the coefficient of thermal expansion mismatch between 3C–SiC and Si (about 8%) as well as the lattice mismatch (about 20%) between the two materials [71]. Partly to resolve these difficulties, and to achieve a simpler and more robust manufacturing process, the development of a poly-SiC oxide technology has been achieved and preliminary MEMS structures realized [60].

1.7 SILICON CARBIDE AS A BIOMEDICAL MATERIAL

While the durability and physical resilience of SiC is well known and documented, the use of SiC as a biomaterial is less well known, thus providing the motivation for this book. Although the following chapters will provide a detailed discussion on this important subject, at this stage it is useful to review some of the significant biomedical devices that have been reported and use SiC materials. Indeed, SiC has been used in virtually every part of the human body, from a durable coating for bone prosthetics [72] and in dental applications [73], which cover the mechanical/structural biomedical use of SiC, to coatings for BMI devices [32], myocardial heart probes [74], and finally nonfouling coatings for coronary heart stents [43]. A plethora of sensors have been reported, and at this point it is sufficient to review the literature for a sampling of this important work [75]. In all of these instances SiC materials were either in amorphous, bulk crystal, or thin-film (monocrystalline or polycrystalline) form. Thus, in order for the medical community to best make use of this very impressive material, it is incumbent that we begin our discussion of SiC biotechnology with a review of how the various SiC materials are formed and processed into useful devices (Fig. 1.7).

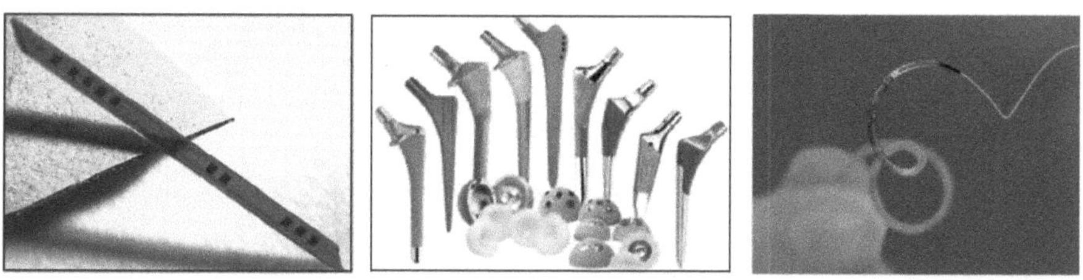

Figure 1.7 ***Reported SiC biomedical devices.*** From left to right, myocardial heart probe [43], bone prosthetic [76], and coronary heart stent [77]. The heart probe is made of single-crystal bulk 6H–SiC, while the bone prosthetic is SiC-coated titanium to eliminate ion uptake in the body. The heart stent is SiC coated to greatly reduce clotting and thus eliminate the need to perform a second surgery to replace the stent. *(Photo courtesy of Christoph Hehrlein [78].)*

1.8 SUMMARY TO THE FIRST EDITION

In this chapter we begin our introduction to SiC as a robust and highly useful material for biomedical devices and applications. A review of the basic material structure was made, followed by the various forms of SiC that are in use today, and finally the many ways that SiC materials are grown/synthesized/formed. An introduction to SiC bioMEMS was made followed by a quick review of the important SiC biomedical device applications that have been reported in the literature. Based on this introduction, the reader should be able to review the subsequent chapters of this book with a feeling of the overall SiC biotechnology landscape, and thus gain an in-depth and comprehensive feeling of why SiC is ideally poised to lead the way to the development of the next generation of biomedical devices, especially for long-term chronic implantation applications. We hope you enjoy the following discussion as much as we enjoyed putting it together. In fact, once completed, this book motivated us to accelerate the development and incorporation of SiC in advanced next-generation biomedical devices, the result being the realization of the second edition you are now reading.

ACKNOWLEDGMENTS

I would first like to thank all the former and present members of my research group, the USF SiC Group, for their hard work and dedication that led to the research that forms the underpinning of this chapter and book. Specific members stand out, in particular the bio-SiC team of Dr Camilla Coletti, Dr Norelli Schettini, and Dr Christopher Frewin for their pioneering work in the area of SiC biomaterials, which formed the backbone of the first edition of *Silicon Carbide Biotechnology*. This second edition would not have been possible without the device-related work performed by Dr Shamima Afroz, Dr Alexandra Oliveros, Dr Joe Register, and Dr Maysam Nezafati. One person stands out above all and that is my good colleague and friend Dr Chris Frewin, who has shared my vision from the beginning and recently launched a new startup company, Crystal Cybernetics, LLC, Tampa, FL, to specifically take the great research progress of the USF SiC Group and translate it into a commercial product line. His vision, imagination, and work ethic continue to be an inspiration not only to myself but to many of us who work closely with him. Finally, this journey has been made infinitely more enjoyable by the great colleagues I have had the good fortune to work with over the years, in particular Dr Andrew Hoff, who has been instrumental in transitioning my SiC electronic materials research program into biomedical engineering, Dr Sylvia Thomas, who has been working with me to develop the *in vivo* glucose sensor, and Dr Karl Muffly (USF College of Medicine), who has been instrumental in assisting my students who have been involved in our hemocompatibility studies. I continue to be inspired by my colleague Dr Lennox Hoyte, USF College of Medicine, a world-class robotic surgeon who performed our surgical implants for Dr. Afroz (glucose sensor). I am very grateful to Dr Joseph Pancrazio (UT Dallas), who has proven to be a great mentor and friend as we first worked together on a DARPA project to evaluate biomaterials for neural implants and he continues to impact my research plans in the area of neural implants. I have had the good fortune to work with Dr Mario Gazziro of the Universidade Federal do ABC, Santo André (Sao Paulo), Brazil, who has been my colleague on a Science without Borders project since 2012 and has always proven to be an inspiration to me as he worked tirelessly, even when out of contract, to continue our project objective to develop a wireless neural implant based on 3C–SiC. Finally, and not least, I would like to thank Dr Gabriel De Erausquin (formerly with USF College of Medicine and now at UT-RGV, Brownsville, TX), who, as an MD and PhD, provides an incredible amount of depth and breadth to this work and always takes the time from his overwhelming duties to meet with me to discuss important issues. I am forever indebted to these great colleagues and look forward to continuing our collaborations and interactions.

REFERENCES

[1] Saddow SE, editor. Silicon carbide biotechnology: a biocompatible semiconductor for advanced biomedical devices and applications. UK: Elsevier Ltd; 2011.

[2] Locke CW. Advanced SiC MEMS structures using polycrystalline SiC films. PhD dissertation, University of South Florida; 2011.

[3] Abbati L, Frewin CL, Placidi P, Saddow SE, Scorzoni A. Design and simulation of a 64 channel, high voltage analog interface for stimulation and acquisition of neural signals. In: Fourth IEEE international workshop on advances in sensors and interfaces (IWASI). IEEE; 2011.

[4] Afroz S, Thomas SW, Mumcu G, Saddow SE. Implantable SiC based RF antenna biosensor for continuous glucose monitoring. IEEE Sensor 2013;1–4. DOI: 10.1109/ICSENS.2013.6688379.

[5] Thomas SW, Afroz S, Saddow SE, Muffly K, Hoyte L. *In-vivo* evaluation of subcutaneous histological tissue reaction to SiC based continuous glucose sensor. J Histotechnol 2015, In press.

[6] Oliveros A, Guiseppi-Elie A, Saddow SE. Silicon carbide: a versatile material for biosensor applications. Biomed Microdevices 2013;15(2):353–68.

[7] Register JJ. SiC for advanced biological applications. PhD dissertation, University of South Florida; 2014.

[8] Nezafati M. Silicon carbide corrosion resistance, biocompatibility and hemocompatibility. PhD dissertation, University of South Florida; 2014.

[9] Lagonegro P, Galli C, Smerieri A, Alinovi R, Pinelli S, Nezafati M, Rossi F, Rimoldi T, Attolini G, Negri M, Fabbri F, Cristofolini L, Macaluso G, Saddow SE, Salviati G. Nanowire platforms based on cubic silicon carbide for biomedical applications: cytocompatibility and platelet activation. ACS Biomaterials Science & Engineering, In Press.
[10] Mantripragada VP, Lecka-Czernik B, Ebraheim NA, Jayasuriya AC. An overview of recent advances in designing orthopedic and craniofacial implants. J Biomed Mater Res Part A 2013;101(11):3349–64.
[11] Frewin CL, Locke C, Mariusso L, Weeber EJ, Saddow SE. Silicon carbide neural implants: *in vivo* neural tissue reaction. In: Sixth international IEEE/EMBS conference on neural engineering (NER). San Diego, CA; November 6–8, 2013. p. 661, 664.
[12] Saddow SE, Frewin CL, Nezafati M, Oliveros A, Afroz S, Register J, Reyes M, Thomas S. 3C-SiC on Si: a bio- and hemo-compatible material for advanced nano-bio devices. In: Ninth nanotechnology materials and devices conference (NMDC). IEEE; 2014. p. 49–53.
[13] Salimi A, et al. Electrooxidation of insulin at silicon carbide nanoparticles modified glassy carbon electrode. Electrochem Commun 2009;1.6:1116–9.
[14] Salimi A, et al. Highly sensitive electrocatalytic detection of nitrite based on SiC nanoparticles/amine terminated ionic liquid modified glassy carbon electrode integrated with flow injection analysis. Sensor Actuat B 2014;205:136–42.
[15] Williams EH, et al. Selective streptavidin bioconjugation on silicon and silicon carbide nanowires for biosensor applications. J Mater Res 2013;2801:68–77.
[16] Fradetal L, Stambouli V, Bano E, Pelissier B, Choi JH, Ollivier M, Latu-Romain L, Boudou T, Pignot-Paintrand I. Bio-functionalization of silicon carbide nanostructures for SiC nanowire-based sensors realization. J Nanosci Nanotechnol 2014;14(5):3391–7.
[17] Negri M, Dhanabalan SC, Attolini G, Lagonegro P, Campanini M, Bosi M, Fabbri F, Salviati G. Tuning the radial structure of core–shell silicon carbide nanowires. Cryst Eng Comm 2015;17:1258. www.rsc.org/crystengcomm.
[18] Tehrani Z, Burwell G, Mohd Azmi MA, Castaing A, Rickman R, Almarashi J, Dunstan P, Miran Beigi A, Doak SH, Guy OJ. Generic epitaxial graphene biosensors for ultrasensitive detection of cancer risk biomarker. 2D Mater 2014;1(2), p. 025004.
[19] Catellani A, Calzolari A. Functionalization of SiC(110) surfaces via porphyrin adsorption: ab initio results. J Phys Chem C 2012;116(1):886–92.
[20] Nardi M, Verucchi R, Aversa L, Casarin M, Vittadini A, Mahne N, Giglia A, Nannarone S, Iannotta I. Electronic properties of tetrakis(pentafluorophenyl)porphyrin. New J Chem 2013;37:1036–45.
[21] Rossi F, Bedogni E, Bigi F, Rimoldi T, Cristofolini L, Pinelli S, Alinovi R, Negri M, Dhanabalan SC, Attolini G, Fabbri F, Goldoni M, Mutti A, Benecchi G, Ghetti C, Iannotta S, Salviati G. Porphyrin conjugated SiC/SiOx nanowires for X-ray-excited photodynamic therapy. Sci Rep 2015;5:7606.
[22] Cacchioli A, Ravanetti F, Alinovi R, Pinelli S, Rossi F, Negri M, Bedogni E, Campanini M, Galetti M, Goldoni M, Lagonegro P, Alfieri R, Bigi F, Salviati G. Cytocompatibility and cellular internalization mechanisms of SiC/SiO2 nanowires. Nano Lett 2014;14:4368–75.
[23] Williams EH, Schreifels JA, Rao MV, Davydov AV, Oleshko VP, Lin NJ, Steffens KL, Krylyuk S, Bertness KA, Manocchi AK, Koshka Y. Selective streptavidin bioconjugation on silicon and silicon carbide nanowires for biosensor applications. J Mater Res 2013;28(1):68–77.
[24] Mohd Nasir NF, et al. Fabrication of 3C-silicon carbide membranes: towards development of novel microdevices for biomedical applications. International conference on biomedical engineering (ICoBE). IEEE; 2012.
[25] Iqbal A, Mohd-Yasin F, Dimitrijev S. Design optimization and finite element analysis of 3C-SiC bio-sensors based on dogbone resonator. Appl Mech Mater 2014;595.

[26] Lebedev MA, Nicolelis MAL. Brain–machine interfaces: past, present and future. Trends Neurosci 2006;29.9:536–46.
[27] Pine J. A History of MEA development. In: Baudry M, Taketani M, editors. Advances in network electrophysiology using multi-electrode arrays. New York: Springer Press; 2006. p. 3–23.
[28] Charkhkar H, Frewin C, Nezafati M, Knaack GL, Peixoto N, Saddow SE, Pancrazio JJ. Use of cortical neuronal networks for *in vitro* material biocompatibility testing. Biosensor Bioelectron 2014;53:316–23.
[29] Knaack GL. *In vitro* and *in vivo* biocompatibility testing of silicon carbide for neural interfaces; 2014. Available from: http://hdl.handle.net/1920/8979.
[30] Frewin CL. SiC materials for brain-machine-interface applications. Doctoral dissertation, USF; 2009.
[31] Saddow SE, et al. 3C-SiC on Si: a biocompatible material for advanced bioelectronic devices. ECS Trans 2014;61.7:101–11.
[32] Gazziro M, Zanetti de Florio D, Moreira D, Carvalho A, Simoes E, Soares Jr JN, Saddow SE, Frewin C, Mioni DP, Mateus J, Dal Fabbro PA, Pessatti MP. A 0.18 μm CMOS low-power amplifier suitable to transmission of wireless neural signals. IEEE international symposium on circuits and systems (ISCAS). Lisbon, Portugal; May 24–27, 2015.
[33] Saddow SE, Agrawal A, editors. Advances in silicon carbide processing and applications. Artech House, Norwood, MA USA; 2004.
[34] Guo H, Wang Y, Chen S, Zhang G, Zhang H, Li Z. PECVD SiC as a chemical resistant material in MEMS. Proceedings of the first IEEE international conference on nano/micro engineered and molecular systems. Zhuhai, China; January 18–21, 2006.
[35] Yonenaga I. Thermo-mechanical stability of wide-bandgap semiconductors: high temperature hardness of SiC, AlN, GaN, ZnO and ZnSe. Physica B Condens Matter 2001;308–310:1150–2.
[36] Available from: http://www.forbes.com/2008/02/27/incandescent-led-cfl-pf-guru_in_mm_0227energy_inl.html
[37] Karpova SY, Kulika AV, Zhmakina IA, Makarovb YN, Mokhovc EN, Rammc MG, Ramm MS, Roenkovc AD, Vodakov YA. Analysis of sublimation growth of bulk SiC crystals in tantalum container. J Cryst Growth 2000;211(1–4):347–51.
[38] Cogan SF, Edell DJ, Guzelian AA, Ping Liu Y, Edell R. Plasma-enhanced chemical vapor deposited silicon carbide as an implantable dielectric coating. J Biomed Mater Res Part A 2003;67A:856–67.
[39] Casady JB, Johnson RW. Status of silicon carbide (SiC) as a wide-bandgap semiconductor for high-temperature applications: a review. Solid-State Electron 1996;39(10):1409–22.
[40] In: Harris GL, editor. Properties of silicon carbide, vol. 13. London: INSPEC; 1995.
[41] Available from: http://www.cree.com/about/milestones.asp.
[42] Müller StG, Glass RC, Hobgood HM, Tsvetkov VF, Brady M, Henshall D, Jenny JR, Malta D, Carter CH Jr. The status of SiC bulk growth from an industrial point of view. J Cryst Growth 2000;211(1–4):325–32.
[42] Available from: http://compoundsemiconductor.net/csc/news-details.php?cat=news&id=19732335
[44] Available from: http://www.iiviwbg.com/
[45] Sakwe SA, Stockmeier M, Hens P, Müller R, Queren D, Kunecke U, Konias K, Hock R, Magerl A, Pons M, Winnacker A, Wellmann P. Bulk growth of SiC – review on advances of SiC vapor growth for improved doping and systematic study on dislocation evolution. In: Friedrichs P, Kimoto T, Ley L, Pensl G, editors. Silicon carbide, vol. 1: growth, defects, and novel applications. Wiley-VCH, Weinheim, Germany; 2010.
[46] Ellison A, Magnusson B, Son NT, Storasta L, Janzén E. High temperature CVD growth of SiC. Mater Sci Eng B 1999;61–62:113–20.
[47] Ellison A, Magnusson B, Son NT, Storasta L, Janzén E. HTCVD grown semi-insulating SiC substrates. Mater Sci Forum 2003;433–436:33–8.

[48] Ellison A, Zhang J, Peterson J, Henry A, Wahab Q, Bergman JP, Makarov YN, Vorob'ev A, Vehanen A, Janzén E. High temperature CVD growth of SiC. Mater Sci Eng 1999;B61-B62:113–20.
[49] Syvajarvi M, Yakimova R, Radamson HH, Son NT, Wahab Q, Ivanov IG, Janzen E. Liquid phase epitaxial growth of SiC. J Cryst Growth 1999;197:147–54.
[50] Chen J, Steckl AJ, Loboda MJ. Molecular beam epitaxy growth of SiC on Si(111) from silacyclobutane. J Vac Sci Technol B 1998;16(3):1305–8.
[51] Kordina O, Hallin C, Henry A, Bergman JP, Ivanov I, Ellison A, Son NT, Janzén E. Growth of SiC by "hot-wall" CVD and HTCVD. Phys Status Solidi B 1997;202(1):321–34.
[52] Frewin CL, Coletti C, Riedl C, Starke U, Saddow SE. A comprehensive study of hydrogen etching on the major SiC polytypes and crystal orientations. Mater Sci Forum 2009;615–617:589–92.
[53] Coletti C, Frewin CL, Hoff AM, Saddow SE. Electronic passivation of 3C-SiC(001) via hydrogen-treatment. Electrochem Solid-State Lett 2008;11(10):H285–7.
[54] Larkin DJ, Neudeck PG, Powell JA, Matus LG. Site-competition epitaxy for superior silicon carbide electronics. Appl Phys Lett 1994;65(13):1659–61.
[55] La Via F, Izzo G, Mauceri M, Pistoneb G, Condorelli G, Perdicaro L, Abbondanza G, Calcagno L, Foti G, Crippa D. 4H-SiC epitaxial layer growth by trichlorosilane (TCS). J Cryst Growth 2008;311(1):107–13.
[56] Nishino S, Powell JA, Will HA. Production of large-area single-crystal wafers of cubic SiC for semiconductor. Appl Phys Lett 1983;42:460–2.
[57] Reyes M, Shishkin Y, Harvey S, Saddow SE. Development of a high-growth rate 3C-SiC on Si CVD process. Mater Res Soc Meet Proc 2006;911:79.
[58] Reyes M, Shishkin Y, Harvey S, Saddow SE. Increased growth rates of 3C-SiC on Si(100) substrates via HCl growth additive. Mater Sci Forum 2007;556–557:191–4.
[59] Frewin CL, Locke C, Wang J, Spagnol P, Saddow SE. Growth of cubic silicon carbide on oxide using polysilicon as a seed layer for micro-electro-mechanical machine applications. J Cryst Growth 2009;311(17):4179–82.
[60] Locke C, Frewin C, Abbati L, Saddow SE. Demonstration of 3C-SiC MEMS structures on polysilicon-on-oxide substrates. Mater Res Soc Proc 2010;1246:B08-05.
[61] Deva Reddy J, Volinsky AA, Frewin C, Locke C, Saddow SE. Mechanical properties of single and polycrystalline SiC thin films. Mat Res Soc Symp Proc 2008;1049.
[62] Saddow SE, editor. Silicon carbide biotechnology: a biocompatible semiconductor for advanced biomedical devices and applications. UK: Elsevier Ltd; 2011. Chapter 2.
[63] Yakimova R, Petoral RM Jr, Yazdi GR, Vahlberg C, Lloyd Spetz A, Uvdal K. Surface functionalization and biomedical applications based on SiC. J Phys D 2007;40:6435.
[64] Cogan SF, Edell DJ, Guzelian AA, Ping Liu Y, Edell R. Plasma-enhanced chemical vapor deposited silicon carbide as an implantable dielectric coating. J Biomed Mater Res Part A 2003;67A:856–67.
[65] Hattori K, Mori T, Okamoto H, Hamakawa Y. Photothermal modulation spectroscopy of multilayered structures of amorphous silicon and amorphous silicon carbide. Phys Rev Lett 1988;60:825–7.
[66] Rzany A, Schaldach M. Smart material silicon carbide: reduced activation of cells and proteins on a SiC:H-coated stainless steel. Prog Biomed Res 2001;4:182–94.
[67] Hess A, Parro R, Du J, Scardelletti JDM, Zorman CA. PECVD silicon carbide as a thin film packaging material for microfabricated neural electrodes. In: MRS Spring Meeting 2007. San Francisco, CA; 2007.
[68] Mehregany M, Zorman CA, Rajan N, Chien Hung W. Silicon carbide MEMS for harsh environments. Proc IEEE 1998;86(8):1594–609.
[69] Zorman CA, Parro RJ. Micro- and nanomechanical structures for silicon carbide MEMS and NEMS. Phys Stat Sol B 2008;245(7):1404–24.

[70] Anzalone R, Severino A, Locke C, Saddow SE, La Via F, D'Arrigo G. 3C-SiC hetero-epitaxial films for sensor fabrication. Advances in Science and Technology, vol. 54. Trans Tech Publications: Switzerland; 2008. p. 411–415.
[71] Anzalone R, Camarda M, D'Arrigo G, Locke C, Canino A, Piulso N, Severino A, La Magna A, Saddow SE, La Via F. Advanced stress analysis by micro-structures realization on high-quality hetero-epitaxial 3C-SiC for MEMS application. Mater Sci Forum 2011;679–680:133–6.
[72] Santavirta S, Takagi M, Nordsletten L, Anttila A, Lappalainen R, Konttinen T. Biocompatibility of silicon carbide in colony formation test *in vitro*. A promising new ceramic THR implant coating material. J Biomater Applic 1998;118:89–91.
[73] Hashiguchi K, Hashimoto K. Mechanical and histological investigations on pressureless sintered SiC dental implants. Okajimas Folia Anatomica Japonica 1999;75(6):281–96.
[74] Godignon P. SiC materials and technologies for sensors development. Mater Sci Forum 2005;483–485:1009–14.
[75] Yakimova R, Petoral RM Jr, Yazdi GR, Vahlberg C, Lloyd Spetz A, Uvdal K. Surface functionalization and biomedical applications based on SiC. J Phys D 2007;40:6435–42.
[76] Available from: http://www.zimmer.nl/producten/heup/mis-hip-solutions/mis-antelolateral-hip/
[77] Hamm CW, Hugenholtz PG, Investigators T. Silicon carbide-coated stents in patients with acute coronary syndrome. Cathet Cardiovasc Intervent 2003;60(11/01):375–81.
[78] Hehrlein C. Stent passivation with silicon carbide as a possible alternative to drug-eluting stents – a comprehensive review of pre-clinical and clinical results. Interven Cardiol 2009;4:60–3.

CHAPTER 2

Cytotoxicity of 3C–SiC Investigated Through Strict Adherence to ISO 10993

Christopher L. Frewin*, Maysam Nezafati, Katie Noble** and Stephen E. Saddow****
*The Department of Biomedical Engineering, Erik Johnsson School of Engineering and Computer Science, University of Texas, Dallas, TX, United States of America
**Department of Electrical Engineering, University of South Florida, Tampa, FL, United States of America

Contents

2.1 INTRODUCTION

Humans have always struggled with hardships due to effects of disease, the general rigors and wear due to life, and, for nearly all of our history, we have used the materials around us to improve our life experience [1,2]. In modern times, we have not only come up with cures to many diseases which affected millions of people, such as smallpox, but humans have also dedicated their efforts to using technology for diagnosing disease and physiology through tools like magnetic resonance imaging, or MRI, and electrocardiograms, otherwise known as ECG or EKG [3]. The explosion of technological advances brought by the silicon revolution has led to the emergence of many new avenues to investigate biological activity as well as the ability to provide therapeutics for people suffering from trauma and disease. This technology includes sensors to monitor physiological processes, devices providing therapeutic treatment for diseases, and new methods to diagnose disease [4–7]. Some examples of these devices are biological sensors, which can detect heart attacks before they occur, sense changes in blood sugar on a real-time basis, and intimately link with neurons to provide connectivity or functionality that was once lost. Unfortunately, many of these devices require prolonged contact with the human body in order to perform their designed tasks. This, in turn, leads to many challenges in both the design of the device and the implementation of it within the human body. It is these challenges that often slow the progress in the utilization of implantable devices for the patients that need them. Therefore, we must find methods with which to test these devices, and the materials that comprise them, thoroughly before they are implemented.

Two of the major challenges facing chronic, implantable biomedical devices are the requirements of stable, reliable function and patient safety. Contemporary implantable biomedical devices, such as mechanical prosthetics like knee, hip [8], and dental implants [9], allow for the restoration of a patient's freedom of movement and/or functionality. Emergent applications for implantable biomedical devices involve integration with the nervous system. These implantable neural interfaces (INI) aim to reduce the burden associated with neurological trauma or disease, which have caused partial to total loss of physical or cognitive functionality. Current INIs deliver open-loop therapeutic neural stimulation, assisting numerous individuals suffering from Parkinson's disease and behavioral issues [10–12]. A promising INI application would provide assistance to paralyzed individuals and amputees allowing them increased communicative ability or freedom of movement using state-of-the-art robotic prosthetic limbs controlled by the INI device [13–16].

Biocompatibility, which is defined as the ability of the device to administer intended positive effects on the individual while maintaining desired and reliable operation, is one of the foremost considerations which must be quantified [17]. Many of the chronically implantable biomedical devices have proven to lack adequate biocompatibility and have encountered issues after implantation leading to failure in clinical application due, in part, to their overall long-term reliability *in vivo* [18–22]. Although the underlying causes of deteriorating reliability have proven difficult to isolate as the factors are both biotic and nonbiotic, research has shown that one major factor stems from the materials used in the fabrication of implantable biomedical devices [20,23,24]. In short, the body recognizes these devices as foreign bodies and initiates the inflammatory response. As the devices are much larger than the cells, frustrated phagocytosis begins, in which phagocytic cells attach to the surface of the device and eject lysosomal enzymes onto the surface. The ejections may not only damage the device, but they may also negatively affect heathy tissue in the surrounding area. Finally, if the device is not removed, it will eventually be encapsulated and closed off from the body.

Many new materials are being developed to improve the functionality of implantable devices. These novel materials have often not been examined for biological safety. The future progress for novel biomedical devices requires the application of a standard, reproducible, ethical, and biologically significant testing methodology. Testing an unclassified material directly in humans is dangerous and deemed unethical in civilized society. Biological research, which uses animals, especially higher-order mammals, is also rightfully restricted and controlled. Screening the device, or constituent materials, *in vitro* is the most logical alternative, especially for a material, which has not been examined in previous efforts. Device screening must satisfy several requirements, such as quantifying material properties, corrosion resistance, biocompatibility, hemocompatibility, and durability in a simulated physiological environment [17,25,26]. One method which has been used is to test the reaction of cells, which have been cultured under controlled conditions, to the device outside their natural environment, usually within plastic wells or dishes. This method is primarily what we will focus on in this chapter. We will discuss in the next section the use of cultured cells to investigate if the material is toxic to cellular processes, known as cytotoxicity. We will then discuss a method, which improves on the testing methods specified in international standards. Finally, we discuss the cytotoxic evaluation of a novel material, silicon carbide (SiC), and specifically the cubic polytype (3C–SiC) of this material and the amorphous form of the material using our improved testing protocols.

2.2 *IN VITRO* BIOMEDICAL TESTING METHODS FOR CYTOTOXICITY

2.2.1 International Standards Organization (ISO) 10993

ISO 10993, compiled from numerous previous biological device testing research protocols, provides a general guideline with which to evaluate the biocompatibility of biomedical devices [27]. ISO 10993, which consists of

21 parts, describes sample preparation and reference materials selection [28], chemical stability, and material robustness [29,30], biological evaluation tests, both *in vitro* [31] and *in vivo* [32], and hemocompatibility evaluation [33,34]. The purpose of the standard is to ensure the ability of data comparison and evaluation reliability and safety for biomedical devices.

ISO 10993-5 provides guidelines for *in vitro* cytotoxicity evaluation and has been utilized as one of the initial biomedical device testing protocols by many research groups. ISO 10993-5 details three separate methodologies: testing with extracts derived from biomedical devices, direct contact of the biomedical device with cells, and diffusion of leachable impurities through agar [31]. These methods have been reported in the evaluation of implantable neural interfaces [35], microelectrode arrays [36], biosensors [37], drug delivery devices [38], silica-glass fiber composites for dental applications [39], and for the evaluation of biodegradable implants such as magnesium alloy bone scaffolds [40]. Unfortunately, the methods used in many of these studies have been based loosely on ISO 10993-5, and experimental deviations may result in inconsistencies between groups and uncertainty in the relevance in the resulting data. The purpose of our study was to present and strictly evaluate the guidelines, methodologies, and protocols presented by ISO 10993-5 in conjunction with ISO 10993-12, which details information on control material qualification as well as device preparation protocols. It is our hope that this study will serve as a practical guide on how to perform these highly critical biomedical device compatibility assessments, thereby allowing comparison of reliable and reproducible data across laboratories, especially those involved in INI research and development.

2.2.2 ISO 10993-12 Control Selection and Material Preparation

ISO 10993-12, Sections 4 and 6, relate that controls are a critical aspect of all material and device biocompatibility testing, providing not only sources for comparison between different groups, but also to ensure testing validation [28]. Section 5 dictates that the reference materials used for these controls should be quantitatively qualified before their use in biological testing. This is to be done within the individual's laboratory, as we have done for this study, although it is more desirable to perform a stringent certification requiring control material evaluation by three or more separate entities [28]. We selected our reference control materials from either a list suggested within Annex A.3, or which would closely match the conductive and physical properties of our testing materials, as dictated in Section 6.3 [28]. Gold (Au) and polyethylene Transpore™ medical tape (PE) were selected as negative controls, while copper (Cu) and Tygon® polyvinyl chloride (PVC) fuel and lubricant tubing formulation F-4040-A were utilized as positive control materials. All the reference materials were evaluated, with each sample tested at $N = 09$, using both the methodologies and cytotoxicity tests discussed later in this section. The cellular response to the materials was found to possess the appropriate reliability required by the ISO 10993 standards. The test materials induced cell responses consistent with the ISO 10993 designations, specifically a viability above 70% of culture-treated polystyrene for the negative controls and below 70% for the positive controls [31].

ISO 10993-12, Section 7, implies that biological testing is to be performed either on the final usable device, or, according to Section 9.1, it can be tested in part to evaluate specific reactivity, and that the product that is tested should be processed in the same manner as used in the device [28]. Since our purpose in this study was to demonstrate strict adherence to the ISO 10993-5 protocol and not evaluate a specific biomedical device, we selected material samples that were constituents of existing or emerging INI devices. Silicon (Si) has been the base substrate material for multiple INI architectures and biomedical microelectromechanical (bioMEMS) system research [22,41–43], silicon dioxide (SiO_2) has been a mainstay for electrical isolation in microelectronics for many years [24,44–46], and tungsten (W) was selected as it was one of the first materials used in INI devices and is still used in multiple, modern INI systems [21,47,48].

One hundred millimeter wafers, Si(100) (University Wafer, Boston, Massachusetts, United States, #1910), provided high-purity silicon material. Fifteen hundred-nanometer thick silicon dioxide (SiO_2) samples were created through plasma-enhanced chemical vapor deposition (PECVD) on Si(100) wafers (University Wafer, Boston, #1910). All of our metallic samples consisted of 1500 Å of electron beam physical vapor-deposited (EBPVD) metal (Au, Cu, W) on 200 Å of EBPVD titanium (Ti) deposited on a 100-mm (100) silicon substrate wafer, Si(100) (University Wafer, Boston, #1910). Transpore PE tape (3M, Maplewood, Michigan, United States, #1534-3) is a commercially sold, FDA-approved medical tape composed of a blend of polyethylene, polyester, and acrylic polymers and used in surgical procedures for humans [49]. Tygon PVC fuel and lubricant tubing formulation F-4040-A (Amazon, Seattle, Washington, DC, United States, #TGYF-0140-10) became the positive control due to the reliable cytotoxic reactions reported by Charkhkar et al. [26].

ISO 10993-12, Section 8, states that the preparation and cleaning of samples should be taken into consideration to avoid contamination [28]. All Si substrate samples were diced into 8 × 10 mm rectangular coupons through the use of a dicing saw equipped with a diamond-coated blade. This sample size is easily accommodated within a single well of a 24-well tissue culture (TC)-treated polystyrene plate (CTPSt) (USA Scientific, Ocala, Florida, United States, #5666-2160). The polymer-based samples were cut to 80 mm^2 of exposed surface area. All nonpolymer samples were degreased within baths of acetone, isopropanol alcohol, and deionized water ($\rho > 16$ MΩ/cm), with added ultrasonication, for 10 min in each medium. The PVC was aseptically cleaned with methanol and isopropanol followed by disinfection with 70% ethanol solution. ISO 10993-5 stresses the importance of sterilization in Sections 4.2.1 and 4.3.2, and ISO 19003-12, Section 8.1.2, requires the sterilization of testing materials, which are intended to be sterile before their clinical use [28,31]. The PVC was sterilized via 12-h ethylene oxide (EtO_x) exposure at George Mason University. The silicon substrate samples were sterilized using dry heat sterilization at 170°C for 60 min in the presence of flowing N_2 gas. The PE tape was sold sterile and therefore did not require further sterilization procedures.

2.2.3 L929 Murine Fibroblastoma Cell Culture Protocol

ISO 10993-5, Section 5, suggests the use of a well-established cell line for biological testing, although primary cell lines can be used if specific sensitivity is required [31]. We utilized one of the ISO 10993-5, Section 8, referenced cell lines; NCTC clone 929, strain L (American Type Culture Collection, Manassas, Virginia, United States, CCL-1™) mouse fibroblastoma [31]. The cell culture media consisted of sterilized Dulbecco's Minimum Eagle Medium (DMEM) (Life Technologies, Grand Island, New York, United States, #10313-021), infused with 10% fetal bovine serum (FBS) (Life Technologies, Grand Island, New York, United States, #16000-044), 2.2 mM L-glutamine GlutaMAX-1 (Life Technologies, Grand Island, New York, United States, #35050-061), and 1% antibiotic/antimycotic solution (Sigma-Alderich, St. Louis, Missouri, United States, #P4458-100ML). The L929 cells were cultured in CytoOne T-75, filter cap, 75 cm^2 TC flask (USA Scientific, Ocala, Florida, United States, #CC7682-4875) in a water-jacketed incubator maintained at 37°C, 95% relative humidity, and 5% CO_2. Fifty percent of the culture media was exchanged with fresh media every 72 h. When the cell population exceeded 80% confluence, the cells were trypsinized and utilized in the cytotoxicity assays within this study; otherwise cells were expanded by adding 25% of the cell suspension into a new 75 cm^2 TC flask with fresh cell media.

2.2.4 ISO 10993-5 Extract and Direct Contact Methods

The process of creating extracts containing leachable substances derived from testing materials is detailed in both ISO 10993-5, Section 8, and ISO 10993-12, Section 10 [28,31]. The standard specifies that extraction conditions should be designed to emulate clinical or patient risk-associated conditions [31]. Historical extraction times and temperature are found in ISO 10993-5, Section 4.2.3.2, and ISO 10993-12, Section 10.3.1, a reference to agitation in ISO 10993-12, Section 10.3.5 [28,31]. ISO 10993-5, Section 4.2.2, and ISO 10993-12, Section 10.3.4, detail that the media for the extraction should be chosen with respect to its interaction with polar and nonpolar solvents [28,31]. We selected serum containing culture media as both our polar and nonpolar extraction vehicles. It has been shown by Baek et al. to provide more sensitivity during cytotoxicity testing, and albumin, a serum protein, has been shown to increase a material's degradation rate [50,51].

Extraction concentration consistency is maintained using surface area-to-volume or mass-to-volume ratios as explained in ISO 10993-12, Section 10.3.2 [28]. All of our reference and testing materials possessed a thickness of less than 1 mm, so we used a volume to sample exposed surface area ratio of 3 cm^2/mL as specified in the standard [28]. The samples were placed within individual wells within a 24-well CT plate (USA Scientific, Ocala, #5666-2160), containing 267 μL of sterile media with serum, and then the plate was aseptically sealed with Parafilm M™ Wrapping Film (Fischer Scientific, Pittsburg, Pennsylvania, United States, #S37441). The plates were placed on a shaker at 1 Hz frequency for 24 h within a hotroom at 37°C [28,31]. The derived extracts were used immediately after agitation to ensure that there was little product degradation, settling, or absorption of the leachable materials onto the storage container as determined by ISO 10993-12, Section 10.3.6.

Testing was performed at three different concentrations, with $n = 3$ tested in triplicate, for a total $N = 9$ at each concentration level, as stated by ISO 10993-5, Section 8.2.1. Suspended L929 cells were plated within a 96-well CTPSt well plate (USA Scientific, Ocala, Florida, United States, #CC7682-7596) at 1.5×10^4 cells to achieve 80% of confluence within 24 h, as required by ISO 10993-5, Sections 8.2.2 and 8.2.4. The manufacturer-suggested working media level of 100 μL was added to each well of the 96 plate. The cells were placed within a sterile, water-jacketed incubator for 24 h at 37°C, 95% relative humidity, and 5% CO_2 as dictated by ISO 10993-5, Section 8.2.3 [31]. After 24 h, the cells were observed to reach subconfluence, at which time 100% of the culture media was replaced with extract dilutions at levels of 100, 70, and 50% as stated in ISO 10993-5, Sections 8.2.5 and 8.2.7 [31]. The media within three wells containing subconfluent L929 cells was replaced with normal serum containing media to provide a blank, which will be used as a population baseline to allow for statistical comparison between the replicate tests. In accordance with ISO 10993-5, Sections 8.2.8 and 8.2.9, the 96-well plate was returned to the incubator for an additional 24 h at 37°C, 95% relative humidity, and 5% CO_2. At the conclusion of this time period, the plates were ready for cytotoxicity evaluation using either the MTT [3-(4,5-dimethylthiazol-2-yl)-2,5-diphenyltetrazolium bromide] or LIVE/DEAD® assays in accordance with ISO 10993-5, Section 8.5 [31]. Fig. 2.1a displays a simplified illustration denoting the main steps involved in the method.

Testing by direct contract methodology, ISO 10993-5, Section 8.3, is much like the testing protocols described for testing by extract, ISO 10993-5, Section 8.2, with only one major exception [31]. The material under test is placed directly on top of a bed of semiconfluent cells covering a total surface area approximately 10 times larger than the sample (ISO 10993-5, Sections 8.3.4 and 8.3.6) [31]. We plated 1.8×10^5 of the L929 cells on a CytoOne™ six-well CTPSt plate (USA Scientific, Ocala, Florida, United States, #CC7682-7506) (8.33% actual sample/plate well area) to achieve subconfluence (ISO 10993-5, Section 8.3.2) [31]. Each well received the manufacturer-recommended working volume of 2 mL fresh cell culture media, and the plate was incubated at 37°C, 95% relative humidity, and 5% CO_2 (ISO 10993-5, Section 8.3.3) [31]. At approximately 24 h, the subconfluence of the cell layer was verified (ISO 10993-5, Section 8.3.4) and the original media was replaced by 2 mL of fresh media with serum (ISO 10993-5, Section 8.3.5) [31]. The sterilized testing and control samples (ISO 10993-5, Section 8.3.7) were carefully placed in the center of the well directly on the subconfluent cell layer (ISO 10993-5, Section 8.3.6). The plate was returned to the incubator and maintained at 37°C, 95% relative humidity, and 5% CO_2 (ISO 10993-5, Section 8.3.7) [31]. As with the extraction method, the plates were ready for cytotoxicity evaluation after 24 h had elapsed (ISO 10993-5, Section 8.3.8) [31]. Fig. 2.1(b) illustrates the direct contact methodology.

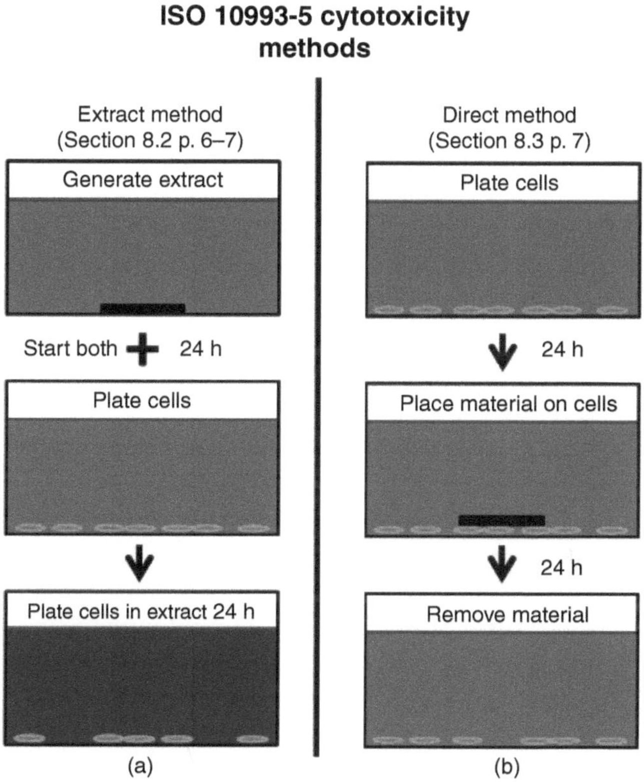

Figure 2.1 ***A simplified cartoon representation of two of the main cytotoxicity methodologies presented in ISO 10993-5.*** (a) The extract method involves plating cells to semiconfluence and exposing them to a solution, which has been exposed to a device under test. (b) The direct method is where the device under test is placed directly on a bed of semiconfluent cells. At the conclusion of these testing methods, the cells are then evaluated for viability.

ISO 10993-5, Section 5, details that the cells will be evaluated by quantitative or qualitative means, with preference given to quantitative methods [31]. ISO 10993-5, Annex A–D, lists established quantitative methods for determining cytotoxicity [31]. We used the MTT cytotoxicity assay protocol, detailed in Annex C, with our material extracts, direct contact, and the direct seeding methodologies [31,52]. MTT power (Life Technologies, Grand Island, New York, United States, #M-6494) was added to sterile phosphate buffered saline (PBS at a concentration of 1 mg/mL (ISO 10993-5, Annex C.2.2.4.3) [31]. After removing the media from the testing well, 156 μL/cm^2 (extrapolated from ISO 10993-5, Annex C.2.3.4) of MTT solution was added to each well followed by incubation for 2 h at 37°C [31]. The MTT solution was then carefully decanted from each well into individual, identifiable, 1.5-mL conical tubes (USA Scientific, Ocala, Florida, United States, #1415-8700), allowing the capture of purple formazan crystals dislodged during the media extraction. The conical tubes were placed in a centrifuge for 2 min at 1200 rpm and then the remaining MTT solution was decanted from the conical tube leaving a pellet of purple formazan crystals. A total of 313 $\mu L/cm^2$ (extrapolated from ISO 10993-5, Annex C.2.3.4) of isopropanol alcohol was added to each sample, control, and baseline well and allowed to incubate for 30 min at 37°C to allow the dissolution of the purple formazan crystals [31]. The isopropanol/formazan solution was then transferred to the 1.5-mL conical tubes and incubated for an additional 30 min at 37°C.

A 96-well plate was loaded with 100 μL of untreated isopropanol solution into the first column and 100 μL of deionized water in the first and last rows. One hundred milliliters of the isopropanol solution was then taken from each sample, control, and baseline well and loaded into the remaining wells of the 96-well plate. The plate was read within an Epoch Microplate Spectrophotometer (BioTec, Winooski, Vermont, United States) at the detected absorption wavelength of 570 nm (reference 650 nm) as specified by ISO 10993-5, Annex C.2.3.4 [31]. The results from these tests were normalized with respect to the CTPSt blank wells and expressed as the sampling distribution of the mean (x) and a standard deviation (σ), as reported in the next section.

As of the 2009 revision of ISO 10993-5, three of the four assays listed in the Annex examples used colorimetric assay techniques to evaluate cellular viability. In these assays, the color intensity increases in proportion to the presence of living cells. One of the four assays directly counted cell colonies and morphologies, while another, the neutral red dye, helped the user evaluate morphology. New cytotoxic assays have been developed since the release of ISO 10993-5 that may allow better quantification of cellular cytotoxic responses.

The LIVE/DEAD Assay Kit for mammalian cells (Life Technologies, Grand Island, New York, United States, #L-3224) is a powerful technique that not only quantifies cellular viability, like the ISO 10993-5 Annex colorimetric assays MTT, XTT, and neutral red, but provides additional information by clearly delineating between live, apoptotic, and dead cells and facilitates enhanced morphological evaluation of cells and cell-to-cell interaction. The assay contains two main permanent fluorescent dyes: calcien AM (CaAM) fluorescently marks living cells in the green spectrum; while ethidium homodimer-1 (Eth-1) identifies dead or apoptotic cells [52–54]. The LIVE/DEAD cytotoxicity assay was prepared by adding 5 μL of the 1-mg/mL CaAM kit solution with 7.5 μL of 2-mM Eth-1 kit solution to 10 mM PBS. This solution was then added to 10 mL of PBS. The PBS/CaAM/Eth-1 solution is then added to the wells containing cells to cover the cells completely, and the plates were incubated at 37°C for 15 min.

The cells were viewed using a Zeiss Image.M2m fluorescent microscope (Carl Zeiss Microscopy, LLC., Thornwood, New York, United States), with live cells expressing fluorescence when viewed through a filter appropriate for green fluorescent protein (470 nm/emission 525 nm) (Carl Zeiss Microscopy, LLC., Thornwood, New York, United States, filter set 38), and dead or apoptotic cells through a filter ideal for rhodamine (560 nm/emission 630 nm) (Carl Zeiss Microscopy, LLC, Thornwood, New York, United States, filter set 45). Digital images were captured with an AxioCam MRm Rev.3 digital camera (Carl Zeiss Microscopy LLC.), and processed with the Zeiss Zen software (Carl Zeiss Microscopy, LLC.). To facilitate quantification of the cytotoxicity, we recorded the total area occupied by living cells as measured by Zen software pixel selection algorithms, eliminating the area occupied by apoptotic cells. The total area occupied by all of the cells was normalized with respect to the area of the cells occupying the CTPSt blank wells. The results were reported as a sampling distribution of the mean (x) and the standard deviation (σ).

2.2.5 Results and Discussions

The extract method allows for the observation of the cellular cytotoxic reaction for nearly any material or biological implant, regardless of shape or physical construction. This method has been highly favored in many laboratories due to this fact, and our own experience indicated that it was perhaps the easiest of the evaluated methods to perform. Another major advantage is that the concentration-dependent effects of a material can be readily quantified, unlike direct contact, which relies on a specified size as dictated by ISO 10993-12. However, it should be taken into consideration that extracts derived from a device constructed with robust, chemically inert materials would not be expected to generate a significant particulate concentration, although this method can identify potential issues with byproducts from the materials fabrication process. As such, these materials and devices should instead be evaluated by the direct cellular contact protocols to ensure the correct evaluation of cytotoxicity. It should also be noted that ISO 10993-12 requires media agitation, but it does not elaborate on the

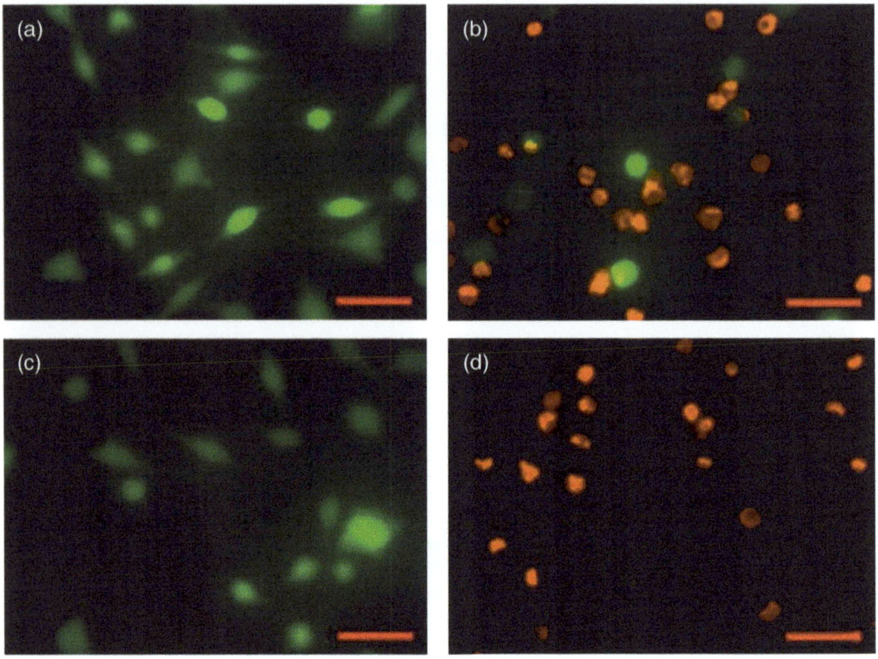

Figure 2.2 Selected fluorescent optical micrographs obtained from a Zeiss Image.M2m fluorescent microscope displaying L929 cells stained with the LIVE/DEAD assay displaying the differences between extract concentrations. (a) Fifty percent concentration of Au extract, (b) 50% concentration of Cu extract, (c) 100% concentration of Au extract, and (d) 100% concentration of the Cu extract. Live cells appear as green while dead or apoptotic cells are orange. The red scale bars represent 50 μm.

exact parameters to use during extract preparation. This factor could be a source of variation leading to discrepancies in experimental results between laboratories.

Fig. 2.2 displays selected fluorescent micrographs after the application of the LIVE/DEAD assay demonstrating the effects of the extract concentration. The top row shows a concentration level of 50, and 100% for the lower row. Note that unlike the MTT assay, the LIVE/DEAD technique allows the user to observe differences in the morphology of the cells, with the cells on the positive control assuming a very rounded morphology, which normally is an indication that cytoskeleton attachment to the well surface is very weak [55]. On the other hand, the morphology of the cells for the negative control shows an extended, oval morphology with visible extensions of lamellipodia and filopodia, which are indications of cellular attachment to the surface, and also resemble the cells in the baseline CTPSt well.

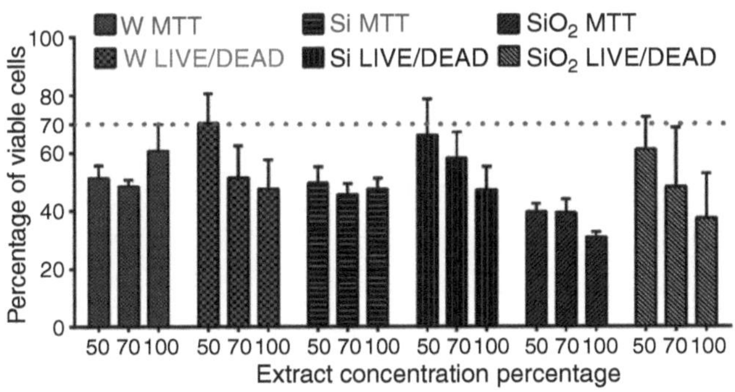

Figure 2.3 ***Histograms reporting the quantitatively measured results from the ISO 10993-5 extract methodology using either MTT or LIVE/DEAD assays.*** The testing results were normalized to the CTPSt blank wells, and cellular viability reported as the mean (c) with the error bars representing the standard deviation (σ). The dotted line at 70% represents the minimum threshold required by ISO 10993-5 for an acceptable cytotoxic reaction.

Both assays allow the quantitative evaluation of cellular viability through a relative measurement with either the baseline or negative control. Fig. 2.3 displays a histogram reporting results from the ISO 10993-5 extract method using W, Si, and SiO_2 as the materials under test. The controls for these tests produced cellular activity within ISO 10993-5 expected parameters (not shown). Two-way analysis of variance (ANOVA) performed along the sample parameter at a 95% confidence resulted in $p < 0.0001$ for both the MTT and LIVE/DEAD assays; however, the $p = 0.0005$ when evaluated versus concentration for the LIVE/DEAD assay. The majority of the tested materials were found to have relative levels of viability below 70% of the L929 cells on CTPSt using both the MTT and LIVE/DEAD assays [31]. The only exception was that a 50% concentration of W extract using the LIVE/DEAD assay was measured to have a mean (χ) value equal to 70.3% with a standard deviation (σ) of 10.3%. Bonferroni posttest analysis comparing the two methods shows that there are significant differences between the means of all three materials at a concentration of 50% ($p < 0.01$), but only W possesses a significant difference at 70% concentration ($p \leq 0.001$).

The direct contact test enables examination of not only cellular reaction to particulates or leachable substances, but also the effects of the presence of bulk material in physical or close contact with the tested cells. Fig. 2.4 demonstrates one major issue with the direct method. Materials possessing a great amount of mass, such as Au, inordinately affect the layer of semi-confluent cells upon which it was placed, as is seen in Fig. 2.4a. The cells

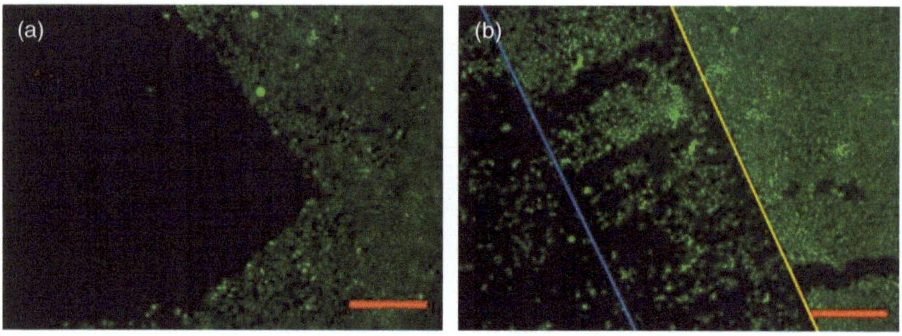

Figure 2.4 ***Fluorescent optical micrographs displaying the effects of the negative control material, Au, on L929 cells utilizing the LIVE/DEAD assay where issues with the direct contact methodology become apparent.*** (a) The digital micrograph showing a loss of cells in an area directly correlated to the location occupied by the sample. (b) A well showing similar effects that were noticed in (a); however, the sample seemingly moved from its original position. The blue line denotes a border of the sample's original position, and the yellow line marks the second boundary. The red scale bars in the lower right-hand corner represent 500 μm.

under the material were noticed to be greatly reduced in number. The loss of viability was most likely due to the weight of the sample upon the bed of cells and not cytotoxic effects, as is denoted by a quantitative comparison of the cells against the cells in the CTPSt baseline well. Additionally, Fig. 2.4b supports this observation. This micrograph shows two areas of differing cellular loss. The area on the left bounded by the blue line shows a similar reduction in cellular viability as seen in Fig. 2.4a, but in a directly adjacent area, bounded by the yellow line, has evidence of a much greater cellular viability but is marked with great variability. Although we have only circumstantial evidence, the two well-defined boundaries indicate that the material experienced unknown perturbations during its time in contact with the cellular bed and these perturbations on the material were likely the cause of damage to the cells. This conclusion was drawn from the fact that the cells in the well outside the material's position for the negative controls possessed viability similar to the baseline well.

Fig. 2.5 shows histograms of the sampling distributions of the mean and standard deviations of the mean obtained from the quantitative analysis of the direct contact L929 cytotoxicity assays. The skew due to the material perturbations was eliminated by disregarding the affected area during the comparison of both the measured sample and the baseline, thereby maintaining a 1:1 comparison. One-way ANOVA at a 95% confidence interval indicates that there are significant differences between the means of the

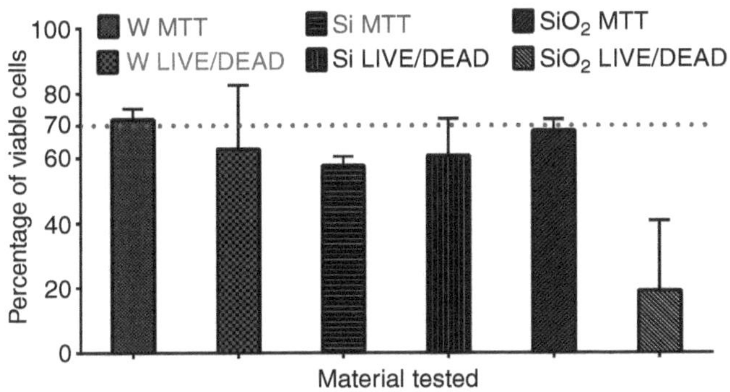

Figure 2.5 ***A histogram representing the quantitatively measured results from the ISO 10993-5 direct contact method using either the MTT or LIVE/DEAD assays.*** The testing results were normalized to the L929 cell viability on CTPSt blanks and reported as the mean (c) with the error bars representing the standard deviation (σ). The dotted line denotes the 70% minimum viability level required by the ISO standard for a material display sufficient cytotoxicity.

materials at $p < 0.0001$ for both the MTT and LIVE/DEAD assays. As was seen with the extract method, W, Si, and SiO_2 did not show the requisite viability level of 70% as required by ISO 10993-5 when examined by either MTT or the LIVE/DEAD assay, with one exception. W displays a mean of 71.9% and a standard deviation of 3.4% using the MTT assay analysis. Bonferroni posttest analysis comparing the two methods shows that there was no significant difference between the means of W and Si, but a significant ($p \leq 0.001$) difference between the means of SiO_2. It is also noted that the mean values of W and SiO_2 tested by MTT assay are larger than the same materials tested by LIVE/DEAD assay. This measurement was unexpected as the MTT assay tests the entire population of the living cells in the well, and there were notable decreases in cellular viability due to the mechanical removal of L929 cells by the material sample.

2.2.6 A Need for More Efficient Methodologies

The ISO 10993-5 standard provides multiple methodologies, which are designed to efficiently evaluate the cytotoxic effects of biomedical devices *in vitro*. Two of the methodologies listed in the standard, testing cells with an extract derived from a material and through the direct contact of the material surface with a bed of semiconfluent cells, were selected for comparison in this study. In addition to these protocols, we evaluated two different assay methods. The MTT colorimetric assay is one of the methods listed in the appendix of the ISO 10993-5, and the LIVE/DEAD assay uses chemical fluorescent stains, which are sensitive to either living or dead cells. We have also incorporated the establishment of reference materials and standards for comparisons established in ISO 10993-12. We established the use of gold and Transpore PE medical tape as negative controls, while copper and F-4040-A Tygon tubing served as positive controls. Finally, we evaluated materials which have been used in biomedical implants, tungsten, silicon, and silicon dioxide. It was our goal to determine if two of the ISO 10993-5 methodologies produced similar results and if the LIVE/DEAD assay was a suitable testing method for accurately determining cytotoxicity by comparing the results with the ISO 10993-5 appendix example, MTT assay.

We have seen that there are distinct differences between the two different assays that may affect a researcher's decision as to their utility. The MTT assay was relatively easy to perform, although it requires multiple procedures to achieve, and it produces a single, quickly acquired quantitative mean value for an entire population of cells; however, it does not facilitate the examination of cellular morphology, which is a requirement of ISO 10993-5. On the other hand, the LIVE/DEAD assay is easily prepared and administered in a single application, and allows the examination of cellular morphologies or groupings, but requires extensive postprocessing to acquire quantitative cellular counts. Also, at higher microscope magnifications, the microscope can only capture very small samplings of the overall cellular population, requiring the user to decrease the level of the magnification, which naturally affects cellular morphological resolution. Alternatively, they may invest in capturing a large amount of images to increase the sampling or invest in a motorized stage to increase the power of their statistical analysis.

The source for the variation may be related to the actual procedures followed to perform the assays. There is an actual recorded image of the amount of cells present at the time of the LIVE/DEAD assay. However, image processing techniques and errors may add some variability to the testing results. In contrast, the MTT assay protocol contains multiple experimental stages where errors could introduce variability within the test. It was noticed through light microscopy that all the cells did not metabolize the MTT into purple formazan equally, with many cells having different sized crystalline clusters emerging from their cellular membranes. Another source of error could occur when the MTT solution was removed from the well, extracting some of the purple formazan crystals in the process. We reduced this error through collection of the formazan crystals by centrifuging the

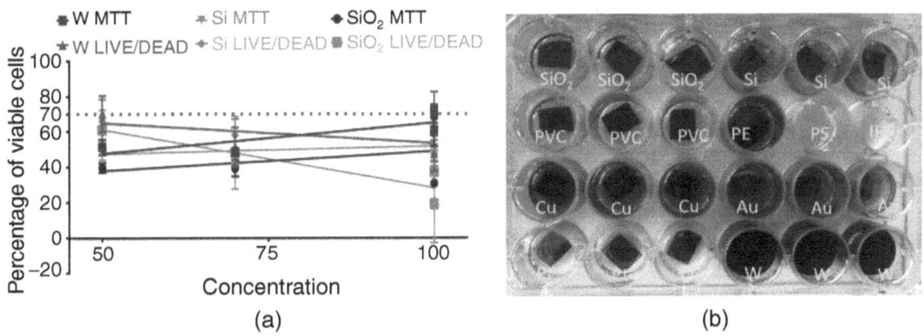

Figure 2.6 (a) A plot of both the LIVE/DEAD and MTT assay relative viabilities over the three concentrations and two ISO 10993-5 methods. The points represent the mean viabilities (χ), the error bars representing the standard deviation (σ), and the lines represent the best-fit equation between the points. The three LIVE/DEAD assays resulted in a negative correlation between concentration and viability, whereas all MTT assays show a positive correlation. (b) A digital image of various sterile metals, ceramics, and polymers tested with MTT assay. The colorimetric changes for various materials even without the presence of cellular activity may be a factor in the error in the means for the MTT and LIVE/DEAD assays.

removed MTT solution as well as allocating additional processing time to allow for the complete dissolution of the crystals. However, cellular metabolism of the MTT is a biotic variable, which is not easily controlled and may have been a source of experimental error. The largest variation we noticed was from a direct reaction initiated between the MTT assay and material by-products, which was verified by performing an MTT assay on sterile materials in the absence of any cells (Fig. 2.6b) producing purple formazan. This was especially evident when using metallic materials, like Cu and W, and led us to the conclusion that this assay method is unsuitable for metallic test samples.

The Bonferroni posttest analysis shows that there were statistically significant differences in the methods used for this work. Fig. 2.6a shows a correlation plot combining both extract and direct methodologies and the two assay methods. Although there is little statistical difference between the viability of the 100% extract and direct contact methodologies, an interesting difference is displayed between the MTT and LIVE/DEAD assays. LIVE/DEAD shows a negative correlation between the viability and concentration for all of the materials, whereas MTT displays a positive correlation. The negative reaction of cells to the presence of a material should increase with the removal of the material, therefore MTT's positive correlation indicates that another mechanism is likely leading to variability between the assays.

Fig. 2.6b shows a digital photograph of the results of the MTT assay performed on sterile control and testing materials without the presence of cells. The procedures used for this test were discussed in the methods section. The photograph shows that yellow MTT was transformed into purple formazan, or other by-products, in the presence of many of the materials. The results indicate that the presence of a material when performing the MTT assay could create false positives and decrease the validity of this cytotoxic screening assay methodology.

We have verified that, through strict adherence to the ISO 10993-5 protocols, it is possible to determine cellular cytotoxicity using either extracts or through directly contacting the material with the cells. We have shown that MTT reacts with different materials and the results can affect the overall assay leading to false positives. The LIVE/DEAD assay allows both quantification of cellular viability as well as qualification of cellular morphology, but takes longer to evaluate. We have also demonstrated that there are minor inconsistencies in the testing methods that can be improved. The extract method can only test chemical by-products and leachates from a material, while direct contact can damage the cell bed and therefore decrease the overall relative viability for that material.

ISO 10993-5 has a relatively large margin (30%) for producing a favorable result, and if the methods contained within the standard do not produce similar results, many devices and materials which are dangerous could be approved. Our group feels that both tests have many positives in their methodologies and both tests should be applied in the determination of *in vitro* cytotoxicity. In the next section, we used our experience with the ISO methods as well as our previous investigations [56] to create a hybrid cytotoxicity testing methodology, which also addresses some of the shortcomings of the ISO 10993-5 protocols.

2.3 IMPROVED ISO 10993: THE BAMBI METHOD

The Defense Advanced Research Projects Agency (DARPA) desired a screening methodology, which could efficiently and accurately screen novel materials for biocompatibility, especially concerning neural implantation devices. This desire initiated the Biocompatibility of Advanced Materials for Brain Interfaces (BAMBI) project. The ISO 10993 standard, which is composed of 21 *in vitro* and *in vivo* sections, provided accepted and understood methodologies for the testing of biomedical devices [27]. The sections provide protocols for determining the physical and chemical

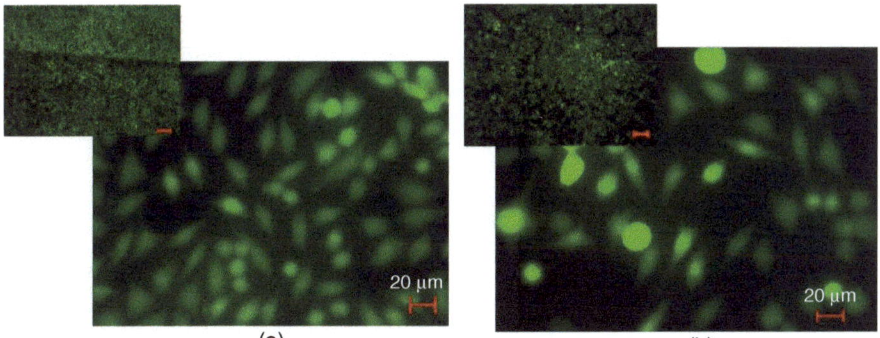

Figure 2.7 ***Fluorescent micrographs of ISO 10993-5 cytotoxicity tests for Au/Si material coupons using LIVE/DEAD assays.*** (a) Shows selected micrographs from the ISO 10993 direct material contact testing protocol, and (b) represents the material extract protocols at 100% concentration. The effects of material perturbation on the cells can be seen in (a). The size of the scale bars for the insets is 500 μm.

properties associated with the device, screening for cytotoxicity and genotoxicity effects, and the utilization of animal models to assess the inflammatory response. In Section 2.2.2, we evaluated the effectiveness of two of the ISO 10993-5 *in vitro* cytotoxicity methodologies, using the recommended L929 mouse fibroblast immortalized cell line. The extract and direct contact methods from the standard were considered as possible protocols for use in BAMBI. In the course of these investigations, we discovered multiple issues with these ISO 10993-5 standard methodologies that could significantly impact testing results. We therefore developed effective means to perform ISO10993-specified tests with complete compliance with the standard (Section 2.2.2).

The ISO 10993 5 direct contact protocol (Fig. 2.7a) evaluates the device or material by placing it in direct, intimate contact on top of a bed of semiconfluent cultured cells (defined by the ISO 10993 as 80% cell/surface coverage) [31]. This method allows the observer to monitor any cytotoxic reaction from the cells due to this intimate contact with the device or test material. In addition one can monitor the reaction of cells surrounding the device or test material to any soluble by-products. However, the major issue with the direct contact protocol was that the test was extremely sensitive to physical perturbations, especially when using dense materials, which is clearly denoted by the physical removal of cells as seen in the lower half of the image for Fig. 2.7a (also see Section 2.2.2). Additionally, materials with a density less than water would not remain in contact with the cell bed and only allow observation of cellular reactions to leachable by-products, which makes the evaluation similar to the extract protocol discussed next.

Another ISO 10993-5 method which was investigated was the extraction protocol (Fig. 2.7b), which immerses a material or device within polar and nonpolar solvents to extract possible leachable products [31]. The solvents containing extracted products are then added to complete cell media to produce an array of varying concentrations, which are then tested via addition to a semiconfluent bed of cells. This method's advantage is that it allows the observation of cellular cytotoxicity at varying levels of exposure. Through our research on this method, we realized that this only exposes cells to possible leachable and chemical by-products. The standard also complicated issues by requiring excitation during extraction generation. Excitation would inherently add kinetic energy into the extraction process, increasing the rate of solubility and chemical reactions at the surface of the tested samples. The major issue is the excitation is not defined, and therefore left to the user, which may cause issues comparing results between laboratories and with the direct method, which does not use excitation.

One popular *in vitro* assay method that has been investigated by many groups, including our own, was designed to monitor cellular affinity and cytotoxicity by culturing cells directly onto the surface of a device or material [25,36,55,57–59]. The issue with this methodology is that cellular attachment to a surface involves complex chemical interactions between extracellular proteins, the surface materials of the biomedical device, the cell's integrin receptors, and protein pathways within the cell itself [60–63]. Although cell attachment to the surface of materials may be important when considering device design functionality and for the selection of device construction materials, especially when they interact within blood or are exposed to the thrombogenic pathway, it must not be mistaken with cellular cytotoxicity. This is especially evident with materials resistant to protein attachment on the surface, such as the negative reference material polyethylene, which is mentioned within ISO 10993-12, as these materials have the possibility of failing this type of cytotoxic testing due to the inability of cells to attach to the device surface.

After our experience in performing and evaluating these three methods, we realized that it was possible to combine the best aspects from each of these methods into one overall testing regime while remaining fully compliant with the ISO 10993 standard. This would not only improve the data quality garnered from a single test, but would also allow us to address the weaknesses we reported for the ISO 10993-5 methods (Section 2.2.2). In this section, we detail a combinational methods protocol, which we call the BAMBI cytotoxicity method after the project it was created for. We detail the BAMBI method results and discuss how they compare with the traditional ISO 10993-5 extract and direct contact protocols.

2.3.1 BAMBI Methodology

The selection of positive and negative materials was conducted in strict adherence to the selection of reference materials outlined in ISO 10993-12 and detailed in Section 2.2.2. In short, we have previously identified gold (Au) and copper (Cu) as suitable negative and positive conductive control materials, respectively. Commercially obtained Transpore PE medical tape and Tygon PVC fuel and lubricant tubing formulation (F-4040-A) have demonstrated suitability as reliable negative and positive controls, respectfully, for insulating materials. All control samples were diced into 8 mm × 10 mm rectangles (0.8 cm^2 surface area), cleaned using our standard processes, and then sterilized as detailed in Section 2.2.2.

The utilization of a cell culture baseline substrate is the strategy used to facilitate measurement of the cytotoxic effects due to unknown tested materials. The baseline surface should promote cellular physiology and, according to ISO 10993-5, allow semiconfluent cell growth [31]. Inexpensive polymers, like polyethylene, polystyrene, and polycarbonate, normally possess positively charged, hydrophobic surfaces inhospitable to cellular adhesion [64]. Exposure to oxygen plasma or coronal discharge breaks the long polymer chain and inserts energetic oxygen-containing products into the surface, so that the surface of the plastic becomes hydrophilic and negatively charged [64–66].

We selected 22-mm diameter OptiClear® polycarbonate cover slides (Hampton Research, Aliso Viejo, California, United States, # HR8-082) over the typical plate well as the baseline for this study. Our reasoning for this change was to facilitate not only the observation of cells within the well, but also to concurrently observe the surface of the tested sample. Many of the observed samples were opaque, with surfaces not easily observed by inverted microscopes. Additionally, the surface of a plate well was difficult to place within the majority of our higher magnification objectives focal distance due to the objectives diameter in comparison with the well diameter. The entire surface of a cover slip could be observed at any magnification as it had no sidewalls.

Culture treatment of the polycarbonate coverslip surface was obtained using a PlasmaTherm PECVD/RIE PT-700 operating at a power of 25 W, a chamber pressure of 100 mTorr, with 300 sccm of O_2 flow for a time of 1 min. We proceeded to then verify that this surface was hospitable to cellular growth as follows. Using our developed culture techniques, L929 cells were cultivated on both bare six-well CT polystyrene culture plates and CT polycarbonate (CTPC) cover slips placed within six-well plates. Using the methods outlined in Section 2.2.4, in conjunction with the LIVE/DEAD Assay Kit for mammalian cells (Life Technologies, Grand Island, #L-3224), we found the level of viability was 99.0 ± 9.59% of the CT polystyrene ($N = 6$), and the morphology of the cells was indistinguishable between the CT polystyrene and CTPC polymer surfaces.

The BABMI cytotoxicity method was designed to address many of the issues that were apparent with the ISO 10993 protocols. The perturbation damage observed with the direct-seeding method (Section 2.2.5) was addressed through the use of adhesives that would not affect the cytotoxicity assays. Histoacryl® (Tissue Seal, Ann Arbor, Michigan, United States), an FDA-approved surgical cyanoacrylate, was evaluated to determine if the material would significantly affect the assay results [67]. Histoacryl (area ~0.8 cm^2) was placed in the center of a six-well plate and allowed to cure at 37°C for 4, 24, and 48 h. Cellular viability was 34.1 ± 3.4%, 93. 9 ± 19.5%, and 100.6 ± 21.3%, respectively. Although there was no statistical difference between 24 and 48 h, we selected a time of 48 h to ensure complete solvent evaporation and curing for future tests using this adhesive.

A 0.1-cm^2 area drop of sterile Histoacryl was placed in the center of the backside of each 8 × 10-mm sample, which was then fixed to the center of a CT (culture treated) OptiClear coverslip. Each assembled fixture was then fixed with a small amount of Histoacryl applied to the outer edge of the coverslip within a CT six-well plate (USA Scientific, Ocala, Florida, United States, #CC7682-7506). The six-well plates were roughly 9.4 cm^2 in area, which meets the 1:10 sample to well area guideline from ISO 10993-5 [58]. Each of the controls, testing materials, and baselines were situated as shown in Fig. 2.8a. The wells were then placed in a sterile environment, at either 37°C for 48 h or 67°C overnight, to ensure complete solvent evaporation and curing of the glue.

Cytotoxicity was tested by plating NCTC clone 929, strain L (American Type Culture Collection, Manassas, CCL-1) mouse fibroblastoma into each of the wells using the cell culture protocols outlined in Section 2.2.3. The amount of cells plated was not constant, and was determined in advance of each testing regimen to ensure that the baseline material, CTPC, expressed semiconfluence. After plating 3 mL of complete cell media DMEM (Life

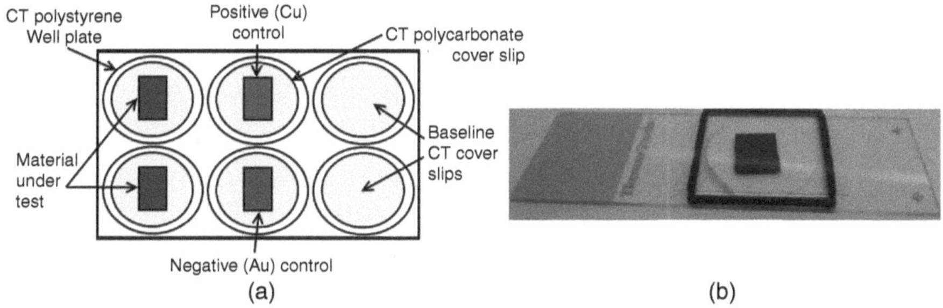

Figure 2.8 ***BAMBI protocol* in vitro *assay method.*** (a) A representation of the typical material placement within a culture treated (CT) polystyrene six-well plate for the BAMBI protocol. (b) A digital photograph of an OptiClear coverslip with mounted 8 × 10-mm sample placed within our "custom-made" live-well microscope slide. The well was filled with PBS/LIVE/DEAD assay solution and covered with a glass microscope slide to preserve the environment and protect the sample and microscope objective. The height of the 3D-printed ABS plastic well was ~550–570 μm, which allowed us to use a 50× magnification objective (Zeiss EC Epiplan 50×, 0.7 numerical aperture, working distance 1 mm) to image the surface of both the OptiClear coverslip and the material.

Technologies, Grand Island, #10313-021), 10% FBS (Life Technologies, Grand Island, #16000-044), 2.2 mM L-glutamine GlutaMAX-1 (Life Technologies, Grand Island, #35050-061), 1% antibiotic/antimycotic solution (Sigma-Alderich, St. Louis, #P4458-100ML) was added to each well. The 3-mL media volume and 0.8-cm^2 sample area are in accordance with the ISO 10993-12 surface area-to-volume concentration of 3 cm^2/mL as we could not perform agitation without disrupting the L929 surface attachment [28]. Although the required ISO 10993 extraction agitation could not be utilized with this testing method, we increased the culture time from 24 to 96 h to allow for the complete media, which has elements of a polar/nonpolar solvent, to diffuse leachable products from the testing material. It should also be noted that the ISO 10993 required agitation specifications were not clearly defined except for the test duration of 24 h [28,31]. The plates were then transferred into a water-jacketed incubator with 5% CO_2, approximately 95% relative humidity, at a temperature of 37°C. The samples were visually monitored on a daily basis to ensure that there were no issues during the incubation.

At 96 h, the six-well plate was removed from the incubator and the LIVE/DEAD cytotoxicity assay prepared. We used 5 μL of the 1 mg/mL calcien AM kit solution, which detects living cells, and combined it with 7.5 μL of 2 mM Eth-1 kit solution, which marks dead or apoptotic cells, within10 mL of pH 7.4 PBS [53,54]. The media was then removed from the well and the cells carefully rinsed with PBS to remove any trace serum, which could react with the assay. Forceps/tweezers were used to carefully dislodge the OptiClear coverslip from the base of the well plate and were placed within our BAMBI live-well microscope slide, as seen in Fig. 2.8b. The BAMBI live-well slide was used not only to study the transparent baseline CTPC material, but also to evaluate the surface of the tested material, which could be opaque in nature.

The BAMBI live-well microscope slide was constructed from a Fisherbrand™ ColorFrost™ Plus Microscope Slide (Fisher Scientific, Pittsburgh, Pennsylvania, United States, #12-550-17), which served as the base substrate. A 24-mm^2 inner diameter, ~570-μm high, 0.5-mm thick acrylonitrile butadiene styrene (ABS) well becomes the wall of the live well and was constructed using a 3D printer. The well was fixed to the slide using Loctite® Super Glue (Henkel Corporation, Westlake, Ohio, United States) and cured at 67°C for 24 h. Our LIVE/DEAD solution was added into the BAMBI live well until the sample was completely immersed. A 24 mm × 30 mm Fisherbrand™ rectangular Cover Glass (Fisher Scientific, Pittsburgh, Pennsylvania, United States #12-545B) was placed over the well to constrain the liquid solution, and the entire assembly was incubated at 37°C for 15 min.

The cells were viewed using a Zeiss Image.M2m fluorescent microscope (Carl Zeiss Microscopy, LLC.), with live cells expressing fluorescence when viewed through a filter appropriate for green fluorescent protein (470 nm/ emission 525 nm) (Carl Zeiss Microscopy, LLC., filter set 38), and dead or apoptotic cells through a filter ideal for rhodamine (560 nm/emission 630 nm) (Carl Zeiss Microscopy, LLC., filter set 45). Digital images were captured with an AxioCam MRm Rev.3 digital camera (Carl Zeiss Microscopy, LLC.).

In our past studies, we were only able to capture a few images representing randomly selected areas within each test well of the attached cells. Our Zeiss Image.M2m microscope has been fitted with a motorized stage and this stage, in conjunction with the Tiles/Positions module for the Zeiss Zen software (Carl Zeiss Microscopy, LLC.), allows us to capture the entire CTPC coverslip area by stitching multiple pictures into a contiguous single image (Fig. 2.9a). We now had the advantage of observing the majority of the cellular population within the plate well. As the material was in a separate focal plane when compared with the CTPC coverslip, it was captured separately as shown in Fig. 2.9b. One issue with tiling the entire surface of the sample and CTPC coverslip concerned the time required to image the entire surface. The Zeiss EC Plan-Neofluar 10× objective (numerical

Figure 2.9 ***Sample BAMBI protocol method data.*** Tiled optical micrographs of L929 mouse fibroblast cells stained with LIVE/DEAD solution and viewed by a Zeiss Image. M2m fluorescent microscope. (a) L929 population attached to the surface of the CTPC cover slip surrounding a Pt sample (dark rectangle in center). The Pt was not imaged, due to its different focal plane (sample outlined in red). The cellular viability was measured with the Zeiss Zen software at 95.4% of the CTPC baseline. The red bar represents 2000 μm. (b) L929 cells attached to the surface of the Pt sample (outlined in yellow) not imaged in (a). The cellular viability of this sample was 62.8% of the baseline CTPC. The bar represents 1000 μm. (c) Viable L929 fibroblastoma cytoplasm area was measured using the Zeiss Zen software for both (a) and (b) and normalized to the L929 cells on the baseline of culture-treated polycarbonate. A scatter plot of each point was created and graphed along with the mean and standard deviation for each tested material (Pt in this example). The dotted line is the 70% threshold for passing the ISO 10993 viability test as discussed previously.

aperture [NA] = 0.30, working distance [WD] = 5.2 mm) limited the capture time to 1 h, which would bring the total test collection time to nearly 8 h including test preparation and incubation times. We used a Zeiss EC Epiplan 50× objective (NA = 0.7, WD = 1.1) at random areas across the CTPC and sample surfaces for higher magnification evaluation of cellular morphological reactions to the tested material's presence.

The microscope images were processed using the Zeiss Zen software. Although the Zen software's Image Analysis module contains cell-counting software, the close knit groups of cells were difficult to automatically separate into individual objects, therefore we devised another method to quantitatively measure the cell's viability. We recorded the total area from the calcien-stained cytoplasm and the ethidium bromide-stained nucleus. The total area measured for the samples would then be normalized with respect to the area measured from the cells on the CTPC cover slips and expressed as a percentage. The resulting percentage allowed the direct comparison of the tested material samples obtained between the different trial dates. The results were then reported as a sampling distribution of the mean (x) and the standard deviation (σ).

2.3.2 BAMBI Method Results

We have created a single method for testing a material's cytotoxicity *in vitro* through the combination of methods specified in the ISO 10993-5 standard. This single method, the BAMBI cytotoxicity method, named after the DARPA program in which it was used, specifically used ISO 10993-5 direct contact and extract protocols. The design was also used to eliminate perturbation damage observed in the direct contact method by using an FDA-approved cyanoacrylic glue, Histoacryl, which did not statistically alter the viability results. The BAMBI method has an additional benefit of allowing the observation of cellular activity not only within the well of a plate, but also to the surface of the tested device or material through the use of a live cell culture well in a reflective microscope. We compared the BAMBI method with L929 mouse fibroblastoma against our ISO 10993-12 established reference materials, Cu, Au, F-4040-A, and PE, and also evaluated three new materials, platinum (Pt), parylene C, and parylene C/Al_2O_3 using this method. The BAMBI method was quantified using the fluorescent tagging system of the LIVE/DEAD assay and the measurements for the tested materials were normalized using measured L929 viability on baseline culture-treated polycarbonate coverslips.

In Fig. 2.10, confidence interval equivalence statistics were used to evaluate the control materials using the BAMBI method compared to both of the ISO 10993-5 direct and extract methods. It was determined that all three methods showed equivalence, rejecting the null hypothesis, for the positive control materials of Cu and F-4040-A. However, the results in Fig. 2.10 show that only the BAMBI and direct contact methods show equivalence when the material tested is Au, while the BAMBI and direct comparison for PE, as well as all comparisons involving the ISO 10993-5 extract method, accepted the null hypothesis and had no direct equivalence. We have proposed that the lack of equivalence is that the BAMBI method was constructed as a conjunction of the direct and extract ISO 10993-5 standard methods, both of which do not show equivalence with each other, and therefore the combinational method will not have equivalence either.

The new BAMBI method, while not equivalent to both the ISO 10993-5 methods for the negative control materials, does improve on some of the issues reported with the standard. It eliminated damage to the cell surface through the Histoacryl glue and allowed extended observation of the cells. The observations of the cellular reactions to the surface of a material can also help develop an overall picture for screening a new novel biomedical device or material. In short, through the use of simulation and careful

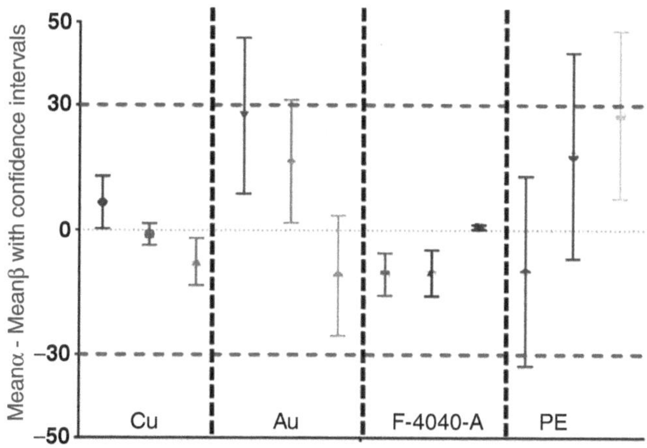

Figure 2.10 ***Confidence interval equivalence tests calculated from the results of dual, single-sided T-tests at 0.1 power on the difference between mean viability levels measured for our control materials copper, gold, Tygon fuel tubing (F-4040-A), and polyethylene.*** ISO 10993-5 direct versus extract (D vs. E) confidence interval equivalence results denoted by the the first error bar shown for each material, BABMI versus ISO 10993-5 extract with the second error bar, and BAMBI versus ISO 10993-5 direct with the third error bar. The dotted lines represent the estimation of equivalence at ±30%, which was determined from the range of ISO 10993-5 acceptable viability level. Based on this statistical information, Cu and F-4050, the two positive controls, display equivalence between all methods, while Au and PE show material and method equivalence issues.

evaluation of agitation, this method combines the ISO 10993-5 methods into one test, reducing screening time for biomedical devices or materials.

2.4 3C–SiC *IN VITRO* EVALUATION

2.4.1 The Advantages of 3C–SiC for Biomedical Devices

Silicon carbide is a compound semiconductor, which has been receiving interest due to its excellent physical properties, radiation resistance, near chemical inertness, and a large bandgap [68]. This material, unlike pure silicon, has many crystalline stacking sequences, denoted as polytypes, which have slightly different physical, electrical, and optical properties. The best-known hexagonal polytypes, 4H–SiC and 6H–SiC, have been used in commercial power devices and as substrates for gallium nitride devices [68]. One of the polytypes, 3C–SiC, has not been normally considered for power applications due to its lower bandgap (2.3 eV), but it has found applications in the fabrication of MEMS devices, as it can be grown with less expensive heteroepitaxial chemical vapor deposition (CVD) processes on silicon substrates. Our epitaxial growth process on 100 mm Si substrates has previously reported using a hot-wall CVD reactor with silane and ethylene precursors [69]. 3C–SiC shares the near chemical inertness and physical robustness seen with the more developed hexagonal polytypes, and many biomedical device applications do not require large power requirements. Also, unlike metals and insulators, 3C–SiC is a semiconductor with modifiable electrical conductivity, enabling us to create complex electrical interactions with the physiological environment.

One way to construct a "smart" device is to construct it using electrical components that control current flow. Semiconductor materials are the backbone of these types of electronics. One of the major components of some of these devices is silicon, which has been the leading electrical material in electronics for many years. Unfortunately, silicon is extremely chemically reactive with many constituents of the physiological environment and does not possess a high level of biocompatibility [70–73]. On the other hand, SiC has shown excellent physical robustness and chemical inertness [68,74]. 3C–SiC has shown excellent *in vitro* biocompatibility with fibroblasts and skin cells [2], neural cell lines [25,75], and has even shown evidence of hemocompatibility [74]. We wished to evaluate 3C–SiC as an ideal material for use in biomedical devices, and, in this section, we detail the *in vitro* evaluation of 3C–SiC using our BAMBI methodology.

2.4.2 Materials and Methods

The development of our 50-mm process for the growth of 3C–SiC has been described in detail through many publications [76–78]. Our new process was reported in Ref. [69]. In short, the CVD reactor was scaled to accommodate 100-mm silicon substrates, and our gas chemistry includes the carrier hydrogen, H_2, with ethylene, C_2H_4, and silane, SiH_4, precursor gases. The deposition process uses a single thermal ramp to the growth temperature with a ramp rate of ~28°C/min under hydrogen carrier gas flow. At a temperature of 1100°C, C_2H_4 is introduced into the gas stream to facilitate surface carbonization, which creates the initial seed layer of SiC, and continues throughout the remaining thermal ramp until the substrate reaches the growth temperature of 1370°C. At 1370°C, SiH_4 is introduced into the gas mixture and the C_2H_4 precursor flow is scaled to achieve a C/Si ratio of 1:2 during the heteroepitaxial growth phase. The process pressure was 375 Torr during the entire growth schedule. The resulting growth speed is approximately 5 μm/h.

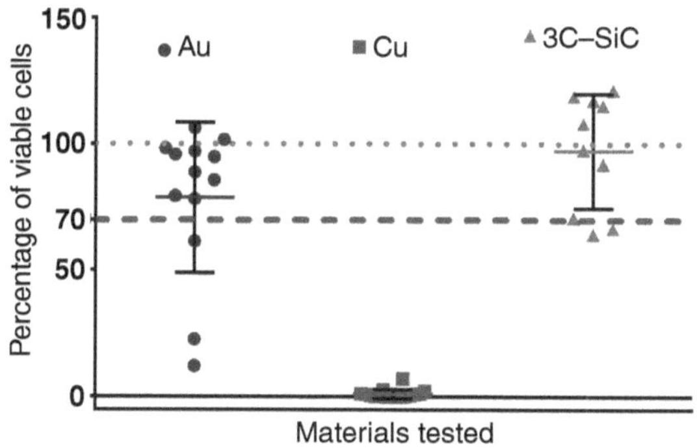

Figure 2.11 ***A scatter plot representing the 3C–SiC cytotoxicity test as evaluated by the BAMBI method.*** The mean ($\bar{x}$) and standard deviation (s) is displayed for the control materials, Cu and Au, along with the tested material, 3C–SiC. These values were normalized against the cells attached to the baseline well, CTPC. The dashed line represents the ISO 10993-5 limitation for acceptable cytotoxicity and the dotted line denotes the mean value for CTPC. As expected Au passes the test indicating that these are valid data. 3C–SiC also passed the test, thus confirming the biocompatible nature of this material as was first reviewed in [74]. Sample raw data shown in Fig. 2.12 for reference.

After growth, a wafer of 3C–SiC was cut into 8 × 10-mm coupons using a dicing saw equipped with a diamond-coated blade to facilitate their use in the BAMBI method. Due to the conductive nature of the material, we chose Au and Cu as our reaction controls. The cleaning methodology in Section 2.2.2 was followed for the control materials as well as 3C–SiC with one exception. 3C–SiC received an additional cleaning step using a piranha solution at a 3:1 sulfuric acid (H_2SO_4) and hydrogen peroxide (H_2O_2) concentration for 30 min to remove any possible carbon surface contaminants. This was followed by a 1:1 buffered hydrochloric acid and water solution for 5 min to remove any metal contamination as well as any sulfuric residues from the piranha clean. All samples were rinsed three times in deionized water and nitrogen dried. The last stage was to sterilize the samples in our dry sterilizer as described in Section 2.2.2.

The samples were then fixed to sterile polycarbonate slides and polystyrene plate wells to prepare them for the BAMBI method protocols detailed in Section 2.3.1. Evaluation was performed with a Zeiss M2M fluorescent microscope as indicated in the same section.

2.4.3 The BAMBI Method Cytotoxicity Testing Evaluation of 3C–SiC

Fig. 2.11 displays the analysis for 3C–SiC as tested using our BAMBI method. 3C–SiC displays a mean viability level, which was 97.4% of the baseline

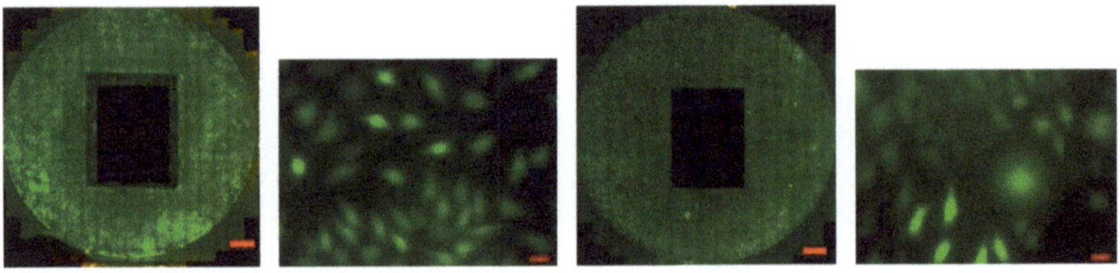

(a) (b) (c) (d)

Figure 2.12 ***Optical micrographs of L929 mouse fibroblast cells stained with LIVE/DEAD solution and viewed by a Zeiss Imager.M2m fluorescent microscope.*** (a) The tiled micrographs for the L929 population attached to the surface of the CTPC cover slip surrounding a 3C–SiC sample (dark rectangle in center). (b) A 50× magnified micrograph obtained from a random area in the well surrounding the 3C–SiC. Cells are elongated and have multiple lamellipodia/filopodia extensions, which are indications of excellent surface attachment. (c) A stitched micrograph of the L929 fibroblastoma attached to the baseline well from the same test for (a). (d) A 50× micrograph of the L929 fibroblastoma in the CTPC baseline well. The cells have a similar morphology with the cells in (b). Scale = 2000 μm in (a) and (c), 50 μm in (b) and (d), respectively.

CTPC, which is well above the value required for passing the cytotoxicity test as recommended by ISO 10993-5. The Au and Cu control materials resulted in a negative and positive cytotoxic cellular reaction as expected and seen in the previous testing protocols.

Fig. 2.12 displays selected stitched micrographs for the entire CTPC cover slips used in the BAMBI method test for 3C–SiC (Fig. 2.12a) and the baseline well (Fig. 2.12c). The wells display the semiconfluence required by ISO 10993-5 and show no obvious differences in cellular grouping. At 50× magnification, we are able to closely compare the morphology of individual cells against each other. The cells in the well alongside the 3C–SiC samples (Fig. 2.12b) possessed a similar morphology to the cells in the baseline CTPC (Fig. 2.12d), further verifying that this material may be compatible for use with biological processes.

2.5 SUMMARY AND THE FUTURE OF 3C–SIC BIOMEDICAL TESTING

3C–SiC is a material well known for its physical robustness, near chemical inertness, and excellent electrical and optical properties. This material could be a major step forward in the creation of biomedical devices, especially concerning long-term implantable devices. To date, our group has investigated this material, starting with skin and connective cell lines [1,2], glia and neurons [25], and static hemocompatibility [56]; however, we had not completely followed the ISO 10993-5 protocols to the letter. We have now used our BAMBI method, which was developed in strict accordance with the ISO cytotoxicity testing protocols. Not surprisingly, the tests proved that 3C–SiC may be a material that could exist within the physiological environment without eliciting undo cytotoxic effects.

3C–SiC displayed a mean value near that of culture-treated polycarbonate plastic. Culture-treated plastics are the gold standard in cellular culture and for many cell lines produce excellent cellular adhesion and proliferation. It should be noted that the 3C–SiC samples, unlike the polycarbonate, were not specially treated in any way. In conclusion, once again, 3C–SiC has shown that it does not initiate cytotoxic reactions with cells *in vitro* and should be considered for further biological testing.

ACKNOWLEDGMENTS

This work was sponsored by the Defense Advanced Research Projects Agency (DARPA) Microsystems Technology Office under the auspices of Dr Jack Judy through the Space and Naval Warfare Systems Center, Pacific Grant/Contract no. N66001-12-1–4026—Biocompatibility of Advanced Materials for Brain Machine Interfaces (BAMBI). We wish to thank Edwin J. Weeber, PhD, for the use of his laboratory facilities during the development stage of the ISO 10993 protocol procedures. We would also like to thank Justin King, Luis Maruisso, Courtney Penn, Jennifer Volger, and Kevin Kohler for their assistance in the development for the LIVE/DEAD cytotoxicity protocols and development of the Zeiss fluorescence microscopy image analysis protocols. The authors wish to thank Dr Christopher Locke and Dr Meralys Reyes for growing the 3C–SiC used in this study.

REFERENCES

[1] Coletti C, Jaroszeski MJ, Hoff AM, Saddow SE. Culture of mammalian cells on single crystal SiC substrates. MRS Proceedings, vol. 950. Cambridge University Press; 2006. p. 0950-D04.

[2] Coletti C, Jaroszeski MJ, Pallaoro A, Hoff AM, Iannotta S, Saddow SE. Biocompatibility and wettability of crystalline SiC and Si surfaces. Conf Proc IEEE Eng Med Biol Soc 2007;2007:5850–3.

[3] Goodacre CJ, Bernal G, Rungcharassaeng K, Kan JY. Clinical complications with implants and implant prostheses. J Prosthet Dent 2003;90:121–32.

[4] Richards-Grayson C, Shawgo RS, Johnson AM. A bioMEMS review: MEMS technology for physiologically integrated devices. Proc IEEE 2004;92:6–21.

[5] James T, Mannoor MS, Ivanov DV. BioMEMS – advancing the frontiers of medicine. Sensors 2008;8:6077–107.

[6] Veiseh O, Doloff JC, Ma M, Vegas AJ, Tam HH, Bader AR, et al. Size- and shape-dependent foreign body immune response to materials implanted in rodents and non-human primates. Nat Mater 2015;14:643–51.

[7] Khan W, Muntimadugu E, Jaffe M, Domb AJ. Implantable medical devices. In: Domb AJ, Khan W, editors. Focal controlled drug delivery: advances in delivery science and technology. New York: Springer; 2014. p. 33–59.

[8] Long M, Rack HJ. Titanium alloys in total joint replacement – a materials science perspective. Biomaterials 1998;19:1621–39.
[9] Hashiguchi K, Hashimoto K. Mechanical and histological investigations on pressureless sintered SiC dental implants. Okajimas Folia Anat Jpn 1999;75:281–96.
[10] Weaver FM, Follett K, Stern M, Hur K, Harris C, Marks WJ Jr, et al. Bilateral deep brain stimulation vs best medical therapy for patients with advanced Parkinson disease: a randomized controlled trial. J Am Med Assoc 2009;301:63–73.
[11] Denys D, Mantione M, Figee M, van den Munckhof P, Koerselman F, Westenberg H, et al. Deep brain stimulation of the nucleus accumbens for treatment-refractory obsessive-compulsive disorder. Arch Gen Psychiatry 2010;67:1061–8.
[12] Redgrave P, Rodriguez M, Smith Y, Rodriguez-Oroz MC, Lehericy S, Bergman H, et al. Goal-directed and habitual control in the basal ganglia: implications for Parkinson's disease. Nat Rev Neurosci 2010;11:760–72.
[13] Wessberg J, Stambaugh CR, Kralik JD, Beck PD, Laubach M, Chapin JK, et al. Real-time prediction of hand trajectory by ensembles of cortical neurons in primates. Nature 2000;408:361–5.
[14] Lebedev MA, Nicolelis MAL. Brain-machine interfaces: past, present and future. Trends Neurosci 2006;29:536–46.
[15] Velliste M, Perel S, Spalding MC, Whitford AS, Schwartz AB. Cortical control of a prosthetic arm for self-feeding. Nature 2008;453:1098–101.
[16] Nicolelis MAL, Lebedev MA. Principles of neural ensemble physiology underlying the operation of brain–machine interfaces. Nature 2009;10:530–40.
[17] Williams DF. On the mechanisms of biocompatibility. Biomaterials 2008;29:2941–53.
[18] Stensaas SS, Stensaas LJ. Reaction of cerebral-cortex to chronically implanted plastic needles. Acta Neuropathol 1976;35:187–203.
[19] Rousche PJ, Normann RA. Chronic recording capability of the Utah Intracortical Electrode Array in cat sensory cortex. J Neurosci Methods 1998;82:1–15.
[20] Polikov VS, Tresco PA, Reichert WM. Response of brain tissue to chronically implanted neural electrodes. J Neurosci Methods 2005;148:18.
[21] Prasad A, Xue QS, Sankar V, Nishida T, Shaw G, Streit WJ, et al. Comprehensive characterization and failure modes of tungsten microwire arrays in chronic neural implants. J Neural Eng 2012;9:056015.
[22] Woolley AJ, Desai HA, Otto KJ. Chronic intracortical microelectrode arrays induce non-uniform, depth-related tissue responses. J Neural Eng 2013;10:026007. 11 pp..
[23] McConnell GC, Butera RJ, Bellamkonda RV. Bioimpedance modeling to monitor astrocytic response to chronically implanted electrodes. J Neural Eng 2009;6:055005.
[24] Winslow BD, Christensen MB, Yang WK, Solzbacher F, Tresco PA. A comparison of the tissue response to chronically implanted Parylene-C-coated and uncoated planar silicon microelectrode arrays in rat cortex. Biomaterials 2010;31(35):9163–72.
[25] Frewin CL, Jaroszeski M, Weeber E, Muffly KE, Kumar A, Peters M, et al. Atomic force microscopy analysis of central nervous system cell morphology on silicon carbide and diamond substrates. J Mol Recognit 2009;22:380–8.
[26] Charkhkar H, Frewin C, Nezafati M, Knaack GL, Peixoto N, Saddow SE, et al. Use of cortical neuronal networks for *in vitro* material biocompatibility testing. Biosens Bioelectron 2014;53:316–23.
[27] ISO, 10993-1. Biological evaluation of medical devices. Part 1: evaluation and testing, ed. Geneva, Switzerland: International Organization for Standardization; 2009.
[28] ISO, 10993-12. Biological evaluation of medical devices. Part 12: sample preparation and reference materials, ed. Geneva, Switzerland: International Organization for Standardization; 2009.
[29] ISO, 10993-14. Biological evaluation of medical devices. Part 14: identification and quantification of degradation products from ceramics, ed. Geneva, Switzerland: International Organization for Standardization; 2001.

[30] ISO, 10993-15. Biological evaluation of medical devices. Part 15: identification and quantification of degradation products from metals and alloys, ed. Geneva, Switzerland: International Organization for Standardization; 2000.
[31] ISO, 10993-5. Biological evaluation of medical devices. Part 5: tests for *in vitro* cytotoxicity, ed. Geneva, Switzerland: International Organization for Standardization; 2009.
[32] ISO, 10993-6. Biological evaluation of medical devices. Part 6: tests for local effects after implantation, ed. Geneva, Switzerland: International Organization for Standardization; 2009.
[33] Seyfert UT, Biehl V, Schenk J. *In vitro* hemocompatibility testing of biomaterials according to the ISO 10993-4. Biomol Eng 2002;19:91–6.
[34] ISO, 10993-4. Biological evaluation of medical devices. Part 4: selection of tests for interactions with blood, ed. Geneva, Switzerland: International Organization for Standardization; 2002.
[35] Rubehn B, Stieglitz T. *In vitro* evaluation of the long-term stability of polyimide as a material for neural implants. Biomaterials 2010;31:3449–58.
[36] Green RA, Ordonez JS, Schuettler M, Poole-Warren LA, Lovell NH, Suaning GJ. Cytotoxicity of implantable microelectrode arrays produced by laser micromachining. Biomaterials 2010;31:886–93.
[37] Koschwanez HE, Reichert WM. *In vitro*, *in vivo* and post explantation testing of glucose-detecting biosensors: current methods and recommendations. Biomaterials 2007;28:3687–703.
[38] Kotzar G, Freas M, Abel P, Fleischman A, Roy S, Zorman C, et al. Evaluation of MEMS materials of construction for implantable medical devices. Biomaterials 2002;23:2737–50.
[39] Meriç G, Dahl JE, Ruyter IE. Cytotoxicity of silica–glass fiber reinforced composites. Dental Mater 2008;24:1201–6.
[40] Zhang S, Zhang X, Zhao C, Li J, Song Y, Xie C, et al. Research on an Mg–Zn alloy as a degradable biomaterial. Acta Biomater 2010;6:626–40.
[41] Wise KD, Angell JB, Starr A. An integrated-circuit approach to extracellular microelectrodes. IEEE Trans Biomed Eng 1970;17:238–47.
[42] Kuperstein M, Whittington DA. A practical 24 channel microelectrode for neural recording *in vivo*. IEEE Trans Biomed Eng 1981;28:288–93.
[43] Normann RA, Campbell PK, Jones KE. Three-dimensional electrode device, USA Patent 5,215,088; 1993.
[44] Edell DJ, Toi VV, McNeil VM, Clark LD. Factors influencing the biocompatibility of insertable silicon microshafts in cerebral cortex. IEEE Trans Biomed Eng 1992;39:635–43.
[45] Nordhausen CT, Rousche PJ, Normann RA. Optimizing recording capabilities of the Utah intracortical electrode array. Brain Res 1994;637:27–36.
[46] Wise KD, Sodagar AM, Yao Y, Gulari MN, Perlin GE, Najafi K. Microelectrodes, microelectronics, and implantable neural microsystems. Proc IEEE 2008;96:1184–202.
[47] Hubel DH. Tungsten microelectrode for recording from single units. Science 1957;125:549–50.
[48] Sankar V, Patrick E, Dieme R, Sanchez JC, Prasad A, Nishida T. Electrode impedance analysis of chronic tungsten microwire neural implants: understanding abiotic vs. biotic contributions. Front Neuroeng 2014;7:1–12.
[49] 3M™ Transpore™ White Surgical Tape – 1534, 3M 21-5837-6; 2012.
[50] Scheideler L, Füger C, Schille C, Rupp F, Wendel HP, Hort N, et al. Comparison of different *in vitro* tests for biocompatibility screening of Mg alloys. Acta Biomater 2013;9:8740–5.
[51] Baek HS, Yoo JY, Rah DK, Han DW, Lee DH, Kwon OH, et al. Evaluation of the extraction method for the cytotoxicity testing of latex gloves. Yonsei Med J 2005;46: 579–83.

[52] Mosmann T. Rapid colorimetric assay for cellular growth and survival: application to proliferation and cytotoxicity assays. J Immunol Methods 1983;65:55–63.
[53] Haugland RP, MacCoubrey, IC, Moore PL. Dual-fluorescence cell viability assay using ethidium homodimer and calcein AM, ed: Google Patents; 1994.
[54] Slepian MJ, Massia SP, Whitesell L. Pre-conditioning of smooth muscle cells via induction of the heat shock response limits proliferation following mechanical injury. Biochem Biophys Res Comm 1996;225:600–7.
[55] Kumari TV, Vasudev U, Kumar A, Menon B. Cell surface interactions in the study of biocompatibility. Trend Biomater Artif Organs 2002;15:37–41.
[56] Saddow SE, Coletti C, Frewin CL, Schettini N, Oliveros A, Jaroszeski M. Single-crystal silicon carbide: a biocompatible and hemocompatible semiconductor for advanced biomedical applications. Mater Sci Forum 2011;679–680:824–30.
[57] Coletti C. Silicon carbide biocompatibility, surface control and electronic cellular interaction for biosensing applications, PhD, Electrical Engineering, University of South Florida, Tampa, FL, USA; 2007.
[58] Hsu SH, Liu BS, Lin WH, Chiang HC, Huang SC, Cheng SS. Characterization and biocompatibility of a titanium dental implant with a laser irradiated and dual-acid etched surface. Biomed Mater Eng 2007;17:53–68.
[59] Anthony HDG, Bowen CR, Taylor J, Robbins J. Neuronal cell biocompatibility and adhesion to modified CMOS electrodes. Biomed Microdevices 2009;11:1091–101.
[60] Braun D, Fromherz P. Fluorescence interferometry of neuronal cell adhesion on microstructured silicon. Phys Rev Lett 1998;81:4.
[61] Frame MC, Fincham VJ, Carragher NO, Wyke JA. v-Src's hold over actin and cell adhesions. Nat Rev Mol Cell Biol 2002;3:233–45.
[62] Zeck G, Fromherz P. Repulsion and attraction by extracellular matrix protein in cell adhesion studied with nerve cells and lipid vesicles on silicon chips. Langmuir 2003;19:1580–5.
[63] Low SP, Williams KA, Canham LT, Voelcker NH. Evaluation of mammalian cell adhesion on surface-modified porous silicon. Biomaterials 2006;27:4538–46.
[64] van Kooten TG, Spijker HT, Busscher HJ. Plasma-treated polystyrene surfaces: model surfaces for studying cell-biomaterial interactions. Biomaterials 2004;25:1735–47.
[65] Underwood PA, Steele JG, Dalton BA. Effects of polystyrene surface chemistry on the biological activity of solid phase fibronectin and vitronectin, analyzed with monoclonal antibodies. J Cell Sci 1993;104:793–803.
[66] Shen M, Horbett TA. The effects of surface chemistry and adsorbed proteins on monocyte/macrophage adhesion to chemically modified polystyrene surfaces. J Biomed Mater Res 2001;57:336–45.
[67] Frewin CL, Locke C, Mariusso L, Weeber EJ, Saddow SE. Silicon carbide neural implants: *in vivo* neural tissue reaction. Sixth international IEEE/EMBS conference on neural engineering (NER); 2013. p. 661–664.
[68] Kordina O, Saddow SE. Silicon carbide overview. In: Saddow SE, Agarwal A, editors. Advances in silicon carbide processing and applications. 1st ed. Boston, MA: Artech House, Inc.; 2004. p. 2–3, 7–8, 18.
[69] Reyes M, Frewin C, Ward PJ, Saddow SE. 3C-SiC on Si hetero-epitaxial growth for electronic and biomedical applications. ECS Trans 2013;58:119–26.
[70] Carlisle EM. Silicon as an essential trace element in animal nutrition. Ciba Found Symp 1986;121:123–39.
[71] Birshall JD, Espie AW. Biological implications of the interaction (via silanol groups) of silicon with metal-ions. In: Evered D, O'Connor M, editors. Silicon biochemistry, vol. 121. Chichester, UK: John Wiley & Sons; 1986. p. 140–59.
[72] Hench LL, Wilson J, Biocompatibility of silicates for medical use. In: Evered D, O'Connor M, editors. Ciba foundation symposium – silicon biochemistry, vol. 121. Chichester, UK: John Wiley & Sons; 1986. p. 231–253.

[73] O'Neil C, Jordan P, Bhatt T, Newman R. Silica and oesophageal cancer. In: Evered D, O'Connor M, editors. Ciba foundation symposium – silicon biochemistry, vol. 121. Chichester, UK: John Wiley & Sons; 1986. p. 214–230.
[74] Saddow SE, editor. Silicon carbide biotechnology: a biocompatible semiconductor for advanced biomedical devices and applications. Amsterdam: Elsevier; 2011.
[75] Frewin CL. The neuron-silicon carbide interface: biocompatibility study and BMI device development, PhD, Electrical Engineering, University of South Florida, Tampa, FL, USA; 2009.
[76] Myers RL, Shishkin Y, Kordina O, Saddow SE. High growth rates (>30 mu m/h) of 4H-SiC epitaxial layers using a horizontal hot-wall CVD reactor. J Crystal Growth 2005;285:486–90.
[77] Harvey S, Reyes M, Shishkin Y, Saddow SE. High growth rate of single crystal 3C-SiC on 2 inch Si (001) wafers. In: Electronic Materials Conference (EMC), Penn State University, Pennsylvania, USA; 2006.
[78] Locke C, Kravchenko G, Waters P, Reddy JD, Du K, Volinsky AA, et al. 3C-SiC films on Si for MEMS applications: mechanical properties. Mater Sci Forum 2009; 615–617:633–6.

CHAPTER 3

Study of the Hemocompatibility of 3C–SiC and *a*-SiC Films Using ISO 10993-4

Maysam Nezafati*, Christopher L. Frewin and Stephen E. Saddow***
*Department of Electrical Engineering, University of South Florida, Tampa, FL, United States of America
**The Department of Biomedical Engineering, Erik Johnsson School of Engineering and Computer Science, University of Texas, Dallas, TX, United States of America

Contents

3.1 INTRODUCTION

As a result of the evolution in health science, the living condition for human beings has become longer and of higher quality. The discovery of new medicines, the invention of biomedical devices, the widespread use of prosthetic implants, as well as new cures and treatment methods, are examples of this progress. Many diseases and traumas that were not curable or correctable in the past can now be treated through applications of pharmaceuticals or the application of biomedical devices. In reference to the latter therapeutic pathway, Williams et al. asserted that biomaterials, or the materials used in the construction of a biomedical device, which are in close and persistent contact with the human body, should not cause deleterious harm to the physiological environment [1]. This is an extremely important point in that biomedical devices, and by extension the materials from which they are composed, must not only be highly compatible with the human body, but must maintain this compatibility for the lifetime of the treatment. For highly complex prosthetic devices and biosensors, the device contact time can be the equivalent of the lifetime of the patient. In realistic terms, this is normally considered to be 15–20 years as a minimum, but the real goal is to have this time limit approach 75 years.

Modern biomedical devices offer a wide range of therapeutic solutions to improve the life of humans suffering from disease or trauma. Examples of this broad range of treatments include diabetic management, dental/bone prosthetic implants, and, more recently, robotic prosthetic replacements, which enable the restoration of physical motility after the loss of a limb or brain trauma. Although these devices function very differently, they share one thing in common: they must function reliably inside the human body, that is, *in vivo*, for long periods of time. Therefore, implant and prosthetic device materials must possess certain critical characteristics, such as mechanical strength, surface hardness, wear resistance, chemical stability, corrosion resistance, biocompatibility, and hemocompatibility. In chapter: Cytotoxicity of 3C–SiC Investigated Through Strict Adherence to ISO 10993, the *in vitro* cytotoxicity of single crystalline cubic silicon carbide, 3C–SiC, in full compliance with ISO 10993, was examined and presented. This chapter focuses on expanding that evaluation to include hemocompatibility in accordance with ISO 10933-4 [2]. This chapter will also introduce the noncrystalline form of SiC, amorphous silicon carbide, denoted as *a*-SiC, which can be deposited at much lower temperatures, and therefore may be used as a coating in many more applications than those of 3C–SiC.

We begin this chapter using the BAMBI method of cytotoxic comparison, which was detailed in chapter: Cytotoxicity of 3C–SiC Investigated Through Strict Adherence to ISO 10993, to evaluate *a*-SiC. The next section focuses on the ISO 10993-4 hemocompatibility study of both of these SiC materials using static and dynamic methodologies. The chapter concludes with a discussion of the results and the future works necessary to fully characterize the hemocompatibility of SiC.

3.2 *IN VITRO* BIOMEDICAL TESTING METHODS FOR CYTOTOXICITY

Silicon carbide (SiC), unlike pure group IV–IV semiconductors like silicon (Si) and diamond/graphene, is interesting in that this material can stack with different crystalline orientations and sequences, which in turn produces many different "flavors" called polytypes. Just like different flavors of ice cream, SiC polytypes share many innate similarities, such as chemical resistance, physical resilience, and modifiable conductivity. They differ in some very minor, but important, ways in that they differ in bandgap energy, optical absorption and transmission, fracture resistance, as well as other fundamental properties [3,4]. As with everything in nature, there are advantages and disadvantages to this overall diversity. These properties allow SiC to be tailored to specific applications but, on the other hand, these polytypes can mix during growth and create defects in the otherwise pure crystal.

In the last chapter, we expounded on one of the polytypes of SiC, the single cubic form, 3C–SiC. 3C–SiC has often been the neglected polytype of SiC in that it has a lower bandgap than the two major hexagonal polytypes, 4H–SiC and 6H–SiC. However, this form of SiC offers some advantages in that it can be grown heteroepitaxially on Si substrates, which enables lower cost device processing and electronics options not possible with the hexagonal polytypes [3,4]. Unfortunately, the single crystal semiconductor quality form of SiC requires a large amount of activation energy to create the network of strong, covalent bonds [5]. This activation energy usually requires temperatures well above 800°C [6]. Many materials are not able to withstand this temperature, making SiC a difficult material to incorporate into many types of devices.

Incorporation of SiC with other materials requires different techniques than what is used in the Si industry, for example. Film deposition can still be used, but it requires much lower temperatures that are only possible with the amorphous form of SiC (*a*-SiC). Like the crystalline forms of SiC, *a*-SiC is extremely chemically resilient and physically robust but, unlike the semiconducting crystalline forms, it is an inherent insulator, which possesses a high dielectric constant, *k* [7]. Deposition temperatures have been reported to be as low as 100°C through the use of plasma-enhanced chemical vapor deposition (PECVD), enabling this material to be deposited on higher temperature polymers [8].

At these lower temperatures, *a*-SiC has been deposited on myriad polymers and metals and has been proposed for use as a chemically resistant coating for many types of biomedical devices. The American Food and Drug Administration has seen that *a*-SiC could possibly have an advantage as a hemocompatible surface, and has approved Rithron-XR™ cardiac stents, produced by the German firm Biotronik GmbH, into commercial clinical use [9–11]. The hemocompatibility of *a*-SiC was not only also tested in artificial heart valves [12], but it has also been examined as a biocompatible material for use as a wear-resistant coating on titanium for bone replacements [13], and as a protective dielectric coating on cortical electrode polymer insulation, which is discussed in more detail in chapter: Amorphous Silicon Carbide for Neural Interface Applications [14,15]. However, there is little quantitative information reported on the biocompatibility of *a*-SiC, as most of the focus has been on its hemocompatibility. In this section, we report the cytotoxicity of nondoped *a*-SiC films deposited on various semiconductor material surfaces with the question "will *a*-SiC effectively protect and enhance the capability of materials like silicon and silicon dioxide, which have been reported as cytotoxic materials?" (see chapter: Cytotoxicity of 3C–SiC Investigated Through Strict Adherence to ISO 10993).

3.2.1 Testing Materials

3.2.1.1 (100) Silicon

Silicon wafers (University Wafer, Inc.) with the crystallographic orientation of (100) and thickness of 500 μm were used as test materials. Since this type of wafer is used as the substrate for the entire semiconductor and conductor materials in this work it can be considered as a baseline material.

3.2.1.2 Cubic Silicon Carbide

Thin films of 3C–SiC on (100) silicon were grown via chemical vapor deposition at the University of South Florida [16]. The process involves using a horizontal hot-wall, low pressure reactor at the growth rate of ~5 μm/h at 1350°C. Quantitative analysis was performed via atomic force microscopy (AFM) using a Park Systems XE-100 AFM in tapping mode (Si_3N_4 probes) (see Fig. 3.1). The scan rate was 0.2 Hz for all of the AFM characterization reported here. Fig. 3.1 shows 10 × 10 and 5 × 5 μm AFM micrographs of

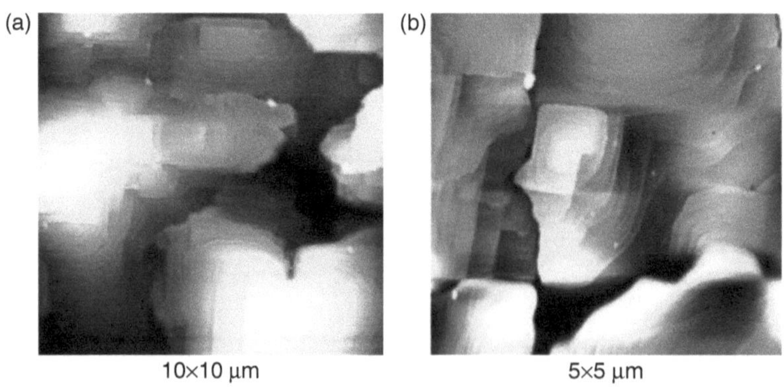

Figure 3.1 ***AFM micrographs of (a) 10 × 10 μm and (b) 5 × 5 μm scans of 3C–SiC.*** AFM measurement was performed in tapping mode with a 0.2 Hz scan rate. The micrographs are normalized to a Z height interval of (+10, −10) nm. Note the surface displays atomic steps, which is an indication of high-quality crystal growth.

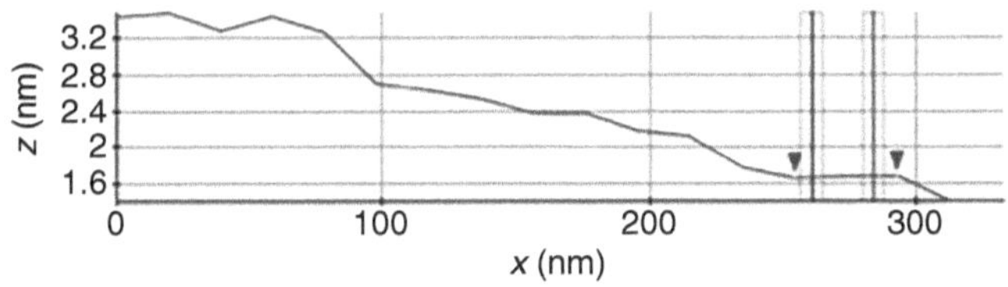

Figure 3.2 ***Line scan of the 5 × 5 μm AFM micrograph of 3C–SiC on Si sample of Fig. 3.1.*** The r_q value was 7.2 ± 0.5 nm RMS. From this line scan the terrace width is ~41 nm and the step height (not shown) about 4.5 nm.

the 3C–SiC samples, again for reference, as the sample morphology changes across the wafer but the reported RMS roughness value was ~7 nm across the wafer. A representative r_q value for the sample was ~7.2 ± 0.5 nm RMS taken from Fig. 3.1 (5 × 5 μm scan). AFM data elaboration was performed using the XEI software (Park Systems). In these measurements the average value for the terrace width was about 41 nm and the step height was about 4.57 Å. Fig. 3.2 shows this AFM data elaboration.

3.2.1.3 Amorphous Silicon Carbide

Four different variants of *a*-SiC were prepared by J. Register of the USF SiC Research Group using the USF-NREC Plasmatherm 700, PECVD tool; details of the *a*-SiC film deposition process that he developed and that were tested in this chapter can be found in [17]. For convenience, a brief summary of the process developed is included here.

To develop this capability, we began by benchmarking a PECVD *a*-SiC process that could be performed at USF. The characterization runs were done at the Nanotechnology Research and Education Center cleanroom located within the University of South Florida's Tampa campus using a Plasmatherm 700 PECVD system. Fifty-millimeter (2-in.) silicon (100) wafer substrates were coated with films at 250°C using methane (CH_4) and silane (SiH_4) gas precursors, with argon (Ar) as a bulk dilution gas. PECVD *a*-SiC film uniformity is particularly sensitive to the deposition pressure. The literature indicated that maximum film uniformity was achieved at a pressure of 900 mT [8]. Our pressure was held at 900 mT while varying the precursor gas ratio (CH_4/SiH_4) from a ratio of 6.6:30. The resulting films were verified as *a*-SiC, with characterization of stress, chemical resistance, and optical index included in [17].

The early films exhibited large stress gradients, leading to film/substrate delamination and indicating the process was significantly detuned. The measured film stress versus film deposition conditions are shown in Fig. 3.3. Film stress was estimated by measuring wafer curvature changes using a Dektak 150 optical profilometer, which calculates stress using the modified

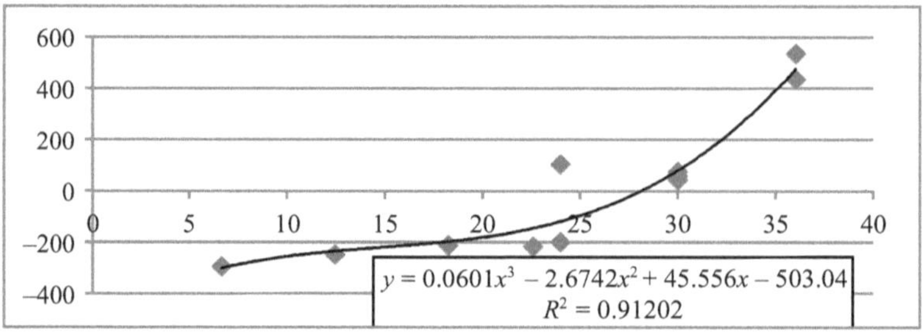

Figure 3.3 ***Average film stress (MPa) versus CH_4/SiH_4 gas ratio by flow volume.*** A polynomial fit was applied to the data as shown in the text box with an R^2 value of 0.9 indicating a strong fit [17].

Stoney's equation [18], prescan curvature measured before the deposition, and postscan data measured along the same path [17]. Standard 2-in. (100) silicon wafers were RCA cleaned and marked with polyimide tape running across the length of the substrate. The polyimide tapeline was used for film thickness verification and as a geometric reference for the profilometer probe measurement location. We varied methane concentration while holding silane (SiH_4) flow rate at 15 sccm, the temperature at 250°C, and the pressure at 900 mT during the deposition. Argon acted as the dilution gas and had a flow of 500 sccm. After each run, the stress was profiled and the results were compiled into Fig. 3.3. The test resulted in very low stress for $CH_4/SiH_4 \sim 30$. The final optimized process parameters are shown in [17]. It was found that by adding an Ar sputtering step before the PECVD deposition clean, pinhole-free, thin films could be realized, as demonstrated through AFM in Fig. 3.4. The thickness of the film was maintained for both tests of 230 nm.

In order to evaluate both the *a*-SiC film itself and the impact of the underlying material to the biological environment, four film stacks with *a*-SiC as the final layer were fabricated for bio- and hemocompatibility testing as shown in Fig. 3.5. The *a*-SiC films were deposited using the process in Table 3.1. The SiO_2 layers were PECVD deposited to a thickness of 800 nm. 3C–SiC film thickness was ~6.2 μm grown using hot-wall CVD on a 100 mm, Si(100) wafer using our established process and described in Section 3.3.2 [16].

3.2.2 *In vitro* BAMBI Cytotoxicity Assay for *a*-SiC

One of the first tests performed on a novel material or biological device application is the cytotoxicity test, which is performed to see if the device can safely exist *in vivo* without causing cellular apoptosis. In chapter: Cytotoxicity of 3C–SiC Investigated Through Strict Adherence to ISO 10993, we discussed

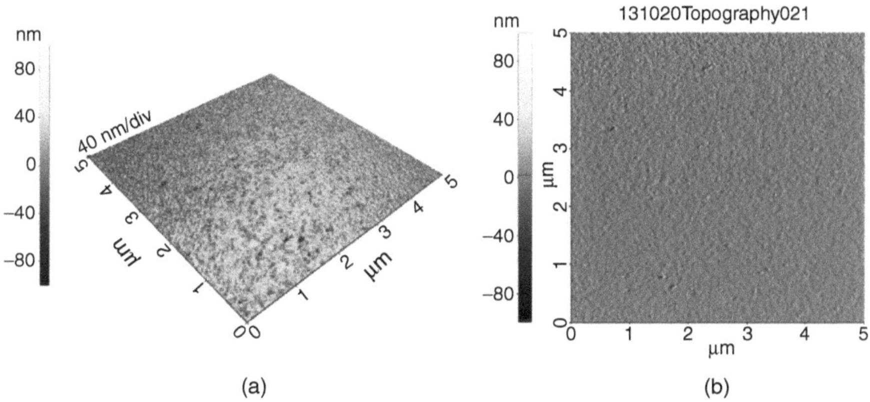

Figure 3.4 ***AFM image of a 230-nm thick* a-*SiC film.*** (a) A 5 × 5 µm AFM scan of the surface showing 3D topology. (b) 5×5 µm scan of the sample showing reduced roughness and a lack of pinhole defects. Sample RMS roughness of 1.2 nm. *(Data taken in tapping mode using a Park XE-100 AFM by M. Nezafati, USF SiC Group [17].)*

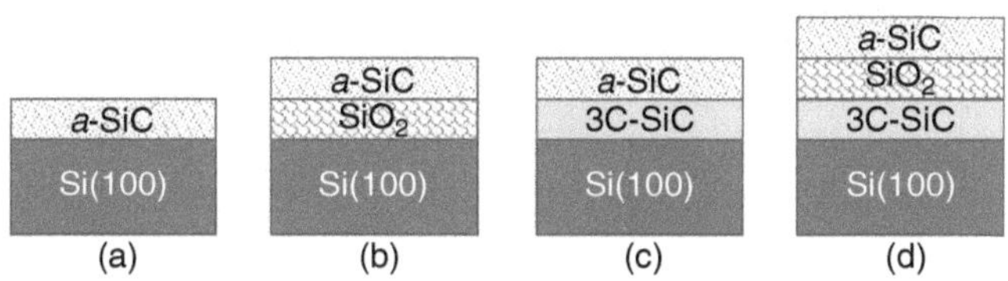

Figure 3.5 ***Films produced to evaluate the bio- and hemocompatibility of* a-*SiC coatings on various semiconductor materials.*** *a*-SiC coating on (a) Si, (b) oxide-coated Si, (c) 3C–SiC, and (d) oxide-coated 3C–SiC. The *a*-SiC coating is 300 nm thick, the oxide coating 800 nm thick, and the 3C–SiC film thickness ~6 µm. Si(100) wafer substrate thickness ~500 µm.

Table 3.1 Process conditions for low stress PECVD *a*-SiC films [17]

Parameter	**Value**
Single RF supply	13.57 MHz
Temperature	250°C
Pressure	900 mT
Silane (SiH_4) flow rate	12 sccm
Methane (CH_4) flow rate	360 sccm
Argon (Ar) flow rate	500 sccm

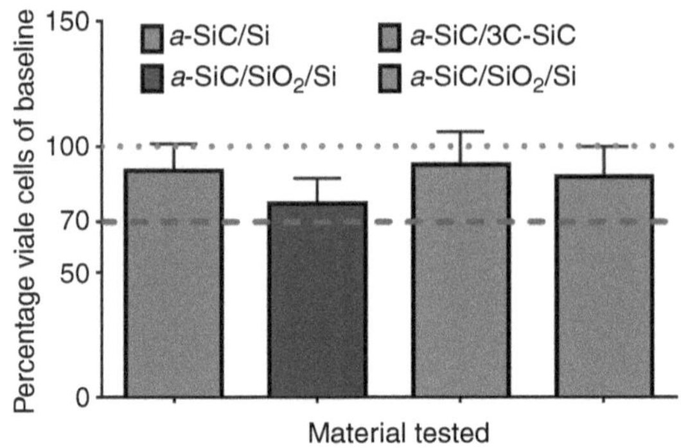

Figure 3.6 ***A bar plot representing*** **a-SiC** ***viability relative to the baseline well of culture treated polycarbonate (CTPC) measured for the BAMBI cytotoxicity test (chapter: Cytotoxicity of 3C–SiC Investigated Through Strict Adherence to ISO 10993).*** The mean $\overline{x}$ and standard deviation (σ) are displayed for the four *a*-SiC stacks. The dashed line represents the ISO 10993-5 limitation for acceptable cytotoxicity, and the dotted line denotes the mean value for CTPC. All stacks passed the cytotoxicity test, further establishing *a*-SiC as a possible material for biological device encapsulation or insulation.

the cytotoxic evaluation using the ISO 10993-4 methods of extract generation or through direct cellular contact with the tested device or material [19]. We surmised that both of these techniques often produced statistically different results, with the results possibly enabling a material or device to pass the test when it should be investigated more thoroughly. Section 2.3 introduced our BAMBI method, which combined these two techniques into one test. We used this BAMBI method to evaluate the cytotoxicity of the four *a*-SiC semiconductor stacks shown in Fig. 3.5.

As displayed in Fig. 3.6, all four stacks pass the BAMBI test with relative viability levels greater than the quantitative ISO 10993-5 cytotoxicity limit of 70% [19]. We have demonstrated that both Si and SiO_2 substrates have repeatedly failed the ISO 10993-5 cytotoxicity tests at a time period of 24 h (Figs. 2.3 and 2.5); however the use of *a*-SiC as a coating allows these materials to pass the BAMBI combined method at a time period of 96 h. This result is encouraging and displays evidence that *a*-SiC provides an effective barrier for the underlying Si and SiO_2 layers to the *in vitro* biological environment.

Although the cellular viability is increased to acceptable levels for all the tested stacks, there is an evident trend displayed in Fig. 3.6. It is observed that the overall level of mean viability is slightly influenced by the substrate material beneath the *a*-SiC coating. The two semiconductors, 3C–SiC and Si, show a slightly higher relative viability when compared to the two SiO_2 substrates. As all the substrates received the same formulation of *a*-SiC (Section 3.2.1.3), and they were all coated at the same time, it could be caused by an effect of the oxide on the SiC coating, either its composition or adhesion. Further analysis is required to understand the mechanism for this observed variation.

One possible explanation for the difference in these films comes from the exploration of hydrogenated amorphous silicon carbide (*a*-SiC:H) by Y. Matsuda et al. in 2012 [20]. Moisture-assisted cracking of stressed thin films had been noticed due to the rupture of O—Si—O bonds [21]. Y. Matsuda et al. noticed that, although their *a*-SiC films did not contain oxygen bonds during deposition, many had a proclivity to form after exposure of the film to moisture, which was linked to film stoichiometry and carbon density [20]. The *a*-SiC that we deposited on SiO_2 would necessarily contain O—Si bonds at the interface, and therefore could lead to fractures and cracking in the film, slowly exposing the SiO_2 beneath. Although both Si and 3C–SiC have SiO_2 as a native oxide, the surfaces would contain OH^- passivation from HF cleaning performed before processing and therefore decrease the proclivity to form these O—Si bonds. In future works, we would need to fully characterize our *a*-SiC using X-ray diffraction or Fourier transform infrared spectroscopy (FTIR) to chemically characterize the *a*-SiC on the various stacks to fully verify this hypothesis. However, it was noted that even if the *a*-SiC is affected by the oxide on the substrate, it still provides an adequate protective shield from the physiological environment for the short term of 96 h and increases the overall biocompatibility of the *a*-SiC-coated oxide films studied here.

In chapter: Cytotoxicity of 3C–SiC Investigated Through Strict Adherence to ISO 10993 and Section 3.2 we have investigated the biocompatibility of various semiconductor materials, namely, Si and SiC. Si and its native oxide have been shown to generate cytotoxicity in the presence of L929 mouse fibroblastoma. On the other hand, 3C–SiC and *a*-SiC have shown viabilities similar to culture treated polymers. We will now explore if these SiC materials are not only biocompatible, but also hemocompatible. If these materials prove to display both desirable characteristics, the combined evidence will warrant the use of animal models to evaluate the inflammatory reactions *in vivo* (chapter: *In Vivo* Exploration of Robust Implantable Devices Constructed From Biocompatible 3C–SiC).

3.3 *IN VITRO* ASSAY TO ASSESS HEMOCOMPATIBILITY OF SiC

Modern biomedical devices have the goal of including distributed sensors, which are able to record pertinent patient biological activity, such as pressure or flow, and ideally transform this information into a closed loop pathway that provides accurate therapeutic treatment, which can include accurate dispensing of medications. In some cases, it is absolutely essential for designed operation that at least one component of the device be able to perform *in vivo*. One example of where traditional medicine is not able to provide sufficient assistance is when the individual has lost a body part, such as limbs or organs. In this case, prosthetic devices have provided some restoration of functionality, with common examples being hip and knee arthroplasty and dental implants. While these types of replacements have been successful, the replacement of lost limbs and organs remains a challenge.

Several investigations are under way to find a viable solution to restoring organ functionality. Restoring lost cognitive functionality after trauma or disease, like assisting individuals with Parkinson's disease or restoring patient motility after the spine was severed through trauma, has been demonstrated through the use of implantable neural interface devices [22–24]. Diabetic patients can be treated by implantable biological microelectromechanical systems (BioMEMS) for drug delivery [25] and patients with arthritis or movement disorders can be cured by joint replacement [26]. Implantable devices that are in contact with the human body must satisfy several requirements for their safe implementation. Paramount to safety are overall device biocompatibility, hemocompatibility, and the durability and reliability of the constituent materials within in the physiological environment of the human body.

ISO 10993 provides guidelines from which to evaluate biomedical devices and biomaterials. It consists of several parts, which describe sample preparation and reference (ie, control) materials selection [27], chemical stability and corrosion [28,29], *in vitro* biological tests, for example, cytotoxicity [19], and testing implants with animal models *in vivo* [30]. In addition, ISO 10993 outlines how to perform a relevant hemocompatibility evaluation [31,32]. For a device to be selected for clinical trial, strict adherence to ISO 10993 provides an abundance of pertinent biological information, hence it is critical to fully understand the standard and how to properly implement it in the laboratory. Chapter: Cytotoxicity of 3C–SiC Investigated Through Strict Adherence to ISO 10993 discussed how to perform proper ISO 10993 compliant *in vitro* cell assays to determine material compatibility with the body. In this chapter we discuss new experiments, performed using similar methods to Schettini et al. [33], to expound upon the hemocompatibility of 3C–SiC, and now include *a*-SiC, which is already becoming a common biocompatible insulating and hermetic material for advanced neural prosthetics (see chapter: Amorphous Silicon Carbide for Neural Interface Applications).

3.3.1 Hemocompatibility

Hemocompatibility is always a challenge given the harsh electrolytic environment of the human vascular system. As was the case for the term "biocompatibility," there is much debate about what hemocompatibility implies. In the literature, which has been focused on the topic of blood compatibility, researchers try to answer the following questions: How can blood compatibility be measured? What materials can be called "hemocompatible?" What is the biological basis of the reaction of materials with blood [34]?

Based on these questions, the focus in this section is on both hemocompatibility methodology and materials evaluation. The minimum requirements for hemocompatibility measurements are mentioned in section four of ISO 10993 [2]. ISO 10993-4 provides test methods to evaluate blood/material interaction [2]. Blood cell adhesion is one measure of the hemocompatibility of a material when considered in conjunction with distal embolization. Also platelet count and platelet aggregation are important for evaluating the hemocompatibility of materials [31,35]. ISO 10993-4 suggested a static state for hemocompatibility tests, which clearly is not close to the real *in vivo* condition. Testing the materials in a dynamic state is a complementary test that can be done on implants that are to be used in contact with blood, which is particularly true for neural implants due to the vasculature of the human brain.

In previous studies, the hemocompatibility of silicon-based semiconductors and *a*-SiC was investigated by Roy et al. in 2009, and by Muthusubramaniam et al. in 2011 [36,37]. The results showed outstanding hemocompatibility for *a*-SiC, supporting its favorable candidacy as the coating of cardiovascular implants [38,39]. Nurdin et al. in 2003 studied the hemocompatibility of SiC in comparison to diamond-like coatings, normally referred to as DLC [40], and Schettini et al. in 2009 showed preliminary hemocompatibility of 3C–SiC based on platelet adhesion and activation studies under static conditions [41]. In this section, we present the hemocompatibility of 3C–SiC and *a*-SiC evaluated in a dynamic state to support the static state tests.

Based on ISO 10993-4, we describe the design of our experiments to evaluate the blood compatibility of materials of interest, both in the static and dynamic state. Indeed, it was with this ISO 10993-4 guidance that Schettini et al. [42] performed static hemocompatibility studies using platelet-rich plasma (PRP), where it was surprisingly discovered that 4H–SiC and 6H–SiC were no more hemocompatible than Si. We have added to this body of knowledge by adding dynamic experiments with PRP flowing in a fluid flow cell in order to understand the impact of fluid motion on the attachment of platelets to various surfaces [43]. Since the static experiments have been described in the literature [42], we will focus here on the addition of dynamic hemocompatibility experiments to this body of knowledge, starting with the classical approach using a flow cell called the Chandler's loop.

3.3.2 ISO 10993-4 Hemocompatibility Evaluation of SiC

3.3.2.1 Chandler's Loop

A useful test configuration was developed by Chandler in 1958 to measure the time period for thrombogenic responses [2]. Chandler's loop system consisted of a tube that was filled to 70% of its capacity with blood and connected with a silastic collar, thus closing the loop. The circuit was tilted at 45 degrees with respect to the horizon. The loop was utilized until either blood clotted in the tube or a clot formed on the test material's surface [44].

In recent years, the Chandler loop system has been modified for anticoagulated blood to measure biomaterial-induced platelet activation [45]. The most noticeable modification in the Chandler's loop is the increase of surface-area-to-blood-volume ratio, which will result in the reduction of background noise. Even with these modifications, the presence of air bubbles at the blood–air interface is a major issue when evaluating the hemocompatibility of unknown materials. This blood–air interface could result in protein aggregation, which may result in thrombotic platelet activation and affect the outcome of the experiment. Even the presence of microscopic air nuclei have resulted in an increase of platelet adhesion, and consequently activation afterwards [46].

Our Chandler's loop consists of a peristaltic pump, 0.2-in. (5.08-mm) diameter Tygon® S-50-HL PVC medical surgical tubing (Saint-Gobain, Paris, France), a sample holder, polymeric gasket, and a vacuum system as shown in Fig. 3.7. The circuit was placed inside a 1-m^3 polyethylene chamber, which included an electrical heater to maintain a temperature of 37 ± 1°C. The 3C–SiC, *a*-SiC, and control sample size was 50 × 75 mm. Experiments were performed in Dr K. Muffly's laboratory in USF Health.

3.3.2.2 Platelet-Rich Plasma Preparation

The University of South Florida has an assurance #A-4100-01 on file, and porcine blood for the experiments was provided by the University of South Florida's Center for Advanced Medical Learning and Simulation

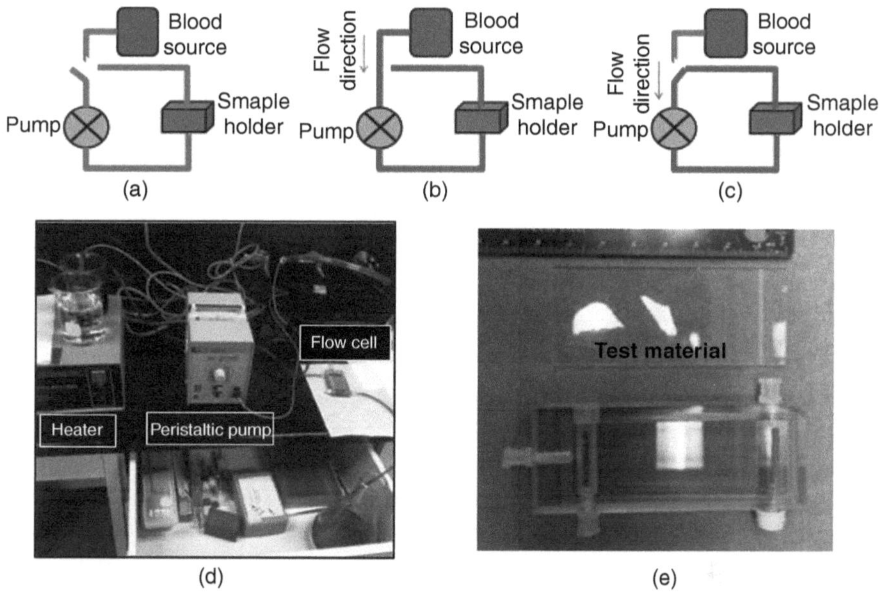

Figure 3.7 ***The schematic of the modified Chandler's circuit design.*** In (a) Stage 0, the circuit is open and the tubes are rinsed with PBS. (b) Stage 1, the circuit is still open but the blood source is connected. (c) Stage 2, the circuit is closed and the tubes are completely filled with PRP. (d) A digital photograph of the dynamic hemocompatibility test setup. The flow cell, represented with a digital photograph in (e), was designed to accommodate 50 × 75 mm samples. *(Experiment performed in Dr K. Muffly's laboratory in USF Health.)*

(USF-CAMLS) labs. The animals for these experiments were maintained in accordance with the USF Institutional Animal Care and Use Committee (IACUC) protocol USF ID:T IS00000216. The blood was collected post-mortem by the CAMLS lab technician in blood bags including acid citrate dextrose (ACD) as an anticoagulant (1:9 ACD:blood volume ratio). The blood bags were stored in a portable refrigerator at 4°C and transported to the experimental apparatus. Fifty-millimeter ACCUSPIN™ tubes were filled with 45 mL of whole porcine blood and centrifuged at 4°C for 10 min at 1000 Gr (standard gravity). As shown in Fig. 3.8a, the blood separated into three phases after centrifuge: 55% of total blood volume is the plasma, which will be above the other layers due to its lower density; the buffy coat is next, which includes leukocytes, and platelets form 1% of whole blood; the erythrocyte (red blood cell) layer is lowest as it has the highest density. Due to the presence of a membrane within the ACCUSPIN tube, the plasma, leukocytes, and platelet phases were not mixed with the erythrocytes,

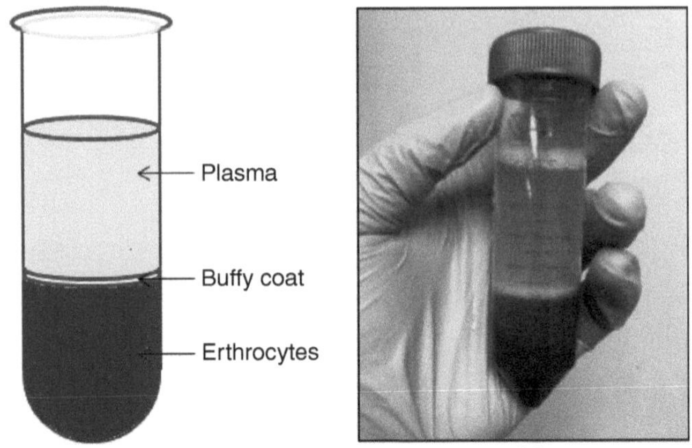

(a) (b)

Figure 3.8 ***Formulation of PRP for static and dynamic hemocompatibility material examinations.*** (a) A schematic representation showing the three phases of the blood after centrifuging. (b) A digital photograph of a 50-mL ACCUSPIN tube showing the three phases consisting of the PRP, buffy coat, and erythrocytes. The PRP is the nearly clear liquid layer on top of the two white and dark gray [red in the web version] layers approximately located at the 15 mL mark on the tube.

forming the PRP. The PRP was removed and placed in separate, sterile, 50 mL conical tubes, and stored at 4°C. The concentration of platelets in the PRP was measured using a hemocytometer at ~150,000 platelets μL^{-1}.

The initial flow rate of the Chandler's loop was measured in open loop configuration (Fig. 3.7b) after loading PRP to a flow rate just below 8 cm^3/s, verified at two separate points located just before and after the sample holder. After verification of the flow rate, 100 mL of fresh PRP was manually loaded into the loop using a syringe and air bubbles were minimized, before the two ends of the tube were connected to have a closed loop flow system (Fig. 3.7c). The pump circulated the PRP across the surface of the tested sample for a time of 15 min, after which time the sample was retrieved and platelets were fixed for florescent microscope analysis.

Platelet activation was evaluated by using fluorescence microscopy as described by N. Schettini of the USF SiC Group during her static hemocompatibility tests of SiC materials [41]. In short, 1 μL/mL rhodamine phalloidin solution, a stain for F-actin protein, which is found in platelets, was dissolved in 1× PBS, the surfaces were imaged using fluorescent microscopy, and the images were analyzed using Image J analysis software [47]. Further evaluation of the material involved analysis of surface modifications through scanning electron microscopy (SEM) and AFM after platelet removal through aggressive piranha cleaning. N. Schettini demonstrated that 3C–SiC had superior hemocompatibility to 4H–SiC and 6H–SiC, as well as Si, via the observation of significantly lower platelet adhesion and activation. Indeed, the work reported by Schettini was the basis for the dynamic work reported here, which was to provide a more realistic model of the physiological environment and reliable measure of a materials compatibility with flowing blood.

3.3.3 Static Hemocompatibility of SiC

The hemocompatibility of Si(100), 3C–SiC, SiO_2, and *a*-SiC-coated Si was evaluated under static conditions in the Chandler's loop with the flow rate equal to 0 cm^3/s. This measurement was made with two goals in sight: first, we wished to study the platelet adhesion to a material's surface without the presence of shear forces; second, we wished to verify our current apparatus with the results of N. Schettini [42,48].

As shown in Fig. 3.9a, the fluorescence micrograph of a selected 3C–SiC surface shows low platelet attachment and no apparent activation. The fluorescent micrographs of the *a*-SiC (Fig. 3.9b) and Si(100) (Fig. 3.9d) shows a

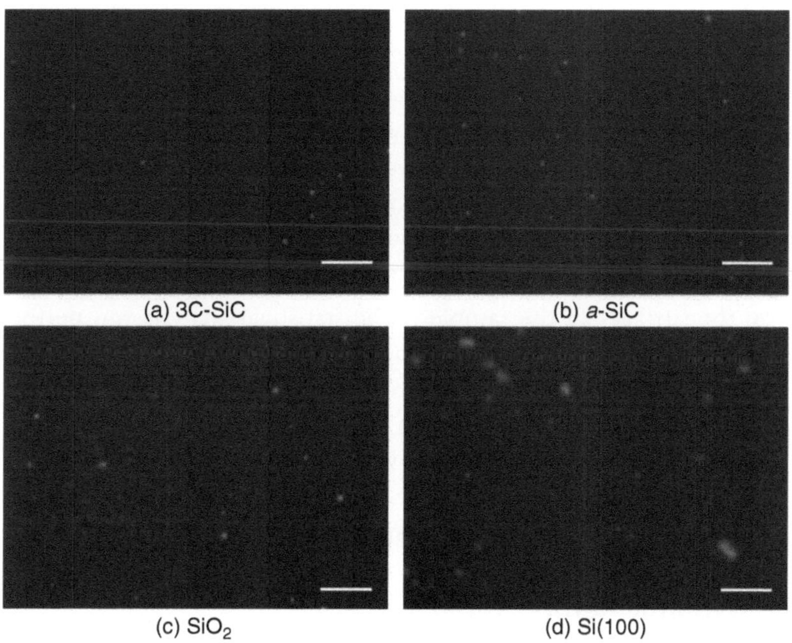

Figure 3.9 ***Static hemocompatibility fluorescence micrographs using rhodamine as a fluorescence tag at 50× of magnification.*** (a) 3C–SiC, (b) *a*-SiC on Si, (c) SiO_2, and (d) Si(100). Scale = 100 μm.

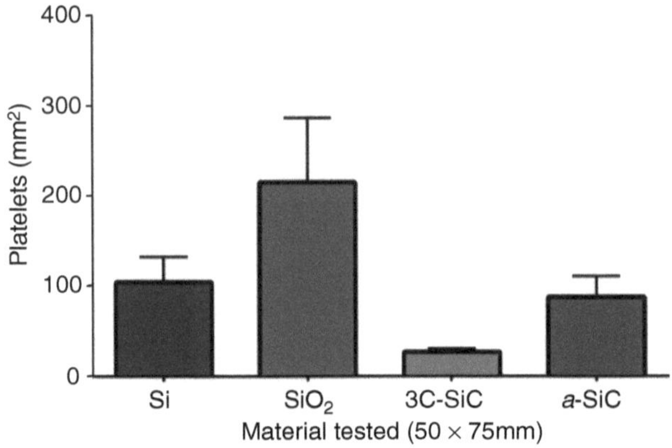

Figure 3.10 ***Static hemocompatibility histogram displaying the mean platelet activation of the Si,*** **a-SiC** ***on Si, SiO_2, and 3C–SiC with the standard deviation representing the error bar.*** Activated platelets per square milimeter used to evaluate the hemocompatibility of the materials.

much larger accumulation of platelets on the surface as well as some platelet activation marked by a spike-like morphological change to the platelets. This result shows that 3C–SiC once again displays evidence of better hemocompatibility than silicon-based material systems, which are already proposed for many biomedical systems. SiO_2 was tested as positive control (Fig. 3.9c), and displayed both platelet adhesion and activation.

The rhodamine phalloidin stains the F-actin protein, which is common to active and resting platelet cytoskeletons [49]. In this work we did not distinguish between these two platelet states as we could only use qualitative morphological changes to ascertain platelet state, so we counted all stained cells on the surfaces of the samples. The statistical analysis was performed on the quantitative mean platelet values, and the results are displayed in the bar chart in Fig. 3.10, which includes the standard deviation as the positive error bar. Analysis of variance (ANOVA) indicated at least two of the materials display significantly different mean values. The mean values were found to be in platelets per area: 3C–SiC displayed $26 mm^{-2}$, whereas *a*-SiC had $\sim 87 mm^{-2}$, Si(100) $\sim 104 mm^{-2}$, and SiO_2 $\sim 215 mm^{-2}$.

3.3.4 Dynamic Hemocompatibility of SiC

The hemocompatibility of the same set of materials tested using the BAMBI method was tested in a dynamic state to simulate the real conditions of the *in vivo* environment. Tests in a static state showed that 3C–SiC and *a*-SiC on

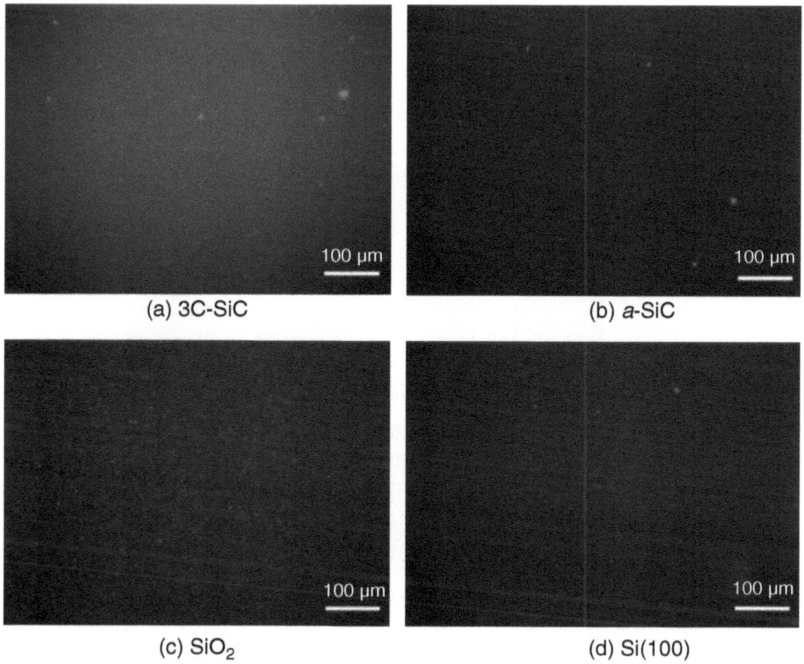

Figure 3.11 ***Dynamic hemocompatibility test fluorescence micrographs using rhodamine as a fluorescence tag at 50× of 10 magnification.*** (a) 3C–SiC, (b) *a*-SiC on Si, (c) SiO_2, and (d) Si(100). Scale = 100 μm.

Si had more resistance to platelet surface adhesion and may increase the performance of biomedical devices. The dynamic tests would enable the examination of the material's performance if placed directly into an active circulatory system.

Fig. 3.11 displays fluorescence micrographs of selected tested surfaces after the dynamic hemocompatibility evaluation with the rate ~8 cm^3/s for a duration of 15 min. Clearly, both the 3C–SiC and *a*-SiC samples displayed a very low attached platelet count. The platelets on Si show an increase when compared to the SiC samples, and SiO_2 is notably covered in attached platelets.

The quantitative mean platelet per area counts are displayed in Fig. 3.12. The standard deviation is displayed as the error bars. Two important points can be understood from the statistical analysis of a dynamic test. The overall platelet activation was slightly higher on *a*-SiC, 52mm^{-2}, when compared to 3C–SiC, which has 15mm^{-2}. Si(100), at 92mm^{-2}, displays a higher platelet activation than both the SiC materials, while SiO_2 has just above 2.5× the attached platelets compared to Si.

One observation that we made is that the overall platelet activation decreased when compared with the static test. We hypothesize that with the

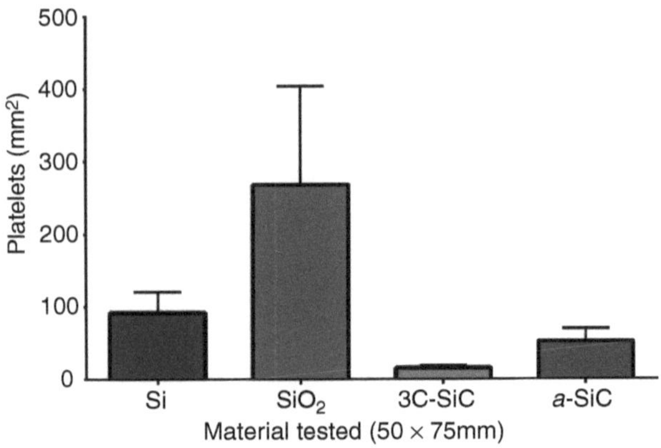

Figure 3.12 ***Dynamic hemocompatibility histogram of platelet activation of the Si, a-SiC on Si, SiO_2, and 3C–SiC using standard deviation as the error bar.*** Activated platelets per square milimeter used to evaluate the hemocompatibility of the materials.

applied shear force from the flow, even within the supposed laminar flow conditions set within the test condition of 8 cm^3/s flow rate, some of the platelets could be removed from the surface. The activation of the platelets decreased by a factor of 40% for both the SiC materials, and 20% for Si under laminar flow. This remains a hypothesis as the platelet attachment to the SiO_2 surface actually increased by 25%, indicating that there are secondary mechanisms at hand beyond physical sheer force application. The latter increase may be due to the properties of the surface of the materials and its proclivity to protein and subsequent cell attachment. However, it is noted that all of the materials show the same trend in both the dynamic and static tests, so material properties may not be the issue. In the future, immunohistochemistry, focused on the platelet activation protein pathway, should be performed in both tests to assist in determination of the thrombogenic quality of the materials [49]. A study of the proteins attached to the surface should also be performed using attenuated total reflectance Fourier transform infrared spectroscopy, ATR-FTIR, accompanied by liquid chromatography/mass spectrometry to analyze the proteins attached to the surface of the materials.

3.4 SUMMARY

In this chapter the biocompatibility and hemocompatibility of 3C–SiC and *a*-SiC were evaluated and both materials found to possess excellent characteristics in both regimes. Biocompatibility was evaluated using the BAMBI method outlined in chapter: Cytotoxicity of 3C–SiC Investigated Through

Strict Adherence to ISO 10993 and the results showed a slight decrease in performance compared to 3C–SiC, but still well within the acceptable threshold for material biocompatibility as stipulated by ISO 10993. Material hemocompatibility under the static and dynamic (ie, under laminar flow) using PRP was studied. A modified Chandler's loop was used as the blood circuit. Platelet surface attachment as a measure of hemocompatibility was studied using fluorescence microscopy. The data obtained from the static test verified the previous work reported using the same method by Schettini in 2009 [48]. It was seen then, and also observed again through our current testing methodology, that 3C–SiC demonstrated an appreciably lower platelet attachment and activation in comparison to conventional semiconductor implant materials such as Si(100) and SiO_2. Although *a*-SiC, a SiC material currently being evaluated clinically as a coating for biomedical devices, displayed better resistance to platelet adhesion than Si(100), it did not perform as well as 3C–SiC.

The results from the dynamic material evaluation under laminar PRP flow displayed the same trend as seen from the samples in static state, but most of the materials showed a 20–40% reduction in overall platelet attachment. SiO_2 was the only exception to this observation and actually displayed a 25% increase. While this preliminary dynamic flow data is encouraging we need to study the performance of the materials to various flow rates and timeframes, as well as incorporate immunohistochemistry and analytical chemistry, to fully understand the hemocompatibility of a particular material.

ACKNOWLEDGMENTS

This work was sponsored by the Defense Advanced Research Projects Agency (DARPA) MTO under the auspices of Dr Jack Judy through the Space and Naval Warfare Systems Center, Pacific Grant/Contract no. N66001-12-1-4026 – Biocompatibility of Advanced Materials for Brain Machine Interfaces (BAMBI). Dr K. Muffly of the USF Morsani College of Medicine is gratefully acknowledged for his technical assistance and the use of his Chandler's loop apparatus and laboratory for the hemocompatibility experiments. The authors also thank the USF CAMLS administration and staff for providing blood supplies, especially Jade Buchanan and Dr Luis Llerena.

REFERENCES

[1] Donaruma LG, Williams DF. Definitions in biomaterials. J Polym Sci Part C 1988;26:414.
[2] ISO, 10993-4. Biological evaluation of medical devices. Part 4: selection of tests for interactions with blood, ed. Geneva, Switzerland: International Organization for Standardization; 2002.
[3] Choyke WJ, Matsunami H, Pensl G. Silicon carbide: recent major advances. Berlin; New York: Springer; 2004.

[4] Saddow SE, Agarwal A. Advances in silicon carbide processing and applications. Boston: Artech House; 2004.

[5] Kordina O, Saddow SE. Silicon carbide overview. In: Saddow SE, Agarwal A, editors. Advances in silicon carbide processing and applications. 1st ed. Boston, MA, USA: Artech House, Inc.; 2004. p. 2–3, 7–8, 18.

[6] Severino A, Frewin C, Bongiorno C, Anzalone R, Saddow SE, La Via F. Structural defects in (100) 3C-SiC heteroepitaxy: influence of the buffer layer morphology on generation and propagation of stacking faults and microtwins. Diamond Rel Mater 2009;18:1440–9.

[7] Matsuda Y, King SW, Bielefeld J, Xub J, Dauskardt RH. Fracture properties of hydrogenated amorphous silicon carbide thin films. Acta Mater 2012;60:682–91.

[8] Iliescu C, Poenar DP. PECVD amorphous silicon carbide (*a*-SiC) layers for MEMS applications. In: Hijikata Y, editor. Physics and technology of silicon carbide devices. Rijeka, Croatia: InTech; 2013. p. 131–48.

[9] Hehrlein C. Stent passivation with silicon carbide as a possible alternative to drug-eluting stents – a comprehensive review of pre-clinical and clinical results. Intervent Cardiol Rev 2009;4:60–3.

[10] Kalnins U, Erglis A, Dinne I, Kumsars I, Jegere S. Clinical outcomes of silicon carbide coated stents in patients with coronary artery disease. Med Sci Monit 2002;8:PI16–20.

[11] USFDA., Rithron-XR Coronary Stent System – P030037; 2005. Available from: http://www.fda.gov/MedicalDevices/ProductsandMedicalProcedures/DeviceApprovalsandClearances/Recently-ApprovedDevices/ucm078628.htm.

[12] Bolz A, Schaldach M. Artificial heart valves: improved blood compatibility by PECVD *a*-SiC:H coating. Artif Organs 1990;14:260–9.

[13] Naji A, Harmand MF. Cytocompatibility of two coating materials, amorphous alumina and silicon carbide, using human differentiated cell cultures. Biomaterials 1991;12:690–4.

[14] Cogan SF, Edell DJ, Guzelian AA, Ping Liu Y, Edell R. Plasma-enhanced chemical vapor deposited silicon carbide as an implantable dielectric coating. J Biomed Mater Res Part A 2003;67:856–67.

[15] Hsu J-M, Tathireddy P, Rieth L, Normann AR, Solzbacher F. Characterization of *a*-SiC(x):H thin films as an encapsulation material for integrated silicon based neural interface devices. Thin Solid Films 2007;516:34–41.

[16] Reyes M, Frewin C, Ward PJ, Saddow SE. 3C-SiC on Si hetero-epitaxial growth for electronic and biomedical applications. ECS Trans 2013;58:119–26.

[17] Register, J. SiC for advanced biological applications. Doctorate of Philosophy, Electrical Engineering, University of South Florida; 2014.

[18] Temple-Boyer P. Residual stress in low pressure chemical vapor deposition SiN_x films deposited from silane and ammonia. J Vac Sci Technol A 1998;16:2003.

[19] ISO, 10993-5. Biological evaluation of medical devices. Part 5: tests for *in vitro* cytotoxicity, ed. Geneva, Switzerland: International Organization for Standardization; 2009.

[20] Matsuda Y, King SW, Bielefeld J, Xu J, Dauskardt RH. Fracture properties of hydrogenated amorphous silicon carbide thin films. Acta Mater 2012;60:682–91.

[21] Wiederhorn SM. Influence of water vapor on crack propagation in soda-lime glass. J Am Ceram Soc 1967;50:407–14.

[22] Follett KA, Weaver FM, Stern M, Hur K, Harris CL, Luo P, et al. Pallidal versus subthalamic deep-brain stimulation for Parkinson's disease. N Engl J Med 2010;362:2077–91.

[23] Kravitz AV, Freeze BS, Parker PR, Kay K, Thwin MT, Deisseroth K, et al. Regulation of parkinsonian motor behaviours by optogenetic control of basal ganglia circuitry. Nature 2010;466:622–6.

[24] Moritz CT, Perlmutter SI, Fetz EE. Direct control of paralysed muscles by cortical neurons. Nature 2008;456:639–42.

[25] Medtronic MiniMed. MiniMed Paradigm® REAL-Time Revel™ System; 2015. Available from: https://professional.medtronicdiabetes.com/paradigm-revel-real-time.

[26] Long M, Rack HJ. Titanium alloys in total joint replacement – a materials science perspective. Biomaterials 1998;19:1621–39.
[27] ISO, 10993-12:2004. Biological evaluation of medical devices. Part 12: sample preparation and reference materials, ed. Geneva, Switzerland: International Organization for Standardization; 2004.
[28] ISO, 10993-14:2001. Biological evaluation of medical devices. Part 14: identification and quantification of degradation products from ceramics, ed. Geneva, Switzerland: International Organization for Standardization: ISO copyright office; 2001.
[29] ISO, 10993-15:2000. Biological evaluation of medical devices. Part 15: identification and quantification of degradation products from metals and alloys, ed. Geneva, Switzerland: International Organization for Standardization; 2000.
[30] ISO, 10993-6. Biological evaluation of medical devices. Part 6: tests for local effects after implantation, ed. Geneva, Switzerland: International Organization for Standardization; 2009.
[31] Seyfert UT, Biehl V, Schenk J. *In vitro* hemocompatibility testing of biomaterials according to the ISO 10993-4. Biomolec Eng 2002;19:91–6.
[32] ISO, 10993-5:2009. Biological evaluation of medical devices. Part 5: tests for *in vitro* cytotoxicity, ed. Geneva, Switzerland: International Organization for Standardization; 2009.
[33] Saddow SE, editor. Silicon carbide biotechnology: a biocompatible semiconductor for advanced biomedical devices and applications. Amsterdam: Elsevier; 2011.
[34] Ratner BD. Blood compatibility – a perspective. J Biomater Sci – Polym Ed 2000;11: 1107–19.
[35] ISO, 10993-4:2002. Biological evaluation of medical devices. Part 4: selection of tests for interactions with blood, ed. Geneva, Switzerland: International Organization for Standardization; 2002.
[36] Muthusubramaniam L, Lowe R, Fissell W, Li L, Marchant R, Desai T, et al. Hemocompatibility of silicon-based substrates for biomedical implant applications. Ann Biomed Eng 2011;39:1296–305.
[37] Roy RK, Choi HW, Yi JW, Moon MW, Lee KR, Han DK, et al. Hemocompatibility of surface-modified, silicon-incorporated, diamond-like carbon films. Acta Biomater 2009;5:249–56.
[38] Armin B, Amon M, Ozbek C, Heublein B, Schaldach M. Coating of cardiovascular stents with a semiconductor to improve their hemocompatibility. Tex Heart Inst J 1996;23(2):162.
[39] Weisenberg BA, Mooradian DL. Hemocompatibility of materials used in microelectromechanical systems: platelet adhesion and morphology *in vitro*. J Biomed Mater Res 2002;60:8.
[40] Nurdin N, François P, Mugnier JKY, Moret M, Aronsson B-O, Descouts P. Haemocompatibility evaluation of DLC and SiC coated surfaces. Euro Cell Mater 2003;5:17–28.
[41] Schettini N. Hemocompatibility assessment of 3C-SiC for cardiovascular applications. PhD, Electrical Engineering, University of South Florida. Tampa, FL; 2009.
[42] Schettini N, Jaeroszeski MJ, West L, Saddow SE. Hemocompatibility assessment of 3C-SiC for cardiovascular applications. In: Saddow SE, editor. Silicon carbide biotechnology: a biocompatible semiconductor for advanced biomedical devices and applications. Amsterdam: Elsevier; 2011. p. 153–208.
[43] Nezafati M. Biomaterial testing methodology for long-term *in vivo* applications: silicon carbide corrosion resistance, biocompatibility and hemocompatibility. PhD, Electrical Engineering, University of South Florida. Tampa, FL; 2014.
[44] Chandler AB. *In vitro* thrombotic coagulation of the blood; a method for producing a thrombus. Lab Invest 1958;7:110–4.
[45] Black J. Biological performance of materials: fundamentals of biocompatibility. 4th ed Boca Raton, FL: CRC Press-Taylor & Francis Group; 2006.

[46] Haycox CL, Ratner BD. *In-vitro* platelet interaction in whole human blood exposed to biomaterial surface – insights on blood compatibility. J Biomed Mater Res 1993; 27:1181–93.
[47] Schneider CA, Rasband WS, Eliceiri KW. NIH Image to ImageJ: 25 years of image analysis. Nat Meth 2012;9:671–5.
[48] Schettini N. Hemocompatibility assessment of 3C-SiC for cardiovascular applications. PhD, Electrical Engineering, University of South Florida. Tampa, FL; 2009.
[49] Barkalow KL, Italiano JE, Chou DE, Matsuoka Y, Bennett V, Hartwig JH. α-Adducin dissociates from F-actin and spectrin during platelet activation. J Cell Biol 2003;161: 557–70.

CHAPTER 4

Graphene Functionalization for Biosensor Applications

Owen J. Guy and Kelly-Ann D. Walker
Centre for NanoHealth, College of Engineering, Swansea University, Swansea, United Kingdom

Contents

4.1 INTRODUCTION

Silicon carbide (SiC), an advanced semiconductor material, is a suitable substrate for graphene growth via a process referred to as sublimation growth [1–4]. During annealing at temperatures between 1200 and 1700°C the SiC surface reconstructs itself, with silicon atoms subliming and leaving behind a layer, or multiple layers, of epitaxial graphene [2]. A key advantage of graphene prepared in this manner over conventional graphene production (exfoliation from highly ordered pyrolytic graphite [HOPG] substrates) is the capacity to grow graphene on flat, large area SiC substrates. These large areas of graphene, up to 100 mm in diameter, can subsequently be processed into devices, using standard semiconductor fabrication techniques. Several papers have reported the development of graphene-based electronic devices, with device performance varying significantly with the growth techniques for graphene [5,6]. Device performance depends on the growth conditions, and consequently on parameters such as number of graphene layers, defect densities, and grain sizes. These factors can vary widely for epitaxial graphene grown on SiC. However, high-quality epitaxial graphene has been achieved and its exceptional electronic and thermal properties make epitaxial graphene particularly suitable for a range of new applications, from electronics to functional nanodevices such as biosensors [7–10]), gas sensors [11], and high-performance transistors [6,12].

Graphene devices and sensors promise to be a disruptive technology in next generation electronics and healthcare diagnostic applications. This is due to graphene's exceptional electronic properties and extreme surface-to-volume ratio, which offer greatly enhanced sensitivity. Graphene devices also provide the potential for portable rapid diagnostic point-of-care sensors compared to conventional labor-intensive, time-consuming, enzyme-linked immunosorbant assays (ELISAs) that are currently used for clinical biomarker analysis. The processes from graphene growth to chemical modification of graphene—in order to develop graphene biosensors—are reviewed in this chapter.

4.2 PRODUCTION OF GRAPHENE

There are numerous methods available for producing graphene, ranging from exfoliation of graphite [13] to chemical vapor deposition (CVD) on transition metal substrates [14]. Each method has its advantages and disadvantages. For example, exfoliation can produce very high-quality graphene, but only for very small sample areas. In contrast, CVD graphene can be grown on large area copper foils, but the quality of CVD graphene is not as high. In addition, CVD graphene usually requires a layer transfer process onto an insulating substrate (often SiO_2 on silicon), in order to be able to use it as an electronic material. The layer transfer process can introduce damage and defects to the graphene layer. This section reviews the various graphene production methods, with particular focus on epitaxial graphene, grown on silicon carbide—since the subject of this book is predominantly focused on SiC.

4.2.1 Mechanical Exfoliation of Graphite

Exfoliation of HOPG, repeatedly cleaved using adhesive tape, was the fabrication method used by Geim and Novoselov et al. in their 2004 seminal paper on graphene [13]. The cleaved graphene layer was eventually deposited onto a silicon dioxide substrate. Single or multiple graphene layers can be identified from their optical contrast with the silicon dioxide substrate visible using an optical microscope [15].

4.2.2 Chemical Exfoliation

Graphene can also be produced via chemical exfoliation or the cleaving of graphite. There are numerous alternative methods, but most rely on the introduction of surface tension using strong acids and oxidizing agents [16], intercalation between graphene layers [17,18], or ultrasonic agitation [19,20]. Using an appropriate solvent, such as *N*-methylpyrrolidone, sonication of graphite causes the graphite to split into platelets. These processes are solution based and often produce a mixture of graphene, few-layer graphene, multilayer graphene, and graphene nanoparticles. Chemical exfoliation can also result in the formation of graphene oxide (GO), rather than graphene, which can subsequently be reduced to yield graphene or reduced graphene oxide (rGO), to use the correct nomenclature [21]. Graphene from rGO contains oxygen impurities and defects, which may be removed by thermal annealing. However, some properties of these chemically derived materials may be inferior to mechanically exfoliated or epitaxial graphene.

Despite its generally inferior properties, manufactured suspensions of graphene particles have already been used in products such as conductive inks [22,23]). Graphene prepared in solutions may be processed into inks that may be printed using scalable techniques [22,23]). Conductive graphene inks may be used for applications such as printed electronics [24], photovoltaics [25], and displays [26], where cost is a driving factor.

Chemically exfoliated graphene is also being developed as a material for use in polymer composites [27].

4.2.3 Supporting Substrates

Though SiO_2 substrates are often used as supporting substrates for exfoliated graphene, the SiO_2 can strongly influence the properties of the graphene layer by introducing charged surface states, impurities, and surface roughness, which modulate the electronic transport properties of the graphene. The observed mobilities of graphene on SiO_2 are lower than theoretical predictions, prompting research into alternative insulating substrates such as hexagonal boron nitride (*h*-BN). *h*-BN is a structural analog of graphene, with B and N atoms in its sublattices in place of C atoms [28]. The lattice mismatch between graphene and *h*-BN is small (1.7 %), which reduces rippling in the graphene overlayer [29]. As *h*-BN has a wide bandgap (5.97 eV), it can be used as a 2D insulating substrate for graphene devices, and exhibits superior dielectric properties to that of SiO_2 ($\varepsilon = 3$–4, $V_{\text{breakdown}} = 0.7\text{V/nm}$). The structure of *h*-BN is atomically flat, and it is free of dangling bonds and surface charge traps. This means that the interference with the conduction electrons of graphene on *h*-BN is substantially lower than that of SiO_2, leading to much higher electron mobility measured for graphene deposited on *h*-BN2 [30]. Thin films of *h*-BN can be produced by mechanical exfoliation, in a similar manner to graphene [31].

4.2.4 Chemical Vapor Deposition

CVD growth is a highly promising method for mass production of large area graphene. It has been known for some time that hydrocarbons will form a carbon deposit on transition metal substrates. Most CVD methods utilize transition metal substrates with hydrocarbon gases as the carbon source. CVD thus has a wide variety of process parameters that can be used to tailor the graphene film produced [14].

Thermal decomposition of hydrocarbon gases occurs on the surface of the transition metal films. Catalytic decomposition of the hydrocarbon gas produces carbon, which is soluble in the transition metal—held near its melting point during the growth process. The exact deposition mechanism is determined by the carbon solubility in the metal and the growth conditions, such as cooling rate, carbon exposure time, carbon concentration, flow rate, and carbon feedstock. The catalytic properties of the metal are important for graphene formation and the choice of substrate can influence uniformity, grain sizes, and number of graphene layers grown.

4.2.5 Metal Substrates

Copper, nickel, ruthenium [32], iridium [33], palladium, platinum, cobalt, and rhenium can all be used as the substrate for CVD graphene [34]. Generally, the quality of the produced graphene is strongly dependent on the crystallinity of the substrate. However, the use of single crystal substrates means single crystal graphene production is currently an expensive process.

4.2.5.1 Growth on Copper

Graphene growth on copper has been reviewed extensively [14,35]. Most reports describe the use of methane gas decomposing on a polycrystalline Cu surface held at around 1000°C. The catalytic activity of copper is enhanced by removing its native oxide using acetic acid or annealing in hydrogen prior to growth. As the growth mechanism is surface related (surface diffusion, rather than bulk diffusion of carbon), growth on copper is self-limiting. Monolayer growth is easily produced, but thicker layers are difficult to achieve.

4.2.5.2 Roll-to-Roll Production

Large area graphene films have been grown on 30-in. copper foils using a roll-to-roll production method [26]. For most applications, it is not desirable to leave the CVD-grown graphene on a metal substrate. The presence of a metal substrate strongly affects the electronic properties of graphene and subsequently, graphene device performance. The graphene films must therefore be transferred onto a polymer support substrate and subsequently passed through a copper etchant, to generate a graphene film supported by the polymer layer. Polymers such as poly(methyl methacrylate) [36] or polydimethylsiloxane are commonly used [37,38]). The graphene/polymer film can then be placed on an insulating substrate, such as SiO_2 on silicon, before removing the polymer film by heating or dissolving in an appropri ate solvent. The layer thicknesses of the film can vary between one and four layers and the process is inherently scalable and suitable for mass production.

4.2.5.3 Growth on Nickel

Growth on copper predominantly yields monolayer graphene. To achieve thicker graphene layers, growth can be conducted on nickel substrates. Atomically smooth micron-size regions have been demonstrated using CVD growth on polycrystalline nickel substrates with a hydrogen–methane gas mixture [39] at 950°C. A $5 \times 5\text{-mm}^2$ area of multilayer graphene was transferred in this way. Continuous, large area single to few-layer graphene films have been reported [36].

Thicker graphene layers can be obtained using nickel substrates due to the higher solubility of carbon in nickel (0.6 wt% at 1326°C) compared to copper [35]. At temperatures above 800°C, carbon and nickel form a solid solution. The solubility of carbon decreases below 800°C. Rapid cooling of the solid solution leads to out-diffusion of carbon from the nickel substrate. Variation in the number of graphene layers grown on nickel is caused by increased diffusion at grain boundaries; this variation can be reduced by using nickel of higher crystallinity.

The nature of the electronic interaction between graphene and substrate varies with the metal, the interaction between the *p* electrons in graphene and the *d* orbitals in the metal is stronger for nickel than in copper [40]. Therefore, once the graphene film has been grown, it must undergo a similar layer transfer process to that used for graphene grown on copper, that is, transfer of the graphene/nickel film onto a polymer substrate, subsequent etching of the nickel, using HCl or $FeCl_3$, and transfer onto an insulating substrate [41].

The supporting polymer is then dissolved, leaving the graphene on the desired substrate. Using a new metal substrate for each batch of graphene is costly. The ability to recycle the substrate is therefore an attractive prospect for mass production. Electrochemical techniques for separating the graphene while leaving the substrate intact have been reported [42–44]).

4.2.6 CVD Growth on SiC

As mentioned in the previous section, CVD graphene growth on metals suffers from the disadvantage that electronic applications require graphene to be transferred to an insulating substrate. Thus, graphene growth on SiC is a particularly attractive alternative as graphene can be grown directly on large area semiinsulating SiC substrates of up to 100 mm diameter without the need for layer transfer processes.

While graphene growth can be achieved by thermal desorption of silicon from SiC (sublimation epitaxial graphene), CVD growth on SiC is another option for growing graphene on semiinsulating SiC substrates. Direct growth of few-layer graphene has been reported on both 6H–SiC and 3C–SiC using propane as the carbon precursor gas [45]. CVD graphene growth could offset any surface quality issues related to the high temperatures (1200–2000°C), for example, step-bunching or similar topographical changes that can occur during high-temperature processes. Strupinski et al. [46] also used propane gas as the precursor for CVD graphene growth on off-axis 4H–SiC substrates. This method was performed at 1600°C under argon flow, to prevent excessive silicon sublimation. The propane gas diffuses through the argon gas layer and deposits an epitaxial graphene layer on the SiC substrate. A uniform layer thickness of 2 and 9 ML was observed on the Si face of SiC(0001), indicating a CVD mechanism rather than any silicon sublimation (as it is known that, in sublimation growth on SiC[0001], the upper limit of the graphene thickness is about 3 ML due to the blocking of Si diffusion by the graphene layer growth). The CVD-assisted epitaxial graphene layers were of good quality, with a mobility of approximately 1800 cm^2/V s at room temperature.

CVD-assisted graphene growth has also been demonstrated on 3C–SiC [45]. Since 3C–SiC can be grown on silicon, this represents a potentially cheaper alternative to all-SiC substrates. 3C–SiC–CVD graphene exhibits preferable stacking and interface properties and the reconstruction is different to that of "furnace grown" 6H–SiC epitaxial graphene. The melting point of silicon (1410°C) prohibits the use of 3C–SiC/Si substrates for high-temperature epitaxial growth, but lower-temperature CVD growth on 3C–SiC/Si is an interesting alternative option.

The quest for cheaper semiinsulating substrates for "transfer-free" graphene growth is ongoing. One promising recent development is the growth of graphene on germanium. Graphene domains have been demonstrated on hydrogen-terminated germanium (110) films [47,48]. The germanium film is obtained by heteroepitaxial growth on a Si(110) wafer using germane as a precursor [49]. After growth, the germanium can be crystallized by thermal annealing at 600°C. Germanium has an extremely low carbon solubility at its melting temperature; consequently, monolayer graphene growth is controllable [50].

4.2.7 Epitaxial Growth on Silicon Carbide

The production of graphite on silicon carbide has been known since 1893 [51]. Recently, this process has generated interest due to the demonstration of single-layer epitaxial graphene, grown on SiC, using a high-temperature sublimation growth process. Single-layer and few-layer graphene have been demonstrated on SiC by various groups [45,52,53]. Growth is achieved by heating SiC to high temperatures (1300–1800°C) in a ultrahigh vacuum or under an inert atmosphere, typically in a radiofrequency induction furnace. The sublimation temperature of silicon is lower than that of carbon. Therefore at around 1500°C, silicon sublimes from the SiC surface, leaving a thin layer of carbon [54]. The number of graphene layers, stacking of layers, and coupling to the substrate varies with crystal face orientation and growth conditions, as reviewed in de Heer [55]. Of the many polytypes of SiC, the hexagonal forms 4H- and 6H- are the most commonly used in epitaxial graphene growth, though, as mentioned earlier, CVD graphene growth has been performed on cubic 3C–SiC. One prohibitive aspect of epitaxial graphene may be the current cost of SiC wafers. However, demand from power electronics research and the LED industry is gradually making SiC wafer costs more competitive. Current suppliers of 4H– or 6H–SiC mass produce 4-in. wafers, and some suppliers offer 6-in. wafers for R&D applications.

4.2.7.1 Si- and C-Face Growth

Hexagonal polytypes of SiC have two possible surface terminations (Si- or C-face). The termination significantly influences the graphene growth process. Here, we refer to "silicon face growth" in regard to epitaxial graphene grown on SiC(0001), and "carbon face growth" in relation to growth on SiC(000-1), respectively. Growth on the Si-face starts with a reconstruction of the SiC surface, progressing with increasing temperature from a Si-rich (3×3) structure to a $(\sqrt{3} \times \sqrt{3})$ R30 surface, ending with a $(6\sqrt{3} \times 6\sqrt{3})$ R30 surface, indicating the formation of graphene.

The first carbon layer formed on the Si-face during annealing is not a true graphene layer, and is often referred to instead as a buffer or interface layer. The atomic bonding of this layer is different to that of graphene, as one-third of the C atoms are covalently bonded to the top SiC layer. The interface therefore contains dangling Si bonds, which disrupt the linear structure of the π bands in the electronic structure of the graphene overlayer, opening a bandgap of 0.26 eV. This disruption is caused by breaking of the symmetry in the A and B sublattices of the graphene structure [56]. This interface effect means that Si-face epitaxial graphene has a lower electron mobility than C-face epitaxial graphene. The effect of the substrate can be ameliorated by hydrogen intercalation [57] or oxidation [58]; both methods producing "quasi-free standing" epitaxial graphene (Fig. 4.1).

C-face grown epitaxial graphene does not have an interface layer between graphene and the substrate, and therefore electron mobilities measured are typically higher in C-face graphene than for Si-face epitaxial graphene. The layer thickness and growth uniformity of Si- and C-face epitaxial graphene are also different. Si-face epitaxial graphene grown in vacuum has small, narrow domains of 30–100 nm in diameter, whereas C-face epitaxial graphene typically has larger, thicker-layer domains (~200 nm in diameter) that are rotationally disordered. Growth on the silicon face is self-limiting, and slows after a few minutes at high temperatures. Carbon face

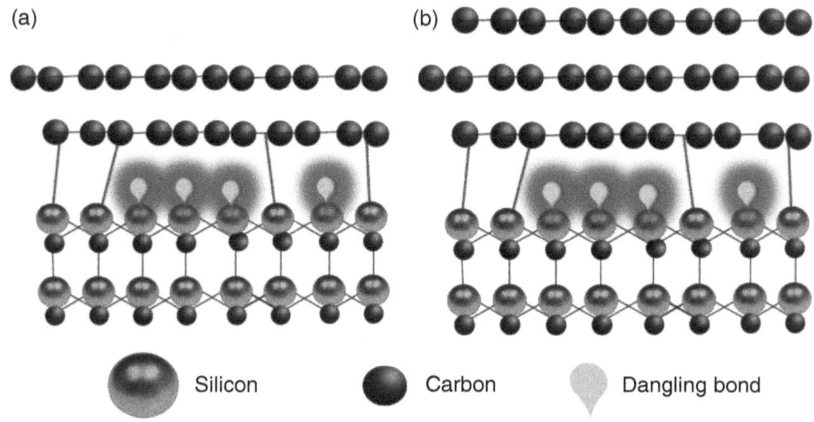

Figure 4.1 ***Structure of graphene on SiC (0001) (Si-face).*** The first carbon layer formed during high-temperature annealing of the Si-face is an interface layer with one-third of the C atoms bonded to the SiC substrate. The interface layer between graphene and SiC influences its electronic properties. (a) Shows a "monolayer" grown on the Si-face; (b) shows a bilayer model.

growth is not limited in this way, allowing for growth of thick (5–100) layers. Monolayer growth on C-face epitaxial graphene is difficult to achieve using standard growth parameters, as structural defects on the SiC surface act as efficient nucleation sites.

Some control over the growth rate on C-face epitaxial graphene can be attained by use of a graphite cap [59], which increases the carbon and silicon partial pressures close to the surface and lowers the silicon sublimation rate. Preparation of the SiC substrate by H etching is important for growth on both faces, but the effect seems to be more pronounced on the C-face due to the absence of an interface layer [60]. High-quality graphene is formed on the C-face when the surface of the SiC substrate consists of well-ordered arrays of parallel, straight step edges—which can be achieved by H etching. Screw dislocations and other defects in the substrate can cause the growth of nanocrystalline graphite on the C-face.

Larger domain sizes are attained on Si-face epitaxial graphene by annealing in an inert atmosphere [53]. Annealing in an inert gas such as Ar lowers the sublimation rate of silicon from the SiC substrate. This allows high-growth temperatures to be used (>1500°C), which increases the detachment, diffusion, and reattachment of surface carbon atoms. The formation of flat terraces is favored at these higher temperatures, producing larger single crystalline regions and therefore higher electron mobility. Inert atmospheric growth is not favorable for C-face epitaxial graphene, as this causes the growth of 3D islands, which grow up to five graphene layers thick before coalescence.

4.2.7.2 Related Growth Techniques on SiC

Graphene has also been produced on SiC using alternative techniques such as molecular beam epitaxy, using a carbon source. This enables the growth of graphene on SiC at much lower temperatures (950°C) [61] than used for conventional epitaxial growth. Graphene can also be formed on SiC using a thin metal layer, such as nickel, deposited on the SiC surface. As the substrate is heated, the nickel extracts and absorbs carbon. As the system is cooled, the carbon precipitates on the nickel surface, forming graphene [62]. An analogous method, using iron films deposited on SiC or diamond, has been used to grow graphene [63].

4.3 GRAPHENE CHARACTERIZATION METHODS

Characterization of graphene materials has progressed rapidly since Geim and Novoselov's seminal work was published in 2004 [13]. Raman spectroscopy in particular has emerged as one of the primary analytical methods for the characterization of graphene. Other valuable analytical techniques include X-ray photoelectron spectroscopy (XPS), FTIR, and electrochemical methods. These methods, and their specific application in graphene characterization, are reviewed in the following sections.

4.3.1 Raman Spectroscopy

Raman spectroscopy and microscopy techniques are now routinely used to characterize graphene, graphene-related materials, and devices fabricated using these materials. Raman spectroscopy enables researchers to understand and quantify a number of important features of graphene. The technique is quick, noninvasive, offers high throughput, and is capable of probing key properties, including: the number of graphene layers; the orientation of layers; strain; doping; disorder and defects; presence of edge boundaries; and surface functional groups [64].

Raman's ability to assess the number of layers [65] is arguably the most prevalent use of Raman analysis of graphene. Research into graphene growth and material development has benefited particularly from Raman characterization. Raman spectroscopy measures vibrational information related to chemical bonds and symmetry of molecules using Raman scattering. The technique examines the inelastic scattering of light to determine properties of the chemical bonds present in a system. Photons scattered from an atom or molecule are predominantly scattered elastically (Rayleigh scattering), but a small fraction are scattered inelastically. Monochromatic light is used to excite the molecular vibrations in a system, and the shift in wavelength between the incident and emitted light is used to characterize the vibrational, rotational, and other low-frequency modes of the atom/molecule. The Raman shift (shift in wavelength) is reported as a wavenumber as shown in Eq. (4.1), with units of inverse length:

$$\Delta\omega = \frac{1}{\lambda_0} - \frac{1}{\lambda_1} \tag{4.1}$$

where $\Delta\omega$ is the Raman shift (expressed in cm^{-1}), $1/\lambda_0$ is the excitation wavelength, and $1/\lambda_1$ is the Raman spectrum wavelength.

The Raman spectrum of graphene has a number of characteristic peaks or bands. The most prominent features in the Raman spectra are: the G band, the 2D band, and, depending on the structural integrity of the sample, the disorder-induced D band. The G band is located at around 1580 cm^{-1} and is indicative of sp^2-bonded carbon compounds. It is attributed to a first-order (one-phonon) process from in-plane C—C bond stretching modes, and is doubly degenerate. Strain of the C—C bonds is therefore evident in a shift of the G band. Thus, the G band shift can be used to measure strain in graphene. The strength of C—C bonds in sp^2-bonded carbon systems results in a relatively high Raman frequency compared to other compounds. The presence of the G band at high Raman frequencies allows for very small shifts to be measured.

As the G band arises from C—C bonds, the intensity of the G peak increases almost linearly with the number of layers. The intensity of the G band is therefore a useful indicator for the number of graphene layers present. The G band is affected by the doping of graphene, which causes modulations in both the peak position and full width of half maximum. The G band shifts linearly as a function of the number of layers, meaning that the G-peak position can also be used as an indicator for the number of layers. The G band is unique from the other features in the graphene spectrum in that the band position is almost independent of excitation laser frequency (it is nondispersive).

The presence of the disorder-induced or D band (located at around 1350 cm^{-1}) is related to edge sites and disordered graphene. The band originates from a second-order Raman scattering process from sp^3 (diamond-like) bonds that involve elastic scattering from a defect, and inelastic scattering by a phonon. In practice, this means that the position and shape of the D band is highly dependent on the excitation frequency used. The ratio of the intensity of the D band to the G band gives a good estimate for the number of defects.

The 2D (or G′) band (located at 2700 cm^{-1}) is present at approximately double the frequency of the D band (second overtone). Like the D band, the 2D band also arises from a second-order scattering process. However, in contrast to the D band, the 2D band arises from inelastic scattering from two phonons, and is therefore present in both disordered and pristine graphene. The shape of the 2D peak is particularly revealing. Single-layer graphene exhibits a sharp and symmetric peak, and the presence of a symmetric peak is used to characterize single-layer graphene. However, characterization is slightly more complicated for two- or few-layer graphene. Both D and 2D bands exhibit dispersive behavior and originate from a double resonance Raman process.

Single-layer exfoliated graphene has a single, sharp 2D peak with roughly four times the intensity of the G band. The structure of the 2D peak increases in complexity with increasing number of graphene layers. Bilayer graphene has a broader peak when compared to that of monolayer graphene due to additional contributions introduced by the splitting of the electronic band structure. This trend continues up to five graphene layers, whereafter the peak is indistinguishable from that of bulk graphite.

Examination of epitaxial graphene on SiC has additional considerations. When present on substrates, such as Si or SiO_2, the strong G and 2D Raman peaks from graphene are clearly visible. However, SiC has peaks present in its Raman spectrum, which overlap the region of the D and G bands (1000–2000 cm^{-1}), from the graphene Raman spectrum. The Raman signal from the surface graphene can be estimated by subtracting SiC reference spectra. The SiC substrate also has a strong influence on the structure and electronic properties of epitaxial graphene, and therefore its Raman spectrum. This influence is remarkably different between the two terminations of SiC [66]. Epitaxial graphene grown on the Si-terminated (0001) face exhibits n-type doping because of the strong interaction of the electrons between the underlying SiC substrate and the graphene layers. Complete coverage is obtained on Si-face growth, displaying Bernal stacking and thickness inhomogeneity. Step bunching causes a compressive strain, which leads to a measurable shift in both the G and 2D bands. C-face (000-1) growth results in a different morphology to that of Si-face epitaxial graphene. In this instance, long isolated graphene ribbons are formed, which are strain relaxed and lightly p-type doped.

Raman spectroscopy also has a lot to offer when investigating the production and structure of graphene. The number of layers and the presence of defects can be ascertained [67], with some notable features present in epitaxial graphene Raman spectra, as follows:

1. The presence of the D band indicates some disorder in graphene, due to the presence of sp^3 bonds. Generally, the intensity of the D band should be lower than that of the G band. The ratio of the intensity of the D band to the G band gives a rough estimate for the number of defects (the ratio of sp^3/sp^2-bonded C).
2. The position of the G band is influenced by the doping and the number of graphene layers. Significant blue shifts of the G band ($\sim$11 cm^{-1}) and 2D band ($\sim$34 cm^{-1}) are observed on epitaxial graphene relative to exfoliated graphene. These are attributed to strain effects from the substrate. Further blue shifts in the 2D band of epitaxial graphene are caused by electron/hole doping. The relative shifts of the 2D and G bands can be used to infer the contribution from both effects [68–70]). Shifts in the position of the G peak can be seen from spectra obtained from different positions of the same graphene sample where strain/doping is inhomogeneous [66].
3. The intensity of the G band increases almost linearly as the number of graphene layers increases [68–70]). Ambiguities can arise when comparing the G band intensity of monolayers and bilayers on epitaxial graphene [66], but the G band intensity still serves as a useful indicator of layer thickness.
4. The intensity ratio of the G and 2D bands of epitaxial graphene differs significantly from that of exfoliated graphene [68–70]).
5. Si-face epitaxial graphene tends to have a narrower G peak than C-face epitaxial graphene, because of its higher crystallinity [68–70]).
6. The 2D-band intensity depends strongly on the Fermi level for monolayers [71]. Therefore, substrate-induced doping effects and strain effects affect the 2D band intensity. The 2D band is further influenced by the number of layers [67]—its position is monotonically blue shifted with increasing layer thickness.
7. The shape of the 2D band is indicative of the number of layers. Monolayer graphene has a 2D band, which can be fitted with a single Lorentzian [65]. The asymmetry of this peak increases with increasing number of layers [64]. It is further influenced by the stacking of the graphene layers (Fig. 4.2).

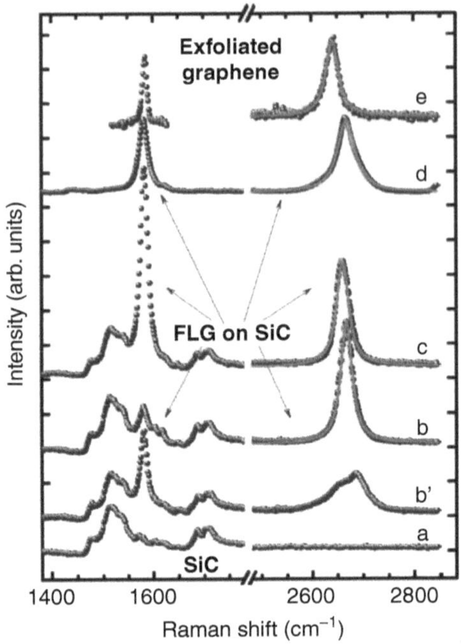

Figure 4.2 ***Raman spectra of a 4H–SiC substrate (a) of a graphitic residue (b,b′) of 5–10 layers of few-layer graphene (FLG) produced on a 4H–SiC substrate (c) of 70–90 layers epitaxial FLG (d), of exfoliated graphene (e).*** The solid lines are Lorentzian fits. Raman spectroscopy is an invaluable tool in the study of graphene. Quantitative information is derived from these Raman spectra, but care must be taken when comparing material on different substrates. *(Reprinted with permission from Ref. [72]. Copyright (2008), AIP Publishing LLC.)*

4.3.2 XPS

XPS is a surface science technique, which is routinely used to reveal information on the elemental composition, empirical formula, chemical state, and electronic state of the elements within a material. XPS analysis relating to the XPS spectra can be obtained by irradiating a sample with X-rays and measuring the kinetic energy of electrons that are emitted from the top 1–10 nm of the material being investigated. Surface atoms have characteristic peaks in the XPS spectrum. The energy and intensity of these peaks enable identification and quantification of all surface elements present (except hydrogen). The surface sensitivity of XPS makes it a valuable tool in the study of graphene—since graphene is intrinsically only a few layers thick. XPS can provide quantitative information such as the number of layers, and identification of impurities present from patterning or processing. XPS has been used to identify metal impurities and the formation of carbides from CVD growth

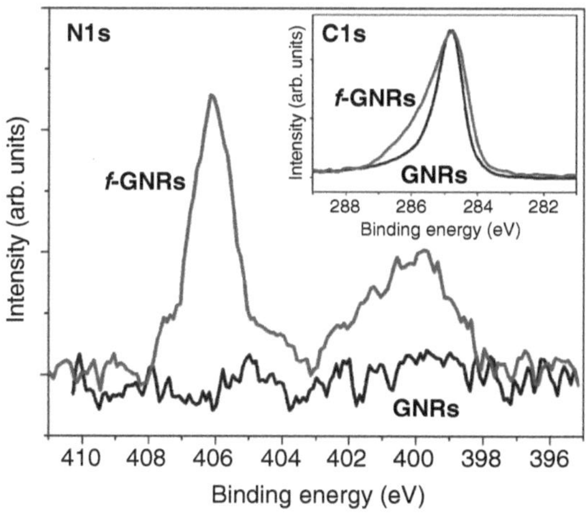

Figure 4.3 ***N1s and C1s XPS spectra of functionalized (f-) and as-prepared GNRs.*** Peaks at ~406 and ~400 eV in *f*-GNRs, respectively, indicate the presence of nitrophenyl groups and lower oxidation states, for example, NH_2 on the surface of the GNRs. *(Reprinted with permission from Ref. [75]. Copyright (2010), American Chemical Society.)*

processes [73], estimating the thickness of epitaxial graphene [74], measuring oxide formation at the interface between epitaxial graphene and the SiC substrate [58], and detecting chemical attachment to graphene [75].

XPS is also an invaluable tool in the analysis of graphene both before and after any chemical modification. Tour and coworkers [75] used XPS to verify attachment of 4-nitrophenyl groups to graphene nanoribbons (GNRs) following treatment with diazonium tetrafluoroborate salts. In these experiments, XPS analysis was conducted on thick films of GNRs on Si/SiO_2 substrates. Fig. 4.3 shows the N1s XPS spectra of the functionalized (*f*-) and control GNR samples. No nitrogen was present in the control GNR sample, suggesting that all of the nitrogen content of the functionalized GNRs was related to the treatment with diazonium salt. No boron or fluorine signals were present in the spectra of the samples, eliminating the possibility of noncovalent physisorption of the diazonium salt or the presence of residual bf_4 counter ions.

The N1s XPS spectrum of functionalized GNRs (Fig. 4.3, labeled *f*-GNRs) has two distinct peaks. The sharp peak at approximately 406 eV corresponds to the nitrogen in the nitro group and confirms the presence of nitrophenyl groups on the GNR surface. The broader peak at around 400 eV is associated with the presence of reduced nitrogen species. It is suggested that these may be due to the transformation of nitro groups via a number of mechanisms: (1) reduction of the nitro groups to amine groups by X-ray irradiation in the XPS chamber [76]; (2) in situ electrolytic reduction of the nitro groups by electrons from neighboring islands of intact graphene; or (3) the 400 eV peak could correspond to diazenyl groups introduced by electrophilic attack of the diazonium cation on aromatic rings (azo coupling reaction).

XPS analysis also revealed a broadening in the peak in the C1s XPS spectrum (inset Fig. 4.3) after nitrophenyl grafting. Sinitskii et al. [75] attributed this effect to either: (1) the presence of C—N bonds from 4-nitrophenyl groups bonded to the GNR sample, which results in the appearance of a new component at 285.7 eV and thus broadening of the main C peak at 284.8 eV; or (2) an increase in the number and contribution of graphene-derived sp^3 C bonds.

4.3.3 Electrical Characterization

Electrical characterization of graphene devices and sensors can be performed using simple current–voltage (*I*–*V*) measurements, Hall-bar measurements, gated field effect transistor (FET) measurements, or capacitance measurements. Electrical sensors are often based on a chemical FET or CHEMFET device structure, where the current passing through a semiconducting channel, between two metal electrodes, is effectively gated by molecular species adsorbing or binding to the channel surface. The adsorption or binding can cause depletion or accumulation of carriers in the channel resulting in a conductivity modulation. Thus, the CHEMFET device can be used to detect molecular-binding events, yielding an electrical signal output. CHEMFETs are discussed further in Section 4.3.4.5. Graphene is an ideal material for the construction of FET biosensors because it is a zero-bandgap semiconductor, and the bandgap can be tuned by surface modification, thus making graphene FETs highly sensitive to charged-binding species [77].

4.3.4 Electrochemical Characterization (Electrochemistry Techniques)

There are several sensing techniques which utilize graphene as the active sensor element including electrochemical, optical (surface plasmon resonance), and electromechanical sensors. However, for the purposes of this chapter, we will concentrate on electrochemical techniques. Graphene is particularly suitable for electrochemical sensing due to its inherently high surface-to-volume ratio, its excellent electronic properties, and its high sensitivity to adsorbed materials.

Graphitic materials, such as glassy carbon, graphite, and carbon nanotubes (CNTs), have been extensively used in electrochemical sensing and thus substitution of these materials with graphene, with its inherently superior properties, is a logical step toward higher sensitivity sensors. Indeed, graphene has been reported to show favorable electrochemical responses when compared with other electrode materials, such as glassy carbon electrodes, graphite, and CNTs. For example, Zhou et al. [78] observed that graphene exhibits a wide electrochemical potential window of about 2.5 V in 0.1 M phosphate-buffered saline (PBS) solution at pH 7. The wide electrochemical window implies that a graphene electrode can be used in electrochemical reactions over a wide voltage range without itself reacting with the electrolyte.

Electrochemical biosensors provide an attractive means to analyze the content of a biological sample due to the direct conversion of a biological event to an electronic signal [79]. Common electrochemical techniques include cyclic voltammetry, chronoamperometry, chronopotentiometry, impedance spectroscopy, and various FET-based methods. Surface modification techniques, the various electrochemical transduction mechanisms, and the choice of the recognition receptor molecules all influence the ultimate sensitivity of the sensor.

Typically in (bio-)electrochemistry, the reaction under investigation would generate a measurable current (*amperometric*); a measurable potential or charge accumulation (*potentiometric*); or demonstrably alter the conductive properties of a medium (*conductometric*) between electrodes [80]. Other types of electrochemical detection techniques include *impedimetric*, which measures impedance (both resistance and reactance) [81,82], and *field effect*, which uses transistor technology to measure current as a result of a potentiometric gating effect [83].

Since electrochemical reactions occur at the surface of the electrode, the electrode material, surface, and dimensions (surface area) are all critical parameters. Electrochemical sensing usually requires a *reference* electrode, a *counter* or auxiliary electrode, and a *working* electrode (the sensing or redox electrode). This section reviews the commonly used electrochemical sensing techniques, which are also applicable to graphene sensors.

4.3.4.1 Amperometry

Amperometry is an electrochemical technique often applied in commercial biosensors. Amperometry takes advantage of the fact that certain chemical species are oxidized or reduced (redox reactions) at inert electrodes; the reactions driven by a constant applied potential. Amperometry is performed using a two or three electrode electrochemical cell, with a working electrode, a reference electrode, and a third counter electrode. The working electrode is usually constructed from a metal such as platinum or gold, but can be made of carbon. The reference electrode, usually Ag/AgCl, provides a fixed potential against which the potential applied to the working electrode is measured and controlled. The third electrode, the counter electrode, can be made of a conductive material, which does not react excessively with the bulk electrolyte solution. Common materials used as counter electrodes are platinum or silver wire. Platinum is often used because of its inert nature. The counter electrode supplies current to the working electrode, and its size should be larger than that of the working electrode. Reactions occurring at the counter electrode surface are unimportant as long as it continues to conduct current well. To maintain a constant current between the counter electrode and the working electrode, the counter electrode will often oxidize or reduce the solvent or bulk electrolyte. There is a small current flow between the counter electrode and reference electrode and for this reason the potential of the reference electrode does not change.

Redox Reactions

If the working electrode is driven to a positive potential, relative to the reference electrode, an oxidation reaction can occur at the working electrode of the electrochemical cell. If a negative potential is used to drive the reaction, this results in a reduction reaction. When applied to biosensors, the working electrode monitors the oxidation or reduction of a species near the surface of the electrode.

Amperometric devices continuously measure current resulting from the oxidation or reduction of an electroactive species in a biochemical reaction. Typically, the current is measured between the working and counter electrodes, at a constant potential, applied between the working and reference electrodes, and this is referred to as amperometry. Amperometric sensing is often used in enzyme-based sensors, where the enzyme catalyzes a redox reaction. For example, the enzyme glucose oxidase catalyzes the oxidation reaction of glucose to gluconic acid, where two electrons are produced per glucose molecule. These electrons reduce the working electrode [84].

Amperometry is also used in immunosensors (antibody-based sensors). In this case most analytes (eg, proteins, hormones) are not electrochemically active and cannot participate directly in redox reactions. Instead, the antibody or antigen is labeled with an electrochemical mediator such as an enzyme, which can undergo a redox reaction.

Potentiometry

Potentiometric devices measure the accumulation of a charge potential at the working electrode, relative to the reference electrode in an electrochemical cell, when zero or no significant current flows between the electrodes. Therefore this technique can be used to provide information about the ion activity (sometimes referred to as the electromotive force) of an electrochemical reaction.

Conductometric Devices

Conductometric devices can be considered as a subset of impedimetric devices, based on measuring capacitance changes. Electrochemical impedance devices are reviewed later (Section 4.3.4.4). One approach, using conductometric devices, is to directly monitor the changes in conductance of an electrode as a result of the immobilization of bioreceptors such as enzymes, complementary antibody–antigen pairs, etc. onto the electrode surface [79]. This technique has developed with semiconductor technology and the advent of nanostructured FET devices. There is an increased interest in conductometric immunosensors in combination with nanostructures, and especially nanowires, for biosensing [85]. There are several reports of conductometric sensing for detection of analytes at very low concentrations, for example, drug detection in human urine [86].

4.3.4.2 Cyclic Voltammetry

Cyclic voltammetry (CV) is the most common, simple, fast, and perhaps the most straightforward electroanalytical technique for acquiring both qualitative and quantitative information about any electroactive species (ie, a species that can be oxidized and/or reduced). This reversible technique is very useful in obtaining information about fairly complicated electrode reactions. CV applies a linear potential sweep to a working electrode at a steady scan rate, first from a negative potential to a positive potential and subsequently in the reverse direction. The resulting current at the electrode surface is measured at each applied potential. CV is, therefore, an amperometric technique.

CV can be used to obtain information about the redox potential and electrochemical reaction rates (eg, the chemical rate constant) of analyte solutions. The scan rate must be sufficiently long to allow chemical reactions to occur to a meaningful degree. A voltage is measured between the reference electrode and the working electrode, while current is measured between the working electrode and the counter electrode.

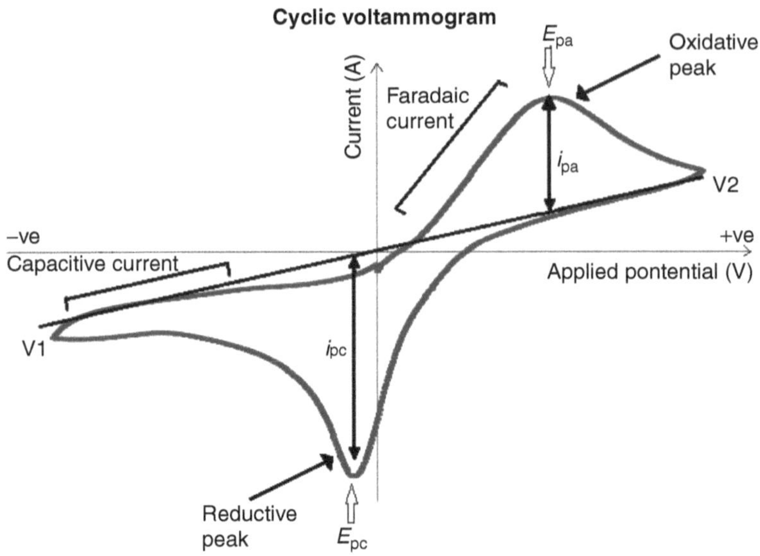

Figure 4.4 ***A schematic diagram of a cyclic voltammogram highlighting the peak cathodic potential (E_{pc}), peak anodic potential (E_{pa}), cathodic current (i_{pc}), and anodic current (i_{pa}).***

The obtained measurements are plotted as current versus voltage in a voltammogram.

A typical voltammogram is illustrated in Fig. 4.4. Generally, the voltammogram will contain an oxidative and reductive peak. As the applied potential increases from *V1* toward *V2*, and toward the electrochemical oxidative potential of the analyte, the current increases from the capacitive (resting) current. This results in a peak, called the peak anodic (oxidative) potential (E_{pa}). The current then decreases as the applied potential surpasses this point. In the reverse cycle, the applied potential sweeps into the negative direction and the measured current becomes increasingly negative until it reaches the electrochemical reductive potential of the analyte. Once the applied potential surpasses this point, a peak polar to the anodic peak is formed and the peak cathodic (reductive) potential (E_{pc}) can be determined. The differences between the cathodic and anodic current from the resting current are denoted as i_{pc} and i_{pa}, respectively.

As the reaction proceeds, the oxidative and reductive CV curves become closer and closer together. This indicates that the reaction is nearing completion.

CV can be conducted using a redox couple, often:

$$Fe_3(CN)_6^{3-} \Leftrightarrow Fe_2(CN)_6^{4-} \tag{4.2}$$

The ferri/ferrocyanide couple provides an ideal electrochemical probe for the study of chemically modified surfaces. This probe is particularly insightful when functionalizing the surface of an electrode. For example, a graphene electrode functionalized with an amino silane (APTES, 3-aminopropyl-triethoxysilane) layer can be characterized by monitoring the redox peak current of the electrode, corresponding to the oxidation potential of Fe(II). Before functionalization with APTES, this peak has a higher current value than after functionalization. This current change indicates successful chemical modification of the surface with APTES as the formation of the C—O—Si bond, between the graphene electrode and the APTES molecule, impedes the transfer of electrons between the electrode surface and the $Fe(CN)_6^{3-/4-}$ electrolyte.

4.3.4.3 Chronoamperometry

Chronoamperometry is a time-dependent technique where a square-wave potential is applied to the working electrode. The current of the electrode, measured as a function of time, fluctuates according to the diffusion of an analyte from the bulk solution toward the sensor surface. Chronoamperometry can therefore be used to measure current–time dependence for the diffusion controlled process occurring at an electrode. This varies with analyte concentration. Chronoamperometry is a sensitive technique which does not require labeling of the analyte or bioreceptor and has been applied in many studies independently or alongside other electrochemical techniques such as CV, for example, Martins et al. utilized chronoamperometry when studying the adsorption of human serum albumin to self-assembled monolayers (SAMs) on gold substrates [87].

4.3.4.4 Electrochemical Impedance Spectroscopy

Impedance is a measure of the ability of a circuit to resist the flow of electrical current. Electrochemical impedence spectroscopy (EIS) uses the application of a small sinusoidal potential to the working electrode in an electrochemical cell, while measuring the resulting current response. By varying the excitation frequency, f, of the applied potential over a range of frequencies, one can calculate the complex impedance; the sum of the real and imaginary impedance components of the system as a function of the frequency (ie, angular frequency *w*). Therefore, EIS combines the analysis of both real and imaginary components of impedance, namely, the electrical resistance and reactance.

In electrochemical sensing, impedance techniques are useful to monitor changes in electrical properties arising from biorecognition events occurring at electrode surfaces. For example, changes in the conductance of the electrode can be measured as a result of protein immobilization and antibody–antigen reactions on an electrode surface [88–90]. The increased binding of molecules at the electrode surface impedes electron transfer from a redox couple to the electrode.

4.3.4.5 CHEMFET

An FET uses an electric field to control the conductivity of a channel between two electrodes (ie, the *source* and *drain*) in a semiconducting material. Control of the conductivity is achieved by varying the electric field potential, relative to the source and drain electrode, at a third electrode, known as the *gate*. Depending on the configuration and doping of the semiconducting material, the presence of a sufficient positive or negative potential at the gate electrode would either attract charge carriers (eg, electrons or holes) or repel charge carriers in the conduction channel. This would either fill or empty the depletion region of charge carriers and thus form or deform the effective electrical dimensions of the conducting channel. Thus, the gate can be used to control the conductance of the channel between the source and drain electrodes.

FETs commonly used in biosensing applications are ISFET (ion-selective field-effect transistor), EnFET (enzyme field-effect transistor), or immunoFET (antibody-based) devices [91–95]. The generic name for chemical sensing FETs is CHEMFETs. In a CHEMFET device, the gating action is related to the interaction of molecules with the electrode channel surface [96]. A reference electrode, also present in the analyte solution, completes the circuit via the gate-voltage bias [91,97]. Binding events at the channel surface influence the presence of accumulated charge carriers near the channel surface. Surface charge accumulation is then proportional to analyte concentration. The change in surface charge produces a modified electrical signal in the form of a change in the channel conductivity [94,98].

Several groups have developed silicon nanowires [99], CNTs [100], or graphene channels [8,9]) CHEMFET sensors. The many emerging applications for analytical CHEMFET sensors [95] and graphene CHEMFET sensors are reviewed in Section 4.7.

4.3.5 Scanning Probe Techniques

The study of graphene using scanning probe microscopy techniques enables examination of a number of crucial properties including morphology and surface potential. Scanning tunneling microscopy (STM) is an important surface science technique and measures the tunneling current between an atomically sharp, metallic tip and a conducting surface. STM is capable of atomic resolution imaging of graphene. It is invaluable in studying the growth of graphene, and is used extensively to understand the processes involved in CVD graphene growth on metals. STM is used to study crystalline structure, boundaries, terraces, and step edges [101,102]. Data from STM measurements are used to provide a better physical understanding of graphene growth processes, which are used in process optimization. STM is also useful for studying the edge states of graphene, and understanding the molecular formation of GNRs [103].

Atomic force microscopy (AFM) is a commonly used technique to examine topographic features of samples. The technique employs a sharp tip, supported by a cantilever, to raster scan over a sample. The tip is held in close proximity (or indeed in contact) with the surface as it traverses the sample. Subtle deflections of the tip caused by changes in the surface topography are measured using a laser, which is reflected off the back of the cantilever to a detector. AFM is capable of providing nanoscale resolution topographical images of both insulators and conductors and is thus capable of discerning the number of layers within a graphene sample. Substrate roughness is commonly monitored using AFM and this has been applied to examine the effect of graphene functionalization with amine functional groups and subsequent biofunctionalization with bioreceptor molecules [10].

Kelvin probe microscopy (KPM) is a noncontact variation of AFM that measures surface potential. KPM has been used in graphene research to probe the electronic uniformity of the surface, provide morphological information, and determine local graphene thickness [104]. The technique has further been used to study the local work function of graphene [105,106], probe chemical variations [107], or examine the effect of interlayer screening of graphene sheets [108]. KPM has been used to measure the local electronic transport properties of functionalized graphene [109], giving insight into the chemical doping effect of the attached functional groups. This is used to understand the active binding regions in graphene, and therefore the mechanisms involved in chemical sensing.

4.3.6 Contact Angle

The contact angle is a quantitative measure of the wettability of a surface or material (wetting of a solid by a liquid). Characteristics such as surface roughness, surface energies, surface chemistry, and surface coatings play a role in a material's wettability. Contact angle measurements are commonly used to study the surface properties of graphene samples and are particularly useful in studying the effects of chemical functionalization of the surface. The contact angle is defined geometrically as the angle formed by a liquid at the three-phase boundary where a liquid, gas, and solid intersect. The contact angle is described by the Young's equation (Eq. 4.3) using interfacial surface tensions between solid and liquid. The contact angle may also be directly measured to calculate the ratio of interfacial surface tensions if the interfacial surface tensions are unknown.

$$Y^{SV} = Y^{SL} + Y^{LV} \cos\theta \tag{4.3}$$

Here θ is the contact angle, γ_{SL} is the solid/liquid interfacial free energy, γ_{SV} is the solid surface free energy, and γ_{LV} is the liquid surface-free energy. Young's equation is used to describe the interactions between the forces of cohesion and adhesion between sample and fluid and measure what is referred to as surface energy.

A water droplet with a contact angle over 90 degrees indicates a hydrophobic surface. This condition is exemplified by poor wetting, poor adhesion, and the solid surface-free energy is low. A droplet with a small contact angle indicates a hydrophilic surface. This condition reflects better wetting, better adhesion, and higher surface energy. Contact angle measurements have been used by Tehrani et al. to study the functionalization of graphene surfaces with surface amine groups [10].

4.3.7 Fourier Transform Infrared Spectroscopy, Ellipsometry

Another chemical analytical technique that can be used to analyze surface chemistry is infrared (IR) spectroscopy. IR spectroscopy interrogates the vibrational characteristics of chemical functional groups in a sample. Different chemical bonds absorb IR radiation at different wavelengths and an IR spectrum displays intensities of the absorptions as a function of wave number. Different functional groups can be identified from characteristic vibrational frequencies of functional groups present within the sample. Functional groups, such as OH, C=O, NH, are readily identifiable in IR spectra.

FTIR instruments employ interferometric techniques for the collection of spectral information and the spectrum is calculated as an inverse Fourier transform of the interferogram. FTIR spectroscopy was used to study the chemical composition of modified HOPG surfaces, and thus used to confirm that functionalization had taken place [10]. FTIR has also been used to assess graphene functionalization in a similar way [76].

Ellipsometry is an optical technique, which characterizes the dielectric properties of thin films. Spectroscopic ellipsometry can be used to analyze the growth of thin films on graphene electrodes [110].

4.4 FUNCTIONALIZATION CHEMISTRIES

The surface chemistry of graphene can be modified in a similar manner to that of HOPG, glassy carbon, or CNTs [111–113]). Chemical functionalization of graphene can be used to tailor the electronic properties of graphene via doping and bandgap engineering effects, produced by chemical modification or adsorption of molecules onto the graphene surface [76,107,114–118]. Chemical modification of graphene can also be used as the basis for graphene sensors. Chemical specificity is induced by binding highly selective "receptor" molecules to the graphene surface. Such "receptors" may include amino acids, enzymes, antibodies, aptamers, or indeed any selective and specific molecule. It is possible to adsorb receptor molecules—and this process is considered in Section 4.4.1. However, the most reliable method for attaching a receptor molecule to graphene is via covalent bonding, usually using a linking molecule or functional group. There are a host of potential linking chemistries and functional groups, which can be derivatized. Examples of graphene functionalization include halogenation, hydroxylation, epoxidation, carboxylation, amination, alkylation, azidation, and many more. These methods are reviewed in Sections 4.4.2 and 4.4.3. A variation of some of these chemical modification processes is electrochemical functionalization (see Section 4.4.3). Electrochemically enhanced functionalization can be used to increase reaction rates and increase the degree of functionalization [10].

Alternative methods for integrating a specific receptor molecule into a graphene sensor utilizes an intermediate polymer layer. The polymer is coated onto the graphene and receptor molecules can then bind with functional groups on the polymer. This method avoids direct chemical modification of graphene, as will be discussed in Section 4.4.4.

Once graphene surfaces have been chemically functionalized, these surfaces can be used in sensing applications. However, it is the addition of selective receptor molecules onto the graphene surface, which introduces specificity into the sensor. Graphene biosensors can be fabricated by covalent immobilization of proteins or other molecular receptors to graphene surfaces through linking groups [10]. Covalent immobilization of receptor molecules onto graphene requires a chemical reaction of the functional groups of the receptor, with functional groups bound to the graphene surface. Such methods are discussed in Section 4.5.

4.4.1 Noncovalent Attachment

Noncovalent interactions primarily involve hydrophobic, van der Waals, and electrostatic forces and require the physical adsorption of molecules on the graphene surface [119]. Perhaps more interesting are the noncovalent attachments which rely on the interaction of π molecular orbitals of a planar molecule, with the sp^2 π-bonded system present in graphene. Noncovalent interactions can be used to modify the chemistry or solubility of graphene, by introducing hydrophilic ligands onto graphene—since pristine graphene is hydrophobic [120].

There are several reported examples of π-stacking of planar molecules on top of graphene using pyrene molecules [121,122]. These pyrenes may themselves be derivatized, for example, with polymer chains or functional groups which modify the surface chemistry of the pyrene/graphene-stacked structure [122].

Fig. 4.5 illustrates the π-stacking of planar molecules on a graphene substrate. Other planar molecules that can form π–π interactions with graphene include porphyrins [123] and phthalocyanines [124–126].

Noncovalent functionalization via π-interactions (also used for CNTs) is an attractive functionalization method, because it offers the possibility of attaching functional groups to graphene without disturbing the electronic sp^2 network [127,128]. Kodali et al. [129] have shown that noncovalently functionalized epitaxial graphene with pyrenebutanoic acid-succinimidyl ester (PYR-NHS) can be prepared without disrupting graphene's electronic structure.

Polymers can also be noncovalently bound to graphene. Sulfonated polyaniline has been used to produce a water-soluble graphene composite [130], while graphene functionalized with polyaniline (PAni) has also been used as

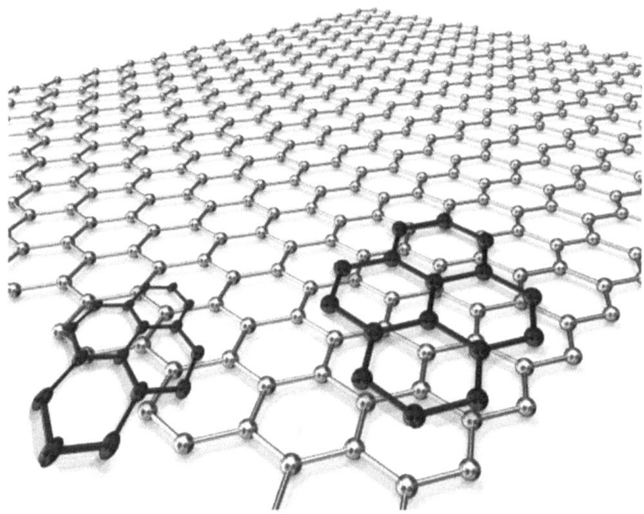

Figure 4.5 ***π-Stacking of a planar pyrene molecule on graphene substrate.***

a high-performance flexible electrode [131,132]). The functionalized material offers enhanced electron transfer from graphene to the PAni film.

Zhang et al. [125,126]) have investigated the theoretical binding of organic donor, acceptor, and metal atoms on graphene sheets. These calculations revealed the effect of the different noncovalent functionalizations on the electronic structure and transport properties of graphene. The adsorption of 2,3-dichloro-5,6-dicyano-1,4-benzoquinone and tetrathiafulvalene induces hybridization between the molecular levels and the graphene valence bands. This transforms the zero-gap semiconducting graphene into metallic graphene. However, there is some debate over whether the current voltage (I − V) response of noncovalent modification by organic molecules is sufficient to significantly alter the transport properties of graphene for sensing applications. Noncovalently bound molecules can be functionalized with side chain groups, which can subsequently be used to attach bioreceptor molecules. These receptors can then be used to interact with target biomarkers. Noncovalent binding in sensor applications is reviewed in detail in Section 4.4.4.

4.4.2 Chemical Methods

4.4.2.1 Acid or Other Wet Treatments

One of the simplest methods for graphene functionalization is graphene oxidation, using mild to strong acid treatments, for example, H_2SO_4, HNO_3. These can produce oxidized surfaces with a variety of possible terminations, for example, CO, C—OH, C—O—C, COOH. These methods have also been used to oxidize HOPG [133] and CNTs [134,135].

Treatment of graphite with strong acids is also a method for producing graphene [136]. The strong acids oxidize the graphite to GO [137] and subsequently induce exfoliation by overcoming the large interlayer cohesive van der Waals energy (5.9 kJ/mol carbon) [138,139]. GO can then be reduced to rGO, also referred to as graphene in some papers. An example of this type of graphene production process is oxidation of graphite using Hummers' modified process [16]. This uses a mixture of sulfuric acid, sodium nitrate, and potassium permanganate in an aggressive chemical process, which disrupts the sp^2-bonded network and introduces hydroxyl or epoxide groups [140,141] in the basal plane, while carbonyl and carboxylic groups, together with lactone, phenol, and quinone, attach to the edges [142].

When developing sensors using oxidized graphene surfaces, one must be careful not to cleave the graphene from its supporting substrate. Controlled oxidation of graphene is thus essential in maintaining the integrity of the graphene layer and also in avoiding complete conversion to GO and hence severe degradation of the layer's electronic properties. The electrochemical oxidation of graphite [143] and CNTs [144–146] is well documented and serves as a starting point for understanding the oxidation of epitaxial graphene. The electrooxidation of graphitic surfaces is known to begin when the surface charge becomes sufficiently positive to surmount the overpotential; at this point oxygen atoms in the electrolyte form covalent bonds to the surface via the conversion of conjugated carbon atoms from sp^2 to sp^3.

Controlled electrochemical oxidation of epitaxial graphene can be performed using nitric acid (HNO_3) under potentiostatic conditions [12]. Selective exposure to the nitric acid allows selective oxidation of the epitaxial graphene substrate. Cyclic voltammetry of epitaxial graphene in 1 M HNO_3, recorded against the Pt counter/reference electrode, is irreversible with an oxidation peak at about +0.6 V. Thus, the application of a potential of +0.8 V to the epitaxial graphene yields controlled oxidation of the epitaxial graphene area that is in contact with the acid.

The hydroxyl, epoxide, and carboxyl groups of oxidized graphene can all be used in further reactions [147,148]. However, the carboxyl termination is the most useful for coupling with amine-terminated molecules, forming an amide bond. This coupling is the basis for attachment of nucleic acids, enzymes, antibodies, and other bioreceptor molecules to carboxyl-terminated graphene.

4.4.2.2 Plasma Methods

Perhaps the simplest, quickest, and most production friendly functionalization method, in terms of scale-up, is plasma functionalization. Gas-phase plasma reactions are very clean, fast, and controllable. Plasma functionalization is used to control hydrophobicity/hydrophilicity of graphene, which is useful for graphene in composites (where graphene needs to bind with polymer chains). Plasma functionalization can also be used to prevent aggregation of exfoliated graphene flakes. Plasma treatment of CNTs and graphite (HOPG) has been widely reported [149,150].

However, plasma methods can be limited (by the gases that are available on commercial plasma systems) in terms of the functional groups that can be attached, for example, oxidation (using an oxygen plasma), capable of producing CO, C—O—C, COOH, and ammonia (NH_2 termination), fluorination/chlorination. An even more critical limitation of plasma functionalization is the potential damage to the graphene surface from plasma processes. Plasma reactions involve reaction of high-energy radicals or ions with the graphene surface. Such reactions inevitably create imperfections in the graphene lattice, which can severely degrade the carrier transport properties. This was reported by Baraket et al. [151], who observed an increased D-peak (sp^3) in the Raman spectra of plasma-functionalized graphene.

As discussed in Section 4.4.1, chemical functionalization methods also induce an increased Raman D peak, and have been observed to degrade the electrical conductivity of graphene devices. If plasma functionalization can be controlled to minimize damage to the graphene, it may be no more detrimental to the electronic properties of graphene than wet chemical modification procedures. Plasma processes can be effective for introducing functional groups onto the graphene surface, for subsequent attachment of bioreceptors in sensor applications [151]

Plasma Chlorination and Fluorination

Chlorination, fluorination, and hydrogenation plasma treatments have all been used to modify graphene [152]. XPS and Raman characterization of treated graphene surfaces revealed how the carbon network in graphene and the electronic properties and band structure of graphene were modified by functionalization. In contrast to pristine conductive graphene, graphane—the product of covalent bonding of hydrogen to graphene—is reported to be highly insulating.

Fluorination of CVD graphene, grown on copper foils, using xenon difluoride (XeF_2) gas, has been reported by Li et al. [153]. Graphene could be fluorinated on either one side or on both sides, by treating the graphene in a XeF_2 etch system. Double-sided fluorination was achieved by transferring the graphene film to an SOI substrate first. This allows the XeF_2 gas to etch away the underlying silicon layer and fluorinate the backside of the graphene film. Exposure of graphene to XeF_2 gas results in fluorine chemisorption, which dramatically changes the optical and electronic properties of the graphene. Most immediately evident is a disappearance of optical contrast with the silicon dioxide substrate upon fluorination, suggesting an insulating nature for fluorinated graphene. Indeed, Robinson et al. observed a bandgap of 2.9 eV in fluorinated graphene—which was electrically insulating [154]. Sherpa et al. performed fluorination of epitaxial graphene using SF_6 plasma treatments and investigated the effect on the work function of fluorine-modified graphene [106]. Fluorination modified the graphene bandgap by up to 0.56 eV.

Photochemical reactions can also be used to produce chlorinated graphene. Uniform 8% coverage with stoichiometry $C_{16}Cl$ was observed, with the basal plane graphene partially chlorinated and exhibiting a nonzero bandgap and a higher sheet resistance than pristine graphene, which can be attributed to increased sp^3 bonding—as evidenced by a D peak (1330 cm^{-1}) in the Raman spectrum of the chlorinated graphene [155,156].

Relative to hydrogen and fluorine plasma treatments, chlorine plasma reactions with graphene have been reported to exhibit the slowest kinetics, with the slowest increase of disorder over reaction time (in minutes), compared to seconds for F and H plasmas. An advantage of chlorination over fluorination and hydrogenation is that chlorination plasma treatment mostly preserves high mobility [157,158]) but also allows control of the graphene Fermi level and the preparation of controllable and effective bonding structures [152]. Indeed, GNRs had p-doping after chlorination and were not extensively structurally damaged by the Cl-plasma treatment.

It is evident that functionalization using Cl, F, and H degrades the electrical conductivity of graphene and that a high degree of functionalization has been shown to destroy the network of graphene for Cl, F, and H treatments, rendering the modified materials insulating [116].

Plasma Amination

Ammonia plasma treatments have been used to controllably functionalize CVD graphene with nitrogen and primary amines [151], thereby tuning the chemical, structural, and electrical properties of the modified graphene. Using low-energy ammonia plasmas avoids generating extensive defects in the graphene basal plane [159]. Mixed Ar/NH_3 plasmas contain reactive species including Ar^+, NH_3^+, NH_2^+ ions along with H_2 and H, N, NH, and NH_2 radicals. Raman and XPS data show the modification of the graphene surface through the emergence of a nitrogen peak in the XPS spectra and corresponding observation of the sp^3 D peak in the Raman spectra of graphene after ammonia plasma treatment. After washing, no change of the C1s and N1s in the high-resolution XPS spectra was observed, confirming that the functional groups were stable and covalently bound. Amine termination of graphene was inferred from contact angle measurements before and after ammonia plasma treatment where the surface changed from hydrophobic to hydrophilic. The trade-off between chemical functionalization of graphene and reduced conductivity, due to the change in hybridization of carbon atoms to sp^3, which creates scattering sites and can act as a carrier trap that reduces carrier density, did not adversely affect the graphene conductivity. This was attributed to n-doping of the graphene films. Sheet resistances of amine-terminated graphene increased by only 6.5 times from 1.2 to 8 kΩ/sq at the highest concentrations. This compares to a 1000-fold increase in sheet resistance of fluorinated or oxygenated graphene. The amine-terminated graphene surface was used to attach single-stranded DNA (ssDNA) probes, which were used to detect complementary target ssDNA. The FET was highly selective to nonspecific ssDNA and had a significantly larger current response than a comparable rGO device.

4.4.2.3 Amination

As well as ammonia plasma treatments, there are several wet chemistry methods for amine functionalization of graphene including diazotization [76], thiol chemistry [160], and surface silanization using aminosilane derivatives, for example, APTES [8,9]). Some of these methods (diazotization being one) directly introduce an amine group onto the graphene surface. However, other methods, such as APTES attachment, require an additional surface treatment of graphene prior to amine termination. For example, APTES is attached to a graphene surface, which has first been hydroxylated. Indeed, APTES can be attached to a variety of substrates provided they contain surface hydroxyl groups. These hydroxyl groups then react with alkoxysilanes to form covalent substrate—O—C bonds. Diazotization, APTES functionalization, and azide are discussed in detail in the following sections.

4.4.2.4 Diazotization

The reduction of diazonium salts to phenyl radicals has been widely employed for grafting aryl groups to the surface of sp^2-hybridized carbon materials including glassy carbon, HOPG, and CNTs [161]. Recently, this type of chemistry has also been applied to chemically functionalize graphene, by several research groups, through spontaneous electron transfer from the graphene layer to the diazonium salt [75,76,162–166]. The aryl diazonium cation accepts an electron from the substrate and generates an aryl radical by releasing a nitrogen molecule. This covalent modification strategy has recently been applied to epitaxial graphene on SiC [167], chemically converted graphene [168], MEG [169,170]), and other sensing platforms [171].

Surface modification with nitrophenyl groups has been achieved through the spontaneous reaction of the diazonium salt (4-nitrophenyl diazonium tetrafluoroborate, denoted as 4-NPD) with a graphene layer, as illustrated in Fig. 4.6. The rate of graphene functionalization can be significantly increased by electrochemical enhancement (see Section 4.4.3). The functionalization changes the electronic structure and transport properties of the epitaxial graphene from near-metallic to semiconducting.

Site-Dependent Functionalization Using Diazonium Molecules

The basal plane and edge sites of graphene layers have been reported to have different reactivity to diazonium functionalization [169,170]). Using spatially resolved Raman spectra obtained from both edge and basal regions

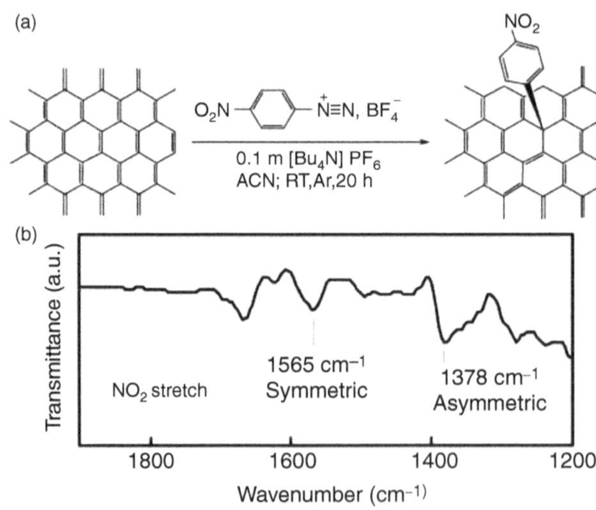

Figure 4.6 (a) Schematic illustration of the spontaneous grafting of aryl groups to epitaxial graphene via reduction of 4-NPD tetrafluoroborate. (b) Transmission FT-IR spectrum of NP-functionalized epitaxial graphene (EG). *(Reprinted with permission from Ref. [76]. Copyright (2009), American Chemical Society.)*

of MEG, Lim et al. revealed that 4-bromobenzene diazonium tetrafluoroborate (4-BBDT) molecules were noncovalently functionalized on the basal region of MEG, while they were covalently bonded to the edges of MEG. A chemical doping effect was induced in the graphene by noncovalently functionalized 4-BBDT molecules on the basal plane region.

4.4.2.5 APTES

Organosilanes are examples of molecules used to form SAMs on a wide variety of hydroxylated surfaces. APTES is one of the most widely used organosilanes for surface modification as it is an economical and efficient method. In biosensing applications, APTES functions as a linker molecule; one end of the APTES molecule binds to the device surface, while the other end immobilizes specific bioreceptor molecules [172].

The reaction mechanism is thought to start with hydrolysis of the ethoxy groups in APTES forming silanols, which subsequently condense with surface hydroxyl groups to form a monolayer of APTES molecules. The surface-bound APTES molecules also cross-link to form a siloxane network with the amino groups protruding from the surface and are capable of binding biological recognition molecules such as DNA, ligands, and antibodies.

However, the process is not always uniform and multiple APTES layers can form at the surface of the device. Organosilane SAM formation is extremely sensitive to the amount of water present within the system, as they are able to readily react with water to form silanols [173]. There may also be nonuniformity in the structure of each layer and therefore the reactivity of APTES layers may differ from one to another. These issues are problematic for biosensing applications due to lack of reproducibility, a trait which is vital for clinical applications [37,38]).

Before the sensor surface can be functionalized with APTES linking molecules, the sensor surface must first be modified to contain hydroxyl groups. This is generally achieved using one of two methods; piranha or Fenton reactions. The piranha solution consists of three parts H_2SO_4 to one part hydrogen peroxide (H_2O_2) at 30% making it extremely oxidizing and highly corrosive. The Fenton reaction takes place between iron(II) sulfate, ethylenediaminetetraacetic acid, and H_2O_2. Both piranha and Fenton reactions generate a hydroxyl-terminated substrate surface. To further increase the amount of hydroxyl groups, the substrate is often rinsed in deionized water ensuring complete hydroxylation. APTES functionalization of the substrate should be immediately after hydroxylation is complete [173].

APTES functionalization of Fenton-hydroxylated epitaxial graphene devices has recently been demonstrated to obtain an amine-terminated surface [8,9]). Antibodies targeted against the pregnancy hormone, human chorionic gonadotropin (hCG), were subsequently grafted to the amine-terminated graphene to produce an amperometric immunosensor. Hydroxylation of the sensor surface using the reaction was first undertaken to yield an —OH terminated graphene layer. Epitaxial graphene samples were then immersed in a solution of hydrogen peroxide and iron(II) sulfate and maintained at pH 3 for 30 min.

Azide Functionalization

Organic azides have the generic formula R–N_3, where the azide functional group can be described by the resonance structures (a) and (b):

(a) (b) (4.4)

Azides are thermally or photochemically unstable and break down, losing nitrogen, to yield nitrene radicals:

$$R\text{–}\ddot{\underset{\cdot}{N}}\cdot \tag{4.5}$$

The nitrene can react with graphene via an electrophilic [2 + 1] cycloaddition reaction a biradical pathway, after intersystem crossing [174]. Nitrene functionalization was observed to open a bandgap epitaxial graphene.

4.4.3 Electrochemically Activated Methods

4.4.3.1 Diazotization

Enhanced reaction rates for diazotization of graphene surfaces also can be achieved using an electrochemical method [10]. Graphene devices were immersed into a solution of 2 mM and 0.1 M tetrabutylammonium tetrafluoroborate in acetonitrile. The electrochemical reaction between graphene and 4-NPD was performed with a conventional three-electrode cell, which included an Ag/Ag+ reference electrode, a platinum wire counter electrode, and the epitaxial graphene channel as the working electrode. Cyclic voltammetry was used to monitor the completion of the reaction. Following attachment of nitrophenyl groups to the graphene surface, reduction of the nitro group to phenylamine was then performed using chronoamperometry. This electrochemically enhanced reduction process has several advantages compared to previously reported diazotizations [76,161,175,176]. It a much faster process (just a few minutes) than other methods, which take several hours [167]; it does not need to be performed in a dark environment and purging argon or nitrogen gas for the exclusion of oxygen is not required [22,23,177]).

The surface-bound amino group can subsequently be used to covalently bind to virtually any biomolecule containing a carboxyl group—forming an amide link. Tehrani used this technique to develop a graphene immunosensor for detection of the oxidative stress marker 8-hydroxydeoxyguanosine (8-OHdG).

4.4.3.2 Electrochemical Oxidation

Electrochemical oxidation of graphene, performed using a strong acid such as HNO_3, is a useful method for introducing functional groups to graphene surfaces. Electrochemical oxidation of pristine epitaxially grown graphene was reported to create numerous surface defects [178]. A similar effect has been observed on the walls of CNTs, leading to an enhanced response in a CNT sensor. It has been suggested that electrochemically oxidized graphene or graphene with large amounts of defects could be used for highly sensitive electrochemical sensing [178]. Electrochemical methods can be used to oxidize selected areas of a graphene device, thereby enabling subsequent attachment of molecules to only a portion of the graphene [12,179].

4.4.4 Polymer Coating of Graphene

4.4.4.1 Larger Molecules

An alternative way of attaching bioreceptor molecules, for example, antibodies to a graphene sensor electrode, is through grafting to a noncovalent intermediate layer such as pyrene [157,158]) or PAni [180]. This avoids the introduction of defects into the graphene substrate and preserves graphene's electrical characteristics [42,43]). Intermediary pyrene or PAni layers allow efficient electron transfer to the graphene electrode.

The conductive polymer PAni has been at the center of many biosensing studies [181]. PAni is a promising platform for electrochemical biosensing devices due to its ease of synthesis and excellent stability. It has also proven to be an effective technique to functionalize nanosensor surfaces, for example, silicon [182]. The formation of PAni from the aniline monomer can be achieved via electropolymerization or chemical polymerization. PAni can be prepared in a number of different morphologies, ranging from nanospheres to fibers. The nano forms of PAni are widely used for sensing applications due to their high surface area, low diffusional resistance, and high current density [183]. PAni functionalized electrodes, prepared by electropolymerization of an aniline monomer solution using a potentiostat or galvanostat, produce a PAni film on the electrode surface—the film thickness being controllable through modulation of the electrochemical polymerization process [184].

Development of biosensors using PAni-coated graphene electrodes requires correct orientation of the antibody on the PAni–graphene support. This is crucial to ensuring that antibody activity is preserved after grafting and that binding sites in each antibody are available for antigen binding. PAni-modified graphene-screen printed electrodes [8,9]) have been used in the development of an hCG (pregnancy hormone) immunosensor. The sensor functionalization method preserves the chemical structure of graphene and allows subsequent binding of the bioreceptor antibody, anti-hCG. A PAni amine layer is obtained by in situ electropolymerization of aniline on the graphene electrode and these amine groups can react with the —COOH groups at the crystallizable fragment (Fc) region of the antibody, activated by carbodiimide chemistry, to ensure successful antibody binding. The antibody is bound to the PAni through its Fc terminal (thus allowing the paratope regions free to bind to the antigen epitopes). The resulting biosensor was used in the detection of hCG in biological samples.

4.5 BIOFUNCTIONALIZATION

Biofunctionalization plays a critical role in the development of biosensors, which have many applications in clinical diagnosis and monitoring of health conditions and diseases. Biofunctionalization of graphene surfaces endows them with highly selective and specific biological recognition capabilities. Following chemical modification of graphene sensor surfaces, using chemical functional linking groups, biofunctionalization can be carried out to attach receptor molecules to the sensor. Attachment of receptor molecules to the sensors results in the ability to efficiently and selectively sense low concentrations of biomolecules or biomarkers.

Various types of receptor molecules can be used, such as enzymes, antibodies, or nucleic acids. Although each type of receptor may work in a slightly different manner, they generally utilize spatial and chemical complementarity to recognize and selectively bind with target molecules. Antibodies are the most common example of such a selective bioreceptor. Antibodies, sometimes called immunoglobulins, are proteins produced by an organism's immune system in response to foreign bodies. They have an

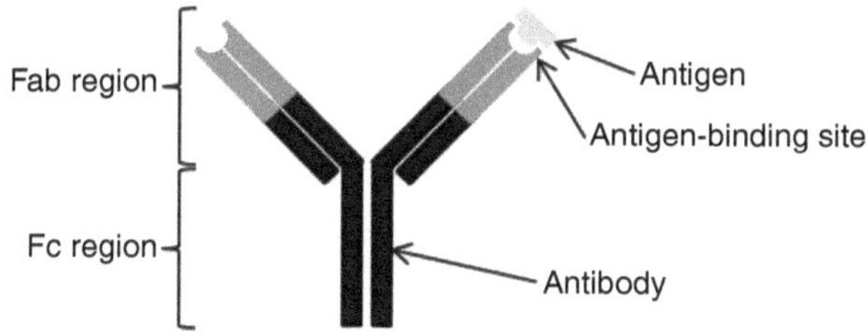

Figure 4.7 ***Illustration showing the antibody's Y-shaped structure.*** The antigen-binding site, or paratope, is situated at the top of the antigen-binding fragment (Fab) region. The complementary antigen's epitope is specific to the paratope and selectively binds to the antibody at this site. The crystallizable fragment (Fc) region is also shown.

extremely high affinity to their target biomolecule and are used routinely in many immuno-based assays, including ELISAs.

Antibodies are large, Y-shaped, multidomain proteins which contain specific binding sites for a complementary antigen, known as paratopes. These binding sites (analogous to a lock) are specific for one epitope (similarly analogous to a key) on a particular antigen as illustrated in Fig. 4.7. The arms of the antibody's Y-shaped structure are the antigen-binding fragments (Fab) region. The paratopes are situated at the end of the Fab fragment. The crystallizable fragment (Fc) region is located at the base of the antibody.

Correct conformation of an antibody is crucial for selective binding to complementary antigens. Antibodies maintain their conformation through disulfide bridges; however, these are highly reactive and easily broken. As such, when covalently bound to the sensor surface, the antibody may become unstable and susceptible to loss of vital traits such as affinity and specificity. This can be problematic in immunosensor applications. Problems may arise from the antibody not binding to the specific epitope or the antibody binding to other species within the sample. However, these effects can be minimized by using smaller, recombinant antibody fragments such as Fab fragments instead of whole antibodies. Not only does this improve specificity but it can also aid sensitivity of the sensor by overcoming steric hindrance effects, resulting in increased antibody coverage of the sensor surface [185].

Other bioreceptor molecules which are commonly employed include aptamers (artificial oligonucleotides), small molecules, cells, peptides, proteins, and nucleic acids. Each of these different bioreceptor types possess specific merits, which lend themselves well to particular applications. For instance, aptamers are short, single-stranded nucleic acid oligomers (DNA or RNA) whose structure is dependent upon their sequence. The structure of aptamers may include a series of loops, bulges, hairpins, etc., which can fold further to form more complex structures. This allows aptamers to form complementary shapes wrapping around all or part of their target or to fit snugly into gaps within the surface of much larger targets. Aptamers are reported to be more stable than antibodies and can be developed in a much shorter timeframe and so offer an advantage where a new biosensor product might require rapid development or where a longer product shelf-life is required.

4.5.1 Selective Antibody Orientation

It is important when antibodies are immobilized on a surface that they maintain their conformation and function. As such it is sometimes necessary to control the orientation of the antibodies. When antibodies are immobilized on a sensor surface, they may randomly adopt a number of orientations. As illustrated in Fig. 4.8, the antibody may attach "end-on" (Fc attached to the surface, Fig. 4.8a), "head-on" (Fab attached to the surface, Fig. 4.8b), sideways (Fig. 4.8c), or they may lie flat on the surface (Fig. 4.8d). Indeed, it is likely that the actual orientation will comprise a combination of these possible positions and result in uncontrolled and unpredictable antibody-binding efficiencies leading to poor sensor sensitivity and reproducibility.

It is therefore desirable to control antibody grafting such that the Fc fragment binds to the sensor surface. This ensures that the antigen-binding sites (on the Fab fragment) protrude away from the electrode surface, thus allowing more efficient binding to target antigens. Oriented antibody immobilization is expected to yield an analytical improvement of the final immunosensor. Several approaches have been developed to achieve this, as reviewed by Wan et al. [186].

One of the simplest processes which can be adopted for controlling antibody orientation uses EDC/NHS chemistry. This method activates the carboxylic group at the Fc terminal via carbodiimide chemistry [187].

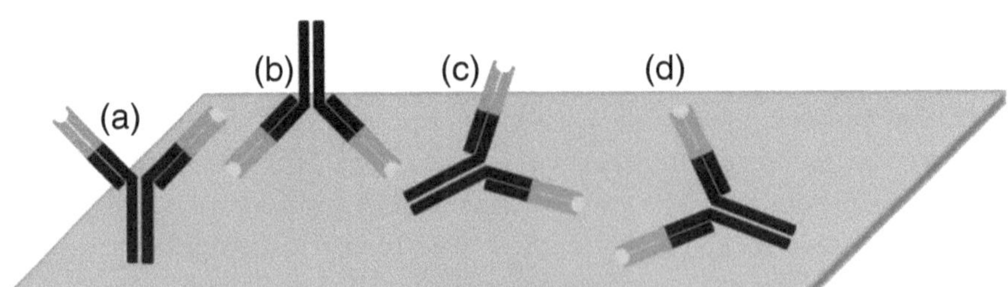

Figure 4.8 ***Illustration of possible different orientations of randomly immobilized antibodies: (a) end-on, (b) head-on, (c) sideways, (d) flat on the surface.***

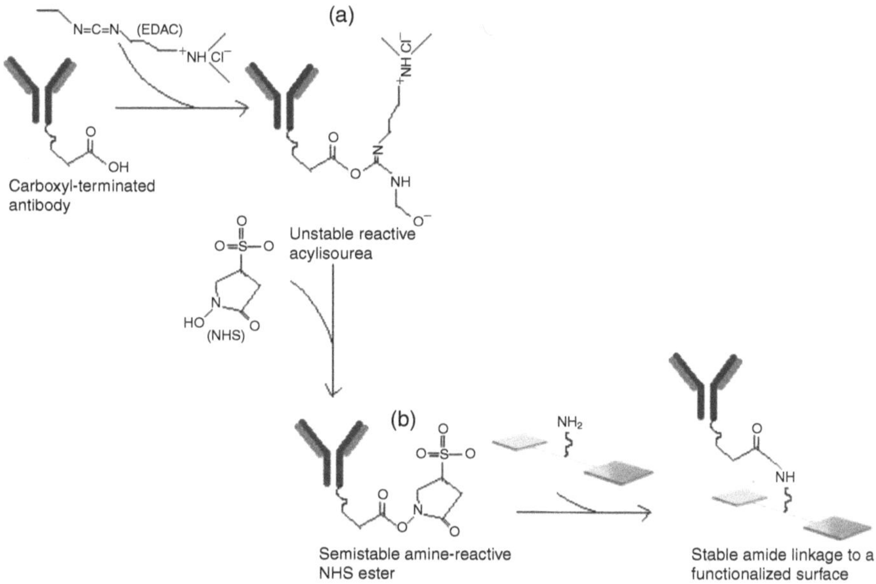

Figure 4.9 ***Activation of an antibody by EDC/NHS chemistry.*** EDC first reacts with the terminal carboxyl of the antibody to form an unstable reactive O-acylisourea (a) before a substitution occurs between EDC and NHS to form a more stable NHS ester (b). This is then reacted with the amine group on the sensor surface to peptide bond the antibody to the sensor surface (c).

4.5.1.1 EDC/NHS Activation of Antibodies

The carboxylic group of the antibody can be activated via 1-ethyl-3-(3-dim ethylaminopropyl)-carbodiimide (EDC, sometimes written as EDAC)/*N*-hydroxysuccinimide (NHS) chemistry, which enables the formation of peptide bonds between an amine-terminated biosensor surface and a carboxylic group on the antibody [188], as illustrated in Fig. 4.9.

EDC/NHS chemistry generates reactive intermediates, NHS ester —COONHS)-terminated antibodies, by reacting the antibodies with the water-soluble carbodiimide. EDC acts to dehydrate the carboxylic groups into an active O-acylisourea intermediate. This is quickly displaced by nucleophilic attack from a primary amine. NHS reacts with the O-acylisourea intermediate to form a semistable NHS ester, which favors conjugation with amines. Exposure of the —COONHS-terminated antibody to an amine-terminated surface results in the formation of a peptide bond, thus covalently attaching the antibody to the biosensor surface [189].

As EDC/NHS chemistry facilitates the conjugation of the carboxyl group to amines without becoming part of the final linkages between molecules, the process is referred to as a zero-length cross-linker [190]. This process is commonly used to attach proteins or peptides to carboxyl/amine-modified surfaces.

The EDC/NHS method has been utilized by Teixeira et al. [8,9]) in the development of an epitaxial graphene sensor for the detection of the hormone hCG—utilized in pregnancy testing. Teixeira et al. employed di-*tert*-butyl dicarbonate (*t*-BOC) to protect the amine groups on the antibodies and thus prevent cross-linking and aggregation between antibodies. The process also ensured that the carboxylic groups present on the antibody bound only to the amine functional groups on the sensor surface.

EDC/NHS activation of the antibodies was then undertaken before the amine-terminated graphene sensor was exposed to activated antibodies. Following attachment of the antibodies to the graphene sensor surface, the *t*-BOC protecting groups were removed to restore the amine groups on the antibody. Exposure for 1 h to trifluoroacetic acid was used to cleave the amide bond between the *t*-BOC and the antibody. The process also hydrolyzes the ester, generated by the activation of the carboxylic groups by EDAC and NHS, back into carboxylic groups. Finally, the authors applied a 5% bovine serum albumin (BSA) in PBS solution for 10 min to the functionalized surface to block nonspecific binding of the target biomarker to any remaining free surface amine groups.

4.5.2 Blocking Nonspecific Binding

Blocking of nonspecific interactions is an essential step in all types of immunoassays. This process minimizes background noise, which may otherwise be introduced through the target analyte binding to components of the sensor other than the active bioreceptor molecules or through any nonspecific binding. Nonspecific binding is mediated through interactions within a sample and/or between a sample and the surface of the immunoassay. Therefore, measures should be taken to block nonspecific binding before sample analysis is undertaken.

The blocking agent typically binds to areas of the sensor surface which are not biofunctionalized and also to antigen-binding sites on antibodies. When the antigen is introduced to the sensor, the blocking agent dissociates from the epitope-binding region of the antibody, leaving the specificity region available for the antigen to bind. This occurs as a result of the high affinity the complementary antigen and antibody pair have for one another, which dominates the lower affinity the antibodies have for the blocking agent [191].

The blocking agent used is generally derived with respect to the antibody used for the analysis. The origin of the blocking serum must not be the same as the origin of the antibody, for example, goat serum should not be selected when a goat-derived antibody is being used. A serum derived from an unrelated species is highly recommended.

The selection of an appropriate blocking system is essential to the development of a specific and sensitive assay. The use of BSA as a blocking reagent is well documented and has been proven to be an effective blocker of nonspecific protein surface binding on medium and high binding surfaces, as well as many preactivated covalent surfaces. BSA is typically used at a 0.5–3% concentration range. BSA is inexpensive and may be stored dry or as a sterile solution at 4°C. However, there are a number of limitations associated with BSA, including lot-to-lot variability—primarily related to the fatty acid content (BSA used as a blocking reagent should be fatty acid free) and lack of molecular diversity (many single molecule blocking reagents lack the diversity to block some surfaces that feature hydrophobic, ionic, and covalent characteristics). Despite these disadvantages, BSA is the most frequently used blocking agent for antibodies or protein bioreceptors across all immunoassay techniques [191].

4.6 EFFECT ON TRANSPORT PROPERTIES

Monolayer graphene is a semimetal with a zero bandgap [192–194]. Therefore, in order to develop effective graphene FETs, a bandgap must be opened in the graphene layer. The challenge is then to maintain the high carrier mobility in this nonzero-bandgap graphene. Several approaches have been developed to open the bandgap of graphene by preparing GNRs, nanomeshes, or by functionalizing graphene [1,114,155,195]. Bandgap opening in graphene on SiC has sparked a lot of interest because of the viability of the growth process. Recently, scanning tunneling spectroscopymeasurements taken on GNRs grown on the terrace structure of SiC have shown a gap of typically 1 eV for 1-nm wide armchair GNRs [196,197]. Graphene grown on SiC tends to be electron doped and the Fermi level lies above the gap. To be viable for electronics, graphene requires hole doping, or Fermi level modulation by applying a gate voltage. In addition, edge disorder causes further problems in the development of nanoribbon transistors, as small fluctuations of the channel width (on the order of nm) can cause a noticeable change in bandgap [198].

Substitutional doping is another promising route for opening a bandgap in graphene. Nitrogen atoms incorporated into graphene have been used to create n-type doping in graphene [199,200]. A bandgap and doping may also be induced by the substrate supporting the graphene layer, or by additional graphene layers. For example, monolayer graphene on SiC has a bandgap of 0.26 eV [201], related to charge transfer from the 4H–SiC substrate to the graphene layers. This charge transfer is suggested to originate from the breaking of sublattice symmetry because of the graphene–substrate interaction [56,202]. Increasing the number of graphene layers on SiC is observed to decrease the bandgap; each subsequent graphene layer reducing the bandgap from 0.26 eV for the layer closest to the SiC, to a negligible bandgap by layer four, due to shielding of the interface charges [56,203].

The number of graphene layers grown on SiC is thus an important consideration in any device or sensor. The first layer of carbon in epitaxially grown graphene grown on the Si-face of SiC has poor electronic properties and acts as a buffer layer [11]. Only subsequent graphene layers are useful for electronic devices or sensors. These subsequent graphene layers are reported to be n-type due to the 4H–SiC substrate–graphene charge transfer [204].

Chemical functionalization of the graphene surface is also a viable method for tuning the electronic properties of graphene, but raises the interesting question: How does chemical functionalization influence the bandgap? Functionalization of graphene can have a significant effect on the bandgap in graphene and on the carrier type of doping. In a similar vein to CNTs, the atomically thin structure of graphene causes the material to be electronically sensitive to the presence of adsorbed atoms and molecules [205,206].

Surface adsorbents such as hydrogen (forming graphane) [207] or fluorine (fluorographene) [19] open up the bandgap in graphene. One must be careful to control the degree of functionalization though, as complete hydrogenation of graphene generates an insulator. Proof-of-concept nitrogen gas sensors have been demonstrated, which are capable of detecting individual adsorption events [205]. Using adsorbed ammonia adsorption on GNRs has been reported to yield n-type doping in GNRs [131,132]).

Single-layer graphene devices, fabricated using graphene grown on 4H–SiC, were found to be extremely sensitive to low concentrations of NO_2 [11]. The as-grown graphene is n-type doped, but exposure to higher concentrations of NO_2 results in a shift of the carrier type from n to p, thought to be related to hole doping from the adsorbed NO_2 molecules. N-type material can be regenerated by desorption of NO_2 from the graphene surface.

The sensitivity of graphene to atoms and molecules adsorbed on the surface is indicative of its suitability for use in chemical sensors. Just a few extra electrons from atoms or molecules adsorbed on the surface of graphene can give rise to notable relative changes in carrier concentration

[208]. Both electron donating and accepting species have been experimentally verified on graphene, such as the strongly electronegative tetrafluorotetracyanoquinodimethane and 7,7,8,8-tetracyanoquinodimethane.

As well as adsorption of molecules onto graphene surfaces, doping and bandgap changes can be induced in graphene via covalent modification of graphene. Experimental data regarding the bandgap induced by covalent modification of graphene are relatively sparse, but covalent functionalization of epitaxial graphene with nitrene has been observed to open a bandgap in graphene [174]. Other methods reportedly used for bandgap tuning include oxidation [209] and diazotization [179].

Chemical modification of epitaxial graphene, using diazotization, has also been shown to affect the conductivity of graphene—more than doubling its room temperature sheet resistance (from 1.5 to 4.2 kΩ/sq) [76]. This is due to the conversion of the carbon atoms in the graphene lattice from sp^2 to sp^3 hybridization, opening a bandgap, which causes a barrier to electron flow.

Greenwood et al. also concluded that grafting of graphene with diazonium 3,5-*bis*-*tert*-butylbenzenediazonium results in bandgap opening, attributed to the breaking of sublattice symmetry of graphene under strain [56,68–70,210,211]. Functionalization using diazonium derivatives can therefore be used to change the electronic properties of epitaxial graphene from metallic to semiconducting. Functionalization of epitaxially grown graphene, where the basal plane of graphene is exposed to diazonium solution, was observed by Loh et al. [169,170]) to resemble chemical modification of the basal plane pyrolytic graphite. The similarity to HOPG was also observed by Tehrani et al. [10], who performed development work on HOPG, before transferring processes to few-layer and multilayer epitaxial graphene.

As well as being used to modify the bandgap and doping of graphene layers, functionalization is used in the development of graphene sensors. For example, binding events such as when a nucleic acid fragment binds to its complementary ssDNA sequence, attached to a CVD graphene device, have been used for highly sensitive detection of DNA hybridization [212]. Graphene sensors are reviewed in Section 4.7.

4.7 APPLICATIONS

Biomolecular analysis has a very important role in a number of application areas including medicine, environmental testing, food and biosecurity, and forensic science. Analytical methods and tools should be capable of providing fast, reliable, and sensitive results, ideally at the point of testing or point-of-care. Biosensors offer a number of advantages over current existing techniques such as mass spectrometry or ELISA tests, which are often labor intensive, require expensive equipment, and in the case of medical testing, results can take a long time to reach the patient resulting in a delay in any required intervention.

Graphene and graphene-related materials have been used to fabricate biosensors for the detection of a number of clinically relevant biomarkers, including glucose [111,112,213–215]), bacteria, DNA [77], and biomarkers relevant to cancer [8–10,216]).

Biosensors comprise a selective interface in close proximity to or integrated with a transducer, which feeds back information about interactions between the surface of the electrode and the local environment. Graphene's exceptional electronic properties offer tremendous potential in this area [217]. Graphene-based biosensors have been demonstrated to exhibit high sensitivities, low detection limits, and long-term stability [8–10]). The highly sensitive electronic response of graphene to its chemical environment allows for the detection of ultralow concentrations of analytes, with limits of detection as low as 50 fM being reported [218]. This level of sensitivity supersedes existing analytical techniques such as ELISAs and could facilitate early diagnosis of medical conditions using noninvasive saliva or urine testing, where levels of biomarkers are typically in much lower concentrations than blood and are generally considered beyond the limit of current technologies.

Following chemical functionalization of graphene (Section 4.4), attachment of "bioreceptor" molecules imparts highly specific and selective recognition capabilities for detection of a particular analyte of interest. Various "bioreceptor" types can be employed (summarized in Section 4.5) depending on the application of use.

Many research groups have focused on the development of graphene-based glucose sensors. Graphene-based glucose oxidase biosensors have been demonstrated for the detection of glucose by several groups and have been shown to exhibit low detection limits, good sensitivity [111,112]) and selectivity, and fast response times [111,112,213–215].

Kang et al. [213] fabricated chitosan/graphene/glucose oxidase modified electrodes for direct electrochemical glucose sensing. Chitosan (a natural biopolymer) was used to immobilize the glucose oxidase enzyme (the bioreceptor) on the electrode. The study reported their sensors displayed a wide linear detection range from 0.08 to 12 mM, with a detection limit of 0.02 mM.

In another study, nitrogen-doped, chemically exfoliated graphene was used in glucose biosensing [214]. Nitrogen plasma treatment of graphene–chitosan on glassy carbon electrodes was undertaken prior to immobilization of glucose oxidase. Detection of glucose was demonstrated with a good response and selectivity to concentrations of glucose as low as 0.01 mM in the presence of other nonspecific substances.

In an initial study using CVD-grown graphene, Huang et al. [215] demonstrated the selective, real-time detection of glucose. CVD graphene FETs were functionalized with 1-pyrenebutanoic acid succinimidyl ester, before being incubated with glucose oxidase. Exposure of the device to glucose resulted in an obvious change in the measured conductance with a detection limit of 0.1 mM being reported. The same study also reported graphene FET sensors for the detection of glutamate. These sensors were functionalized with glutamic dehydrogenase bioreceptor molecules instead of glucose oxidase. In this case a more sensitive detection limit of 5 μM was achieved for glutamate.

Graphene-based immunosensors have also been demonstrated which employ epitaxial or CVD graphene materials. Guy and coworkers [8,9]) developed a chemically modified epitaxial graphene biosensor for the detection of the biomarker hCG. hCG is a key diagnostic marker of pregnancy and has been reported to be an important biomarker in a number of cancers [219]. Patterned multilayer epitaxial graphene devices were functionalized with APTES to produce amine-terminated surfaces. Antibodies targeted against the hCG biomarker were subsequently immobilized on the APTES functionalized devices. A detection limit of 0.62 ng/mL was achieved with these devices, with a linear response to the biomarker reported in the range 0.62–5.62 ng/mL.

The same group recently demonstrated that the epitaxial graphene sensor could be adapted to the detection of other biomarkers by simply changing the antibody bioreceptor [10]. Tehrani et al. [10] used an electrochemical functionalization process to graft phenylamine groups to epitaxial graphene devices. The cancer risk biomarker (related to oxidative stress), 8-OHdG, was then attached to the amine functional groups on the graphene surface. The sensors were capable of detecting 8-OHdG concentrations as low as 0.1 ng/mL (0.25 nM). This result showed that the graphene sensors are five times more sensitive than reported ELISA tests (0.5 ng/mL).

Yang and Gong [216] developed an immunosensor based on exfoliated graphene for the detection of the cancer biomarker prostate-specific antigen (PSA). A detection limit of 0.08 ng/mL was reported with a linear detection range of 0.1–100 ng/mL, well within the relevant clinical PSA concentration ranges. Good selectivity was also demonstrated through monitoring the response of the sensor to PSA in the presence of other nonspecific molecules.

Immunosensors based on CVD graphene have been reported by Eissa et al. [220]. Monolayer CVD graphene supported on a glass substrate was covalently functionalized by electrochemical reduction of carboxyphenyl diazonium salt. The carboxyphenyl-modified graphene was activated using EDC/NHS chemistry prior to attachment of antibodies targeted against ovalbumin. Impedance spectroscopy was used to monitor binding between the surface-immobilized antibodies and ovalbumin. A linear response was observed for ovalbumin concentrations ranging from 1.0 pg/mL to 100 ng/mL, with a detection limit of 0.9 pg/mL. The same group also reported impressive detection limits (0.85 pg/mL) when graphene screen-printed electrodes were employed in immunosensor applications [221].

A number of DNA sensors have been reported in the literature. Dubuisson and coworkers [218] used anodized epitaxial graphene, with inherent carboxyl functional groups from the anodization process, activated using EDC/NHS chemistry. Probe DNA consisting of amine-terminated ssDNA (complementary to the target DNA) was immobilized on the graphene electrodes. Impedance spectroscopy was carried out, which showed that exposure to the target DNA produced a linear response for the measured range 0.05 pM–1 μM, with a detection limit of 0.02 pM. The study also demonstrated the selective nature of the sensor by monitoring its response when exposed to noncomplementary strands of DNA.

Dong et al. [212] fabricated large graphene transistors from CVD graphene on glass substrates. Their study reported that the devices were capable of detecting hybridization of target DNA to surface-bound complementary DNA with a detection limit of 0.01 nM and the ability to distinguish single-base mismatch.

In another study using CVD graphene [222], aptamer-modified graphene FETs were reported for the detection of immunoglobulin E (IgE). Selective detection of IgE was demonstrated, with the detection range of the sensors reported as 0.29–160 nM.

Bonanni and Pumera [223] developed an impedimetric graphene platform for the detection of DNA hybridization and single-nucleotide polymorphism related to the development of Alzheimer's disease. By comparing the performance of different graphene platforms (each with a varying number of graphene layers within a nanoribbon structure) they reported a detection limit in the picomolar range for the detection of single-nucleotide polymorphism.

These numerous reports of graphene biosensors indicate the potential of future health diagnostics based on graphene technology.

REFERENCES

[1] Berger C, Song Z, Li X, Wu X, Brown N, Naud C, Mayou D, et al. Electronic confinement and coherence in patterned epitaxial graphene. Science 2006;312(5777):1191–6.

[2] Castaing A, Guy OJ, Lodzinski M, Wilks SP. Investigation of graphene growth on 4H-SiC. Mater Sci Forum 2009;615–617:223–6.

[3] Guy OJ, Lodzinski M, Teng KS, Maffeis TGG, Tan M, Blackwood I, Dunstan PR, et al. Investigation of the 4H–SiC surface. Appl Surf Sci 2008;254(24):8098–105.

[4] Berger C, Song Z, Li T, Li X, Ogbazghi AY, Feng R, Dai Z, et al. Ultrathin epitaxial graphite: 2D electron gas properties and a route toward graphene-based nanoelectronics. J Phys Chem B 2004;108(52):19912–6.

[5] Schwierz F. Graphene transistors. Nat Nanotechnol 2010;5(7):487–96.

[6] Lin Y-M, Dimitrakopoulos C, Jenkins KA, Farmer DB, Chiu H, Grill A, Avouris Ph. 100-GHz transistors from wafer-scale epitaxial graphene. Science 2010;327:662.

[7] Ang PK, Chen W, Shen Wee AT, Loh KP. Solution-gated epitaxial graphene as pH sensor. J Am Chem Soc 2008;130(44):14392–3.

[8] Teixeira S, Burwell G, Castaing A, Gonzalez D, Conlan RS, Guy OJ. Epitaxial graphene immunosensor for human chorionic gonadotropin. Sens Actuators B Chem 2014;190:723–9.

[9] Teixeira S, Conlan RS, Guy OJ, Goreti M, Sales F. Label-free human chorionic gonadotropin detection at picogram levels using oriented antibodies bound to graphene screen-printed electrodes. J Mater Chem B 2014;2(13):1852.

[10] Tehrani Z, Burwell G, Mohd Azmi MA, Castaing A, Rickman R, Almarashi J, Dunstan P, Beigi AM, Doak SH, Guy OJ. Generic epitaxial graphene biosensors for ultrasensitive detection of cancer risk biomarker. 2D Mater 2014;1(2):025004.

[11] Pearce R, Iakimov T, Andersson M, Hultman L, Lloyd Spetz A, Yakimova R. Epitaxially grown graphene based gas sensors for ultra sensitive NO_2 detection. Sens Actuators B Chem 2011;155(2):451–5.

[12] Ramesh P, Itkis ME, Bekyarova E, Wang F, Niyogi S, Chi X, Berger C, de Heer W, Haddon RC. Electro-oxidized epitaxial graphene channel field-effect transistors with single-walled carbon nanotube thin film gate electrode. J Am Chem Soc 2010;132(41):14429–36.

[13] Novoselov KS, Geim AK, Morozov SV, Jiang D, Zhang Y, Dubonos SV, Grigorieva IV, Firsov AA. Electric field effect in atomically thin carbon films. Science 2004;306(5696):666–9.

[14] Kidambi PR, Ducati C, Dlubak B, Gardiner D, Weatherup RS, Martin M-b, Seneor P, Coles H, Hofmann S. The parameter space of graphene chemical vapor deposition on polycrystalline Cu. J Phys Chem C 2012;116(42):22492–501.

[15] Blake P, Hill EW, Castro Neto AH, Novoselov KS, Jiang D, Yang R, Booth TJ, Geim AK. Making graphene visible. Appl Phys Lett 2007;91:063124.

[16] Hummers WS, Offeman S R.E. Preparation of graphitic oxide. J Am Chem Soc 1958;80(6):1339.

[17] Li X, Zhang G, Bai X, Sun X, Wang X, Wang E, Dai H. Highly conducting graphene sheets and Langmuir–Blodgett films. Nat Nanotechnol 2008;3:538–42.

[18] Widenkvist E, Boukhvalov DW, Rubino S, Akhtar S, Lu J, Quinlan RA, Katsnelson MI, Leifer K, Grennberg H, Jansson U. Mild sonochemical exfoliation of bromine-intercalated graphite: a new route towards graphene. J Phys D Appl Phys 2009;42(11):112003.

[19] Hernandez Y, Nicolosi V, Lotya M, Blighe FM, Sun Z, De S, McGovern IT, et al. High-yield production of graphene by liquid-phase exfoliation of graphite. Nat Nanotechnol 2008;3:563–8.

[20] Blake P, Brimicombe PD, Nair RR, Booth TJ, Jiang D, Schedin F, Ponomarenko LA, et al. Graphene-based liquid crystal device. Nano Lett 2008;8(6):1704–8.

[21] Stankovich S, Dikin DA, Piner RD, Kohlhaas KA, Kleinhammes A, Jia Y, Wu Y, Nguyen ST, Ruoff RS. Synthesis of graphene-based nanosheets via chemical reduction of exfoliated graphite oxide. Carbon 2007;45(7):1558–65.
[22] Huang L, Huang Y, Liang J, Wan X, Chen Y. Graphene-based conducting inks for direct inkjet printing of flexible conductive patterns and their applications in electric circuits and chemical sensors. Nano Res 2011;4(7):675–84.
[23] Huang P, Zhu H, Jing L, Zhao Y, Gao X. Graphene covalently binding aryl groups: conductivity increases rather than decreases. ACS Nano 2011;5(10):7945–9.
[24] Chen J-H, Ishigami M, Jang C, Hines DR, Fuhrer MS, Williams ED. Printed graphene circuits. Adv Mater 2007;19(21):3623–7.
[25] Roy-Mayhew JD, Bozym DJ, Punckt C, Aksay IA. Functionalized graphene as a catalytic counter electrode in dye-sensitized solar cells. ACS Nano 2010;4(10):6203–11.
[26] Bae S, Kim H, Lee Y, Xu X, Park J-S, Zheng Y, Balakrishnan J, et al. Roll-to-roll production of 30-inch graphene films for transparent electrodes. Nat Nanotechnol 2010;5(8):574–8.
[27] Stankovich S, Dikin DA, Dommett GHB, Kohlhaas KM, Zimney EJ, Stach EA, Piner RD, Nguyen ST, Ruoff RS. Graphene-based composite materials. Nature 2006;442(7100):282–6.
[28] Pacilé D, Meyer JC, Girit ÇÖ, Zettl A. The two-dimensional phase of boron nitride: few-atomic-layer sheets and suspended membranes. Appl Phys Lett 2008;92:133107.
[29] Lui CH, Liu L, Mak KF, Flynn GW, Heinz TF. Ultraflat graphene. Nature 2009;462(7271):339–41.
[30] Dean CR, Young AF, Meric I, Lee C, Wang L, Sorgenfrei S, Watanabe K, et al. Boron nitride substrates for high-quality graphene electronics. Nat Nanotechnol 2010;5(10):722–6.
[31] Novoselov KS, Jiang D, Schedin F, Booth TJ, Khotkevich VV, Morozov SV, Geim AK. Two-dimensional atomic crystals. Proc Natl Acad Sci USA 2005;102(30):10451–3.
[32] Sutter PW, Flege J-I, Sutter EA. Epitaxial graphene on ruthenium. Nat Mater 2008;7(5):406–11.
[33] N'Diaye AT, Coraux J, Plasa TN, Busse C, Michely T. Structure of epitaxial graphene on Ir(111). New J Phys 2008;10(4):043033.
[34] Hamilton JC, Blakely JM. Carbon segregation to single crystal surfaces of Pt, Pd and Co. Surf Sci 1980;91(1):199–217.
[35] Mattevi C, Kim H, Chhowalla M. A review of chemical vapour deposition of graphene on copper. J Mater Chem 2011;21(10):3324–34.
[36] Reina A, Jia X, Ho J, Nezich D, Son H, Bulovic V, Dresselhaus MS, Kong J. Large area, few-layer graphene films on arbitrary substrates by chemical vapor deposition. Nano Lett 2009;9(1):30–5.
[37] Kim J, Seidler P, Wan LS, Fill C. Formation, structure, and reactivity of amino-terminated organic films on silicon substrates. J Colloid Interface Sci 2009;329(1):114–9.
[38] Kim KS, Zhao Y, Jang H, Lee SY, Kim JM, Kim KS, Ahn J-H, Kim P, Choi J-Y, Hong BH. Large-scale pattern growth of graphene films for stretchable transparent electrodes. Nature 2009;457(7230):706–10.
[39] Obraztsov AN, Obraztsova EA, Tyurnina AV, Zolotukhin AA. Chemical vapor deposition of thin graphite films of nanometer thickness. Carbon 2007;45(10):2017–21.
[40] Voloshina E, Dedkov Y. Graphene on metallic surfaces: problems and perspectives. Phys Chem Chem Phys 2012;14(39):13502–14.
[41] De Arco LG, Zhang Y, Kumar A, Zhou C. Synthesis, transfer, and devices of single- and few-layer graphene by chemical vapor deposition. IEEE Trans Nanotechnol 2009;8(2):135–8.
[42] Wang Y, Li Z, Wang J, Li J, Lin Y. Graphene and graphene oxide: biofunctionalization and applications in biotechnology. Trends Biotechnol 2011;29(5):205–12.

[43] Wang Y, Zheng Y, Xu X, Dubuisson E, Bao Q, Lu J, Loh KP. Electrochemical delamination of CVD-grown graphene film: toward the recyclable use of copper catalyst. ACS Nano 2011;5(12):9927–33.
[44] Gao L, Ren W, Xu H, Jin L, Wang Z, Ma T, Ma LP, et al. Repeated growth and bubbling transfer of graphene with millimetre-size single-crystal grains using platinum. Nat Commun 2012;3:699.
[45] Michon A, Vézian S, Ouerghi A, Zielinski M, Chassagne T, Portail M. Direct growth of few-layer graphene on 6H-SiC and 3C-SiC/Si via propane chemical vapor deposition. Appl Phys Lett 2010;97:171909.
[46] Strupinski W, Grodecki K, Wysmolek A, Stepniewski R, Szkopek T, Gaskell PE, Grüneis A, et al. Graphene epitaxy by chemical vapor deposition on SiC. Nano Lett 2011;11(4):1786–91.
[47] Nguyen VL, Lee YH. Towards wafer-scale monocrystalline graphene growth and characterization. Small 2015;11:3512–28.
[48] Lee J-H, Lee EK, Joo W-J, Jang Y, Kim B-S, Lim JY, Choi S-H, et al. Wafer-scale growth of single-crystal monolayer graphene on reusable hydrogen-terminated germanium. Science 2014;344(6181):286–90.
[49] Wang G, Zhang M, Zhu Y, Ding G, Jiang D, Guo Q, Liu S, et al. Direct growth of graphene film on germanium substrate. Sci Rep 2013;3:2465.
[50] Scace RI, Slack GA. Solubility of carbon in silicon and germanium. J Chem Phys 1959;30:1551–5.
[51] Acheson EG. Production of artificial crystalline garbonaceous materials. No. 492,767. Patented; 1893.
[52] Hass J, de Heer W, Conrad EH. The growth and morphology of epitaxial multilayer graphene. J Phys Condens Matter 2008;20(32):323202.
[53] Emtsev KV, Bostwick A, Horn K, Jobst J, Kellogg GL, Ley L, McChesney JL, et al. Towards wafer-size graphene layers by atmospheric pressure graphitization of silicon carbide. Nat Mater 2009;8(3):203–7.
[54] Sutter P. Epitaxial graphene: how silicon leaves the scene. Nat Mater 2009;8:171–2.
[55] de Heer WA, Berger C, Wu X, First PN, Conrad EH, Li X, Li T, et al. Epitaxial graphene. Solid State Commun 2007;143(1–2):92–100.
[56] Zhou SY, Gweon G-H, Fedorov AV, First PN, de Heer WA, Lee D-H, Guinea F, Castro Neto AH, Lanzara A. Substrate-induced bandgap opening in epitaxial graphene. Nat Mater 2007;6:770–5.
[57] Riedl C, Coletti C, Iwasaki T, Zakharov AA, Starke U. Quasi-free-standing epitaxial graphene on SiC obtained by hydrogen intercalation. Phys Rev Lett 2009;103(24):246804.
[58] Oida S, McFeely FR, Hannon JB, Tromp RM, Copel M, Chen Z, Sun Y, Farmer DB, Yurkas J. Decoupling graphene from SiC(0001) via oxidation. Phys Rev B 2010;82:041411.
[59] Camara N, Huntzinger JR, Rius G, Tiberj A, Mestres N, Pérez-Murano F, Godignon P, Camassel J. Anisotropic growth of long isolated graphene ribbons on the C face of graphite-capped 6H-SiC. Phys Rev B 2009;80(12):125410.
[60] Luxmi, Fisher PJ, Srivastava N, Feenstra RM, Sun Y, Kedzierski J, Healey P, Gu G. Morphology of graphene on SiC(000-1) surfaces. Appl Phys Lett 2009;95(07):073101.
[61] Moreau E, Ferrer FJ, Vignaud D, Godey S, Wallart X. Graphene growth by molecular beam epitaxy using a solid carbon source. Phys Status Solidi (a) 2010;207(2):300–3.
[62] Pan G, Heath M, Horsell D, Wears ML. Growth of large area graphene from sputtered films. arXiv Preprint arXiv:1209.0489. http://arxiv.org/abs/1209.0489; 2012.
[63] Cooil SP, Song F, Williams GT, Roberts OR, Langstaff DP, Jørgensen B, Høydalsvik K, et al. Iron-mediated growth of epitaxial graphene on SiC and diamond. Carbon 2012;50(14):5099–105.

[64] Malard LM, Pimenta MA, Dresselhaus G, Dresselhaus MS. Raman spectroscopy in graphene. Phys Rep 2009;473(5–6):51–87.

[65] Ferrari AC, Meyer JC, Scardaci V, Casiraghi C, Lazzeri M, Mauri F, Piscanec S, et al. Raman spectrum of graphene and graphene layers. Phys Rev Lett 2006;97(18): 187401.

[66] Tiberj A, Camara N, Godignon P, Camassel J. Micro-Raman and micro-transmission imaging of epitaxial graphene grown on the Si and C faces of 6H-SiC. Nanoscale Res Lett 2011;6(1):478.

[67] Lee DS, Riedl C, Krauss B, von Klitzing K, Starke U, Smet JH. Raman spectra of epitaxial graphene on SiC and of epitaxial graphene transferred to SiO_2. Nano Lett 2008;8(12):4320–4325.

[68] Ni ZH, Chen W, Fan XF, Kuo JL, Yu T, Wee ATS, Shen ZX. Raman spectroscopy of epitaxial graphene on a SiC substrate. Phys Rev B 2008;77:115416.

[69] Ni ZH, Yu T, Lu YH, Wang YY, Feng YP, Shen ZX. Uniaxial strain on graphene: Raman spectroscopy study and band-gap opening. ACS Nano 2008;2(11):2301–5.

[70] Ni Z, Wang Y, Yu T, Shen Z. Raman spectroscopy and imaging of graphene. Nano Res 2008;1(4):273–91.

[71] Basko DM, Piscanec S, Ferrari AC. Electron–electron interactions and doping dependence of the two-phonon Raman intensity in graphene. Phys Rev B 2009;80:165413.

[72] Faugeras C, Nerrìere A, Potemski M, Mahmood A, Dujardin E, Berger C, de Heer WA. Few-layer graphene on SiC, pyrolitic graphite, and graphene: a Raman scattering study. Appl Phys Lett 2008;92:011914.

[73] Dai B, Fu L, Zou Z, Wang M, Xu H, Wang S, Liu Z. Rational design of a binary metal alloy for chemical vapour deposition growth of uniform single-layer graphene. Nat Commun 2011;2:522.

[74] Shivaraman S, Chandrashekhar MVS, Boeckl JJ, Spencer MG. Thickness estimation of epitaxial graphene on SiC using attenuation of substrate Raman intensity. J Electron Mater 2009;38(6):725–30.

[75] Sinitskii A, Dimiev A, Corley DA, Fursina AA, Kosynkin DV, Tour JM. Kinetics of diazonium functionalization of chemically converted graphene nanoribbons. ACS Nano 2010;4(4):1949–54.

[76] Bekyarova E, Itkis ME, Ramesh P, Berger C, Sprinkle M, de Heer WA, Haddon RC. Chemical modification of epitaxial graphene: spontaneous grafting of aryl groups. J Am Chem Soc 2009;131(4):1336–7.

[77] Mohanty N, Berry V. Graphene-based single-bacterium resolution biodevice and DNA transistor: interfacing graphene derivatives with nanoscale and microscale biocomponents. Nano Lett 2008;8(12):4469–76.

[78] Zhou M, Zhai Y, Dong S. Electrochemical sensing and biosensing platform based on chemically reduced graphene oxide supporting information (SI). Anal Chem 2009;81(14):5603–13.

[79] Grieshaber D, MacKenzie R, Vörös J, Reimhult E. Electrochemical biosensors – sensor principles and architectures. Sensors 2008;8(3):1400–58.

[80] Chaubey A, Malhotra BD. Mediated biosensors. Biosens Bioelectron 2002;17(6–7): 441–56.

[81] Mirsky VM, Riepl M, Wolfbeis OS. Capacitative monitoring of protein immobilisation and antibody–antigen interactions on monomolecular alkylthiol films on gold electrodes. Biosens Bioelectron 1997;12(9–10):977–89.

[82] Guiseppi-Elie A, Lingerfelt L. Impedimetric detection of DNA hybridization: towards near-patient DNA diagnostics. Immobilisation of DNA on chips I. Top Curr Chem; Springer Berlin Heidelberg, 2005;260:161–186.

[83] Thevenot DR, Toth K, Durst RA, Wilson GS. Electrochemical biosensors: recommended definitions and classification. Biosens Bioelectron 2001;16(1–2):121–31.

[84] Wilson GS, Hu YB. Enzyme-based biosensors for *in vivo* measurements. Chem Rev 2000;100(7):2693–704.

[85] Patolsky F, Zheng G, Lieber CM. Nanowire-based biosensors. Anal Chem 2006;78(13):4260–9.

[86] Yagiuda K, Hemmi A, Ito S, Asano Y, Fushinuki Y, Chen CY, Karube I. Development of a conductivity-based immunosensor for sensitive detection of methamphetamine (stimulant drug) in human urine. Biosens Bioelectron 1996;11(8):703–7.

[87] Martins MCL, Fonseca C, Barbosa MA, Ratner BD. Albumin adsorption on alkanethiols self-assembled monolayers on gold electrodes studied by chronopotentiometry. Biomaterials 2003;24(21):3697–706.

[88] Katz E, Willner I. Probing biomolecular interactions at conductive and semiconductive surfaces by impedance spectroscopy: routes to impedimetric immunosensors, DNA-sensors, and enzyme biosensors. Electroanalysis 2003;15(11):913–47.

[89] Pei R, Cheng Z, Wang E, Yang X. Amplification of antigen–antibody interactions based on biotin labeled protein–streptavidin network complex using impedance spectroscopy. Biosens Bioelectron 2001;16(6):355–61.

[90] Patolsky F, Zayats M, Katz E, Willner I. Precipitation of an insoluble product on enzyme monolayer electrodes for biosensor applications: characterization by Faradaic impedance spectroscopy, cyclic voltammetry, and microgravimetric quartz crystal microbalance analyses. Anal Chem 1999;71(15):3171–80.

[91] Eggins B. Chemical sensors and biosensors. Analytical techniques in the sciences. West Sussex: John Wiley & Sons; 2002.

[92] Luppa PB, Sokoll LJ, Chan DW. Immunosensors – principles and applications to clinical chemistry. Clin Chim Acta 2001;314(1–2):1–26.

[93] Caras S, Janata J. Field effect transistor sensitive to penicillin. Anal Chem 1980;52(8):1935–7.

[94] Luo X-L, Xu J-J, Zhao W, Chen H-Y. Glucose biosensor based on ENFET doped with SiO2 nanoparticles. Sens Actuators B Chem 2004;97(2–3):249–55.

[95] Schöning MJ, Poghossian A. Bio FEDs (field-effect devices): state-of-the-art and new directions. Electroanalysis 2006;18(19–20):1893–900.

[96] Errachid A, Zine N, Samitier J, Bausells J. FET-based chemical sensor systems fabricated with standard technologies. Electroanalysis 2004;16(22):1843–51.

[97] Schöning MJ, Poghossian A. Recent advances in biologically sensitive field-effect transistors (BioFETs). Analyst 2002;127(9):1137–51.

[98] Estrela P, Stewart AG, Yan F, Migliorato P. Field effect detection of biomolecular interactions. Electrochim Acta 2005;50(25–26):4995–5000.

[99] Patolsky F, Zheng G, Lieber CM. Fabrication of silicon nanowire devices for ultrasensitive, label-free, real-time detection of biological and chemical species. Nat Protoc 2006;1:1711–24.

[100] Balasubramanian K, Burghard M. Biosensors based on carbon nanotubes. Anal Bioanal Chem 2006;385(3):452–68.

[101] Coraux Jo, N'Diaye AT, Busse C, Michely T. Structural coherency of graphene on Ir(111). Nano Lett 2008;8(2):565–70.

[102] Blackwood IP, Teng KS, Maffeïs TGG, Franks JR, Wilks SP. Investigation of annealing effects on the adsorption of Ni on 4H–SiC (0001) surfaces using scanning tunneling microscopy and spectroscopy. J Appl Phys 2005;98:103528.

[103] Cai J, Ruffieux P, Jaafar R, Bieri M, Braun T, Blankenburg S, Muoth M, et al. Atomically precise bottom-up fabrication of graphene nanoribbons. Nature 2010;466(7305): 470–3.

[104] Giusca CE, Spencer SJ, Shard AG, Yakimova R, Kazakova O. Exploring graphene formation on the C-terminated face of SiC by structural, chemical and electrical methods. Carbon 2014;69:221–9.

[105] Filleter T, Emtsev KV, Seyller Th, Bennewitz R. Local work function measurements of epitaxial graphene. Appl Phys Lett 2008;93:133117.
[106] Sherpa SD, Kunc J, Hu Y, Levitin G, de Heer WA, Berger C, Hess DW. Local work function measurements of plasma-fluorinated epitaxial graphene. Appl Phys Lett 2014;104:081607.
[107] Wei Z, Wang D, Kim S, Kim S-Y, Hu Y, Yakes MK, de Heer WA, Sheehan PE, Riedo E. Nanoscale tunable reduction of graphene oxide for graphene electronics. Science 2010;328:1373–7.
[108] Lee NJ, Yoo JW, Choi YJ, Kang CJ, Jeon DY, Kim DC, Seo S, Chung HJ. The interlayer screening effect of graphene sheets investigated by Kelvin probe force microscopy. Appl Phys Lett 2009;95:222107.
[109] Kehayias CE, MacNaughton S, Sonkusale S, Staii C. Kelvin probe microscopy and electronic transport measurements in reduced graphene oxide chemical sensors. Nanotechnology 2013;24:245502.
[110] Gómez-Anquela C, Revenga-Parra M, Abad JM, García Marín A, Pau JL, Pariente F, Piqueras J, Lorenzo E. Electrografting of N′,N′-dimethylphenothiazin-5-ium-3,7-diamine (azure A) diazonium salt forming electrocatalytic organic films on gold or graphene oxide gold hybrid electrodes. Electrochim Acta 2014;116:59–68.
[111] Yang MH, Choi BG, Park H, Hong WH, Lee SaY, Park TJ. Development of a glucose biosensor using advanced electrode modified by nanohybrid composing chemically modified graphene and ionic liquid. Electroanalysis 2010;22(11):1223–8.
[112] Yang W, Ratinac KR, Ringer SP, Thordarson P, Gooding JJ, Braet F. Carbon nanomaterials in biosensors: should you use nanotubes or graphene? Angew Chem 2010;49(12):2114–38.
[113] Loh KP, Bao Q, Ang PK, Yang J. The chemistry of graphene. J Mater Chem 2010;20:2277–89.
[114] Balog R, Jørgensen B, Nilsson L, Andersen M, Rienks E, Bianchi M, Fanetti M, et al. Bandgap opening in graphene induced by patterned hydrogen adsorption. Nat Mater 2010;9(4):315–9.
[115] Boukhvalov DW, Katsnelson MI. Chemical functionalization of graphene. J Phys Condens Matter 2009;21(34):344205.
[116] Elias DC, Nair RR, Mohiuddin TMG, Morozov SV, Blake P, Halsall MP, Ferrari AC, et al. Control of graphene's properties by reversible hydrogenation: evidence for graphene. Science 2009;323:610–3.
[117] Roman T, Diño WA, Nakanishi H, Kasai H, Sugimoto T, Tange K. Hydrogen pairing on graphene. Carbon 2007;45:218–20.
[118] Ruoff R. Graphene: calling all chemists. Nat Nanotechnol 2008;3:10–1.
[119] Kuila T, Bose S, Mishra AK, Khanra P, Kim NH, Lee JH. Chemical functionalization of graphene and its applications. Prog Mater Sci 2012;57(7):1061–105.
[120] Georgakilas V, Otyepka M, Bourlinos AB, Chandra V, Kim N, Kemp KC, Hobza P, Zboril R, Kim KS. Functionalization of graphene: covalent and non-covalent approaches, derivatives and applications. Chem Rev 2012;112(11):6156–214.
[121] Xu Y, Bai H, Lu Ge, Li C, Shi G. Flexible graphene films via the filtration of water-soluble noncovalent functionalized graphene sheets. J Am Chem Soc 2008;130(18):5856–7.
[122] Liu J, Yang W, Tao L, Li D, Boyer C, Davis TP. Thermosensitive graphene nanocomposites formed using pyrene-terminal polymers made by RAFT polymerization. J Polym Sci Part A 2010;48:425–33.
[123] Geng J, Jung H-T. Porphyrin functionalized graphene sheets in aqueous suspensions: from the preparation of graphene sheets to highly conductive graphene films. J Phys Chem C 2010;114(18):8227–34.
[124] Ogawa Y, Niu T, Wong SL, Tsuji M, Wee ATS, Chen W, Ago H. Self-assembly of polar phthalocyanine molecules on graphene grown by chemical vapor deposition. J Phys Chem C 2013;117(42):21849–55.

[125] Zhang X, Feng Y, Tang S, Feng W. Preparation of a graphene oxide-phthalocyanine hybrid through strong π–π interactions. Carbon 2010;48(1):211–6.

[126] Zhang Y-H, Zhou K-G, Xie K-F, Zeng J, Zhang H-L, Peng Y. Tuning the electronic structure and transport properties of graphene by noncovalent functionalization: effects of organic donor, acceptor and metal atoms. Nanotechnology 2010;21(6):065201.

[127] Karousis N, Tagmatarchis N, Tasis D. Current progress on the chemical modification of carbon nanotubes. Chem Rev 2010;110(9):5366–97.

[128] Tasis D, Tagmatarchis N, Bianco A, Preto M. Chemistry of carbon nanotubes. Chem Rev 2006;106(3):1105–36.

[129] Kodali VK, Scrimgeour J, Kim S, Hankinson JH, Carroll KM, de Heer WA, Berger C, Curtis JE. Nonperturbative chemical modification of graphene for protein micropatterning. Langmuir 2011;27(3):863–5.

[130] Bai H, Xu Y, Zhao L, Li C, Shi G. Non-covalent functionalization of graphene sheets by sulfonated polyaniline. Chem Commun 2009;13:1667–9.

[131] Wang D-W, Li F, Zhao J, Ren W, Chen Z-G, Tan J, Wu Z-S, Gentle I, Lu GQ, Cheng H-M. Fabrication of graphene/polyaniline composite paper via in situ anodic electropolymerization for high-performance flexible electrode. ACS Nano 2009;3(7):1745–52.

[132] Wang X, Li X, Zhang L, Yoon Y, Weber PK, Wang H, Guo J, Dai H. N-doping of graphene through electrothermal reactions with ammonia. Science (New York, NY) 2009;324(5928):768–71.

[133] Gewirth AA, Bard AJ. In situ scanning tunnelling microscopy of the anodic oxidation of highly oriented pyrolytic graphite surfaces. J Phys Chem 1988;92(20):5563–6.

[134] Zhang J, Zou Ho, Qing Q, Yang Y, Li Q, Liu Z, Guo X, Du Z. Effect of chemical oxidation on the structure of single-walled carbon nanotubes. J Phys Chem B 2003;107(16):3712–8.

[135] Datsyuk V, Kalyva M, Papagelis K, Parthenios J, Tasis D, Siokou A, Kallitsis I, Galiotis C. Chemical oxidation of multiwalled carbon nanotubes. Carbon 2008;46(6):833–40.

[136] Wu S, He Q, Tan C, Wang Y, Zhang H. Graphene-based electrochemical sensors. Small 2013;9(8):1160–72.

[137] Brodie BC. On the atomic weight of graphite. Phil Trans R Soc Lond 1859;149(1859):249–59.

[138] Chen D, Tang L, Li J. Graphene-based materials in electrochemistry. Chem Soc Rev 2010;39:3157–80.

[139] Zacharia R, Ulbricht H, Hertel T. Interlayer cohesive energy of graphite from thermal desorption of polyaromatic hydrocarbons. Phys Rev B 2004;69:155406.

[140] Cai W, Piner RD, Stadermann FJ, Park S, Shaibat MA, Ishii Y, Yang D, et al. Synthesis and solid-state NMR structural characterization of 13C-labeled graphite oxide. Science 2008;321(5897):1815–7.

[141] Mattevi C, Eda G, Agnoli S, Miller S, Mkhoyan KA, Celik O, Mastrogiovanni D, Granozzi G, Carfunkel E, Chhowalla M. Evolution of electrical, chemical, and structural properties of transparent and conducting chemically derived graphene thin films. Adv Funct Mater 2009;19(16):2577–83.

[142] Ferrari AC, Bonaccorso F, Falko V, Novoselov KS, Roche S, Bøggild P, Borini S, et al. Science and technology roadmap for graphene, related two-dimensional crystals, and hybrid systems. Nanoscale 2015;7(11):4598–810.

[143] Choo H-S, Kinumoto T, Jeong S-K, Iriyama Y, Abe T, Ogumi Z. Mechanism for electrochemical oxidation of highly oriented pyrolytic graphite in sulfuric acid solution. J Electrochem Soc 2007;154(10):B1017–23.

[144] Goldsmith BR, Coroneus JG, Khalap VR, Kane AA, Weiss GA, Collins PG. Conductance-controlled point functionalization of single-walled carbon nanotubes. Science 2007;315(5808):77–81.

[145] Ashraf MK, Bruque NA, Pandey RR, Collins PG, Lake RK. Effect of localized oxygen functionalization on the conductance of metallic carbon nanotubes. Phys Rev B 2009;79:115428.
[146] Kanai Y, Khalap VR, Collins PG, Grossman JC. Atomistic oxidation mechanism of a carbon nanotube in nitric acid. Phys Rev Lett 2010;104:066401.
[147] Widenkvist E, Li J, Jansson U, Grennberg H. Selected area deposition of multiwalled carbon nanotubes from solution. Carbon 2007;45(14):2732–6.
[148] Widenkvist E, Alm O, Boman M, Jansson U, Grennberg H. Functionalization and area-selective deposition of magnetic carbon-coated iron nanoparticles from solution. J Nanotechnol 2011;. Article ID 342368.
[149] Felten A, Bittencourt C, Pireaux JJ, Van Lier G, Charlier JC. Radio-frequency plasma functionalization of carbon nanotubes surface O2, NH3, and CF4 treatments. J Appl Phys 2005;98:074308.
[150] Dongil AB, Bachiller-Baeza B, Guerrero-Ruiz A, Rodríguez-Ramos I, Martínez-Alonso A, Tascón JMD. Surface chemical modifications induced on high surface area graphite and carbon nanofibers using different oxidation and functionalization treatments. J Colloid Interface Sci 2011;355(1):179–89.
[151] Baraket M, Stine R, Lee WK, Robinson JT, Tamanaha CR, Sheehan PE, Walton SG. Aminated graphene for DNA attachment produced via plasma functionalization. Appl Phys Lett 2012;100:233123.
[152] Zhang X. Surface functionalization of graphene devices. MA, USA: Massachusetts Institute of Technology; 2012.
[153] Li X, Cai W, An J, Kim S, Nah J, Yang D, Piner R, et al. Large-area synthesis of high-quality and uniform graphene films on copper foils. Science 2009;324(5932):1312–4.
[154] Robinson JT, Burgess JS, Junkermeier CE, Badescu SC, Reinecke TL, Perkins FK, Zalalutdniov MK, et al. Properties of fluorinated graphene films. Nano Lett 2010;10(8):3001–5.
[155] Li B, Zhou L, Wu D, Peng H, Yan K, Zhou Y, Liu Z. Photochemical chlorination of graphene. ACS Nano 2011;5(7):5957–61.
[156] Wu J, Xie L, Li Y, Wang H, Ouyang Y, Guo J, Dai H. Controlled chlorine plasma reaction for noninvasive graphene doping. J Am Chem Soc 2011;133(49):19668–71.
[157] Zhang F, Chen X, Boulos RA, Yasin FM, Lu H, Raston C, Zhang H. Pyrene-conjugated hyaluronan facilitated exfoliation and stabilisation of low dimensional nanomaterials in water. Chem Commun 2013;49(42):4845–7.
[158] Zhang X, Hsu A, Wang H, Song Y, Kong J, Dresselhaus MS, Palacios T. Impact of chlorine functionalization on high-mobility chemical vapor deposition grown graphene. ACS Nano 2013;7(8):7262–70.
[159] Baraket M, Walton SG, Lock EH, Robinson JT, Perkins FK. The functionalization of graphene using electron-beam generated plasmas. Appl Phys Lett 2010;96:231501.
[160] Bindoli A, Fukuto JM, Forman HJay. Thiol chemistry in peroxidase catalysis and redox signaling. Antioxid Redox Signal 2008;10(9):1549–64.
[161] Bahr JL, Yang J, Kosynkin DV, Bronikowski MJ, Smalley RE, Tour JM. Functionalization of carbon nanotubes by electrochemical reduction of aryl diazonium salts: a bucky paper electrode. J Am Chem Soc 2001;123(27):6536–42.
[162] Englert JM, Dotzer C, Yang G, Schmid M, Papp C, Gottfried JM, Steinrück H-P, Spiecker E, Hauke F, Hirsch A. Covalent bulk functionalization of graphene. Nat Chem 2011;3:279–86.
[163] Shih C-J, Wang QH, Jin Z, Paulus GLC, Blankschtein D, Jarillo-Herrero P, Strano MS. Disorder imposed limits of mono- and bilayer graphene electronic modification using covalent chemistry. Nano Lett 2013;13(2):809–17.
[164] Hong J, Niyogi S, Bekyarova E, Itkis ME, Ramesh P, Amos N, Litvinov D, et al. Effect of nitrophenyl functionalization on the magnetic properties of epitaxial graphene. Small 2011;7(9):1175–80.

[165] Fang M, Zhang Z, Li J, Zhang H, Lu H, Yang Y. Constructing hierarchically structured interphases for strong and tough epoxy nanocomposites by amine-rich graphene surfaces. J Mater Chem 2010;20(43):9635–43.
[166] He H, Gao C. General approach to individually dispersed, highly soluble, and conductive graphene nanosheets functionalized by nitrene chemistry. Chem Mater 2010;22(17):5054–64.
[167] Yan L, Zheng YB, Zhao F, Li S, Gao X, Xu B, Weiss PS, Zhao Y. Chemistry and physics of a single atomic layer: strategies and challenges for functionalization of graphene and graphene-based materials. Chem Soc Rev 2012;41:97–114.
[168] Lomeda JR, Doyle CD, Kosynkin DV, Hwang W-F, Tour JM. Diazonium functionalization of surfactant-wrapped chemically converted graphene sheets. J Am Chem Soc 2008;130(48):16201–6.
[169] Lim CX, Hoh HY, Ang PK, Loh KP. Direct voltammetric detection of DNA and pH sensing on epitaxial graphene: an insight into the role of oxygenated defects. Anal Chem 2010;82(17):7387–93.
[170] Lim H, Lee JS, Shin H-J, Shin HS, Choi HC. Spatially resolved spontaneous reactivity of diazonium salt on edge and basal plane of graphene without surfactant and its doping effect. Langmuir 2010;26(14):12278–84.
[171] Mohd Azmi MA, Tehrani Z, Lewis RP, Walker K-AD, Jones DR, Daniels DR, Doak SH, Guy OJ. Highly sensitive covalently functionalised integrated silicon nanowire biosensor devices for detection of cancer risk biomarker. Biosens Bioelectron 2014;52:216–24.
[172] Li B-R, Chen C-W, Yang W-L, Lin T-Y, Pan C-Y, Chen Y-T. Biomolecular recognition with a sensitivity-enhanced nanowire transistor biosensor. Biosens Bioelectron 2013;45:252–9.
[173] Glass NR, Tjeung R, Chan P, Yeo LY, Friend JR. Organosilane deposition for microfluidic applications. Biomicrofluidics 2011;5(3):36501–17.
[174] Choi J, Kim K-J, Kim B, Lee H, Kim S. Covalent functionalization of epitaxial graphene by azidotrimethylsilane. J Phys Chem C 2009;113(22):9433–5.
[175] Pumera M, Ambrosi A, Bonanni A, Chng ELK, Poh HL. Graphene for electrochemical sensing and biosensing. Trends Anal Chem 2010;29(9):954–65.
[176] Allongue P, Delamar M, Desbat B, Fagebaume O, Hitmi R, Pinson J, Saveant J-M. Covalent modification of carbon surfaces by aryl radicals generated from the electrochemical reduction of diazonium salts. J Am Chem Soc 1997;119(1):201–7. http://pubs.acs.org/doi/abs/10.1021/ja963354s.
[177] Pinson J, Podvorica F. Attachment of organic layers to conductive or semiconductive surfaces by reduction of diazonium salts. Chem Soc Rev 2005;34:429–39.
[178] Pumera M, Sasaki T, Iwai H. Relationship between carbon nanotube structure and electrochemical behavior: heterogeneous electron transfer at electrochemically activated carbon nanotubes. Chem Asian J 2008;3(12):2046–55.
[179] Bekyarova E, Sarkar S, Niyogi S, Itkis ME, Haddon RC. Advances in the chemical modification of epitaxial graphene. J Phys D Appl Phys 2012;45(15):154009.
[180] Hassan HK, Atta NF, Galal A. Electropolymerization of aniline over chemically converted graphene – systematic study and effect of dopant. Int J Electrochem Sci 2012;7(11):11161–81.
[181] Luo X, Killard AJ, Morrin A, Smyth MR. Enhancement of a conducting polymer-based biosensor using carbon nanotube-doped polyaniline. Anal Chim Acta 2006;575(1):39–44.
[182] Deep A, Sharma AL, Kumar P, Bharadwaj LM. Nanostructured polyaniline-silicon substrate for protein biosensing. Sens Actuators B Chem 2012;171–172:210–5.
[183] Basak SP, Kanjilal B, Sarkar P, Turner APF. Application of electrical impedance spectroscopy and amperometry in polyaniline modified ammonia gas sensor. Synth Metals 2013;175:127–33.

[184] Luo X, Killard AJ, Morrin A, Smyth MR. In situ electropolymerised silica-polyaniline core-shell structures: electrode modification and enzyme biosensor enhancement. Electrochim Acta 2007;52(5):1865–70.
[185] Raina M, Sharma R, Deacon SE, Tiede C, Tomlinson D, Davies AG, McPherson MJ, Wälti C. Antibody mimetic receptor proteins for label-free biosensors. Analyst 2015;140:803–10.
[186] Wan Y, Su Y, Zhu X, Liu G, Fan C. Development of electrochemical immunosensors towards point of care diagnostics. Biosens Bioelectron 2013;47:1–11.
[187] Ferreira NS, Sales MG. Disposable immunosensor using a simple method for oriented antibody immobilization for label-free real-time detection of an oxidative stress biomarker implicated in cancer diseases. Biosensor Bioelectron 2014;53:193–9.
[188] Maraldo D, Mutharasan R. Optimization of antibody immobilization for sensing using piezoelectrically excited-millimeter-sized cantilever (PEMC) sensors. Sens Actuators B Chem 2007;123(1):474–9.
[189] Sam S, Touahir L, Salvador Andresa J, Allongue P, Chazalviel J-N, Gouget-Laemmel AC, de Villeneuve CH, et al. Semiquantitative study of the EDC/NHS activation of acid terminal groups at modified porous silicon surfaces. Langmuir 2010;26(2):809–14.
[190] Niemeyer CM. Semisynthetic DNA-protein conjugates for biosensing and nanofabrication. Angew Chem Int Ed Engl 2010;49(7):1200–16.
[191] Walker JM, Rapley R, editors. Molecular biomethods handbook. 2nd ed. Totowa, NJ, USA: Humana Press, Springer Science & Business Media; 2008.
[192] Biswas C, Lee YH. Graphene versus carbon nanotubes in electronic devices. Adv Funct Mater 2011;21(20):3806–26.
[193] Avouris P. Graphene: electronic and photonic properties and devices. Nano Lett 2010;10(11):4285–94.
[194] Sun Z, James DK, Tour JM. Graphene chemistry synthesis and manipulation. J Phys Chem Lett 2011;2(19):2425–32.
[195] Bai J, Zhong X, Jiang S, Huang Y, Duan X. Graphene nanomesh. Nat Nanotechnol 2010;5(3):190–4.
[196] Baringhaus J, Edler F, Tegenkamp C. Edge-states in graphene nanoribbons: a combined spectroscopy and transport study. J Phys Condens Matter 2013;25(39):392001.
[197] Son YW, Cohen ML, Louie SG. Energy gaps in graphene nanoribbons. Phys Rev Lett 2006;97:216803.
[198] Yoon Y, Guo J. Effect of edge roughness in graphene nanoribbon transistors. Appl Phys Lett 2007;91:073103.
[199] Lv R, Li Q, Botello-Méndez AR, Hayashi T, Wang B, Berkdemir A, Hao S Qingzhen, et al. Nitrogen-doped graphene: beyond single substitution and enhanced molecular sensing. Sci Rep 2012;2(586):1–8.
[200] Velez-Fort E, Mathieu C, Pallecchi E, Pigneur M, Silly MG, Belkhou R, Marangolo M, et al. Epitaxial graphene on 4H-SiC(0001) grown under nitrogen flux: evidence of low nitrogen doping and high charge transfer. ACS Nano 2012;6(12):10893–900.
[201] Peng X, Ahuja R. Symmetry breaking induced bandgap in epitaxial graphene layers on SiC. Nano Lett 2008;8(12):4464–8.
[202] Ohta T, Bostwick A, Seyller T, Horn K, Rotenberg E. Controlling the electronic structure of bilayer graphene. Science 2006;313(5789):951–4.
[203] Guinea F. Charge distribution and screening in layered graphene systems. Phys Rev B 2007;75:235433.
[204] Varchon F, Feng R, Hass J, Li X, Ngoc Nguyen B, Naud C, Mallet P, et al. Electronic structure of epitaxial graphene layers on SiC: effect of the substrate. Phys Rev Lett 2007;99:126805.

[205] Schedin F, Geim AK, Morozov SV, Hill EW, Blake P, Katsnelson MI, Novoselov KS. Detection of individual gas molecules adsorbed on graphene. Nat Mater 2007;6:652–5.
[206] Liu L, Shen Z. Bandgap engineering of graphene: a density functional theory study. Appl Phys Lett 2009;95:252104.
[207] Nair RR, Ren W, Jalil R, Riaz I, Kravets VG, Britnell L, Blake P. Fluorographene: a two-dimensional counterpart of Teflon. Small 2010;6(24):2877–84.
[208] Rangel NL, Seminario JM. Graphene terahertz generators for molecular circuits and sensors. J Phys Chem A 2008;112(51):13699–705.
[209] Wu X, Sprinkle M, Li X, Ming F, Berger C, de Heer WA. Epitaxial-graphene/graphene-oxide junction: an essential step towards epitaxial graphene electronics. Phys Rev Lett 2008;101:026801.
[210] Greenwood J, Phan TH, Fujita Y, Li Z, Ivasenko O, Vanderlinden W, Van Gorp H, et al. Covalent modification of graphene and graphite using diazonium chemistry: tunable grafting and nanomanipulation. ACS Nano 2015;9(5):5520–35.
[211] Giovannetti G, Khomyakov PA, Brocks G, Kelly PJ, van den Brink J. Substrate-induced band gap in graphene on hexagonal boron nitride: ab initio density functional calculations. Phys Rev B 2007;76:073103.
[212] Dong X, Shi Y, Huang W, Chen P, Li LJ. Electrical detection of DNA hybridization with single-base specificity using transistors based on CVD-grown graphene sheets. Adv Mater 2010;22(14):1649–53.
[213] Kang X, Wang J, Wu H, Aksay IA, Liu J, Lin Y. Glucose oxidase-graphene-chitosan modified electrode for direct electrochemistry and glucose sensing. Biosens Bioelectron 2009;25(4):901–5.
[214] Wang Y, Shao Y, Matson DW, Li J, Lin Y. Nitrogen-doped graphene and its biosensing. ACS Nano 2010;4(4):1790–8.
[215] Huang Y, Dong X, Shi Y, Li CM, Li L-J, Chen P. Nanoelectronic biosensors based on CVD grown graphene. Nanoscale 2010;2(8):1485–8.
[216] Yang M, Gong S. Immunosensor for the detection of cancer biomarker based on percolated graphene thin film. Chem Commun 2010;46(31):5796–8.
[217] Pumera M. Graphene in biosensing. Mater Today 2011;14(7–8):308–15.
[218] Dubuisson E, Yang Z, Loh KP. Optimizing label-free DNA electrical detection on graphene platform. Anal Chem 2011;83(7):2452–60.
[219] Stenman U-H, Alfthan H, Hotakainen K. Human chorionic gonadotropin in cancer. Clin Biochem 2004;37(7):549–61.
[220] Eissa S, Jimenez GC, Mahvash F, Guermoune A, Tlili C, Szkopek T, Zourob M, Siaj M. Functionalized CVD monolayer graphene for label-free impedimetric biosensing. Nano Res 2015;8(5):1698–709.
[221] Eissa S, Tlili C, L'Hocine L, Zourob M. Electrochemical immunosensor for the milk allergen β-lactoglobulin based on electrografting of organic film on graphene modified screen-printed carbon electrodes. Biosensor Bioelectron 2012;38(1):308–13.
[222] Ohno Y, Maehashi K, Matsumoto K. Label-free biosensors based on aptamer-modified graphene field-effect. J Am Chem Soc 2010;132:18012–3.
[223] Bonanni A, Pumera M. Graphene platform for hairpin-DNA-based impedimetric genosensing. ACS Nano 2011;5(3):2356–61.

CHAPTER 5

SiC Biosensing and Electrochemical Sensing: State of the Art and Perspectives

Salvatore Iannotta, Agostino Romeo, Pasquale D'Angelo and Giuseppe Tarabella
IMEM-CNR Institute of Materials for Electronics and Magnetism – Italian National Research Council, Parma, Italy

Contents

5.1 INTRODUCTION

The research field of bioelectronics, at the fascinating intersection of the worlds of electronics, biology, and biotechnology, has attracted enormous interest in the last few years, due to both the novel perspectives that open up and the potential of applications in terms of devices for diagnostics and healthcare [1]. Bioelectronic devices involve electronic elements such as, for example, electrodes, transistor devices, piezoelectric crystals, magnetoresistance recording media and scanning tunneling microscopy (STM) tips, which are interfaced with biological or biologically derived systems (eg, microorganisms, living tissues, enzymes, antibodies, nucleic acids, engineered proteins, aptamers, etc.). As discussed in other chapters of this book, applications of such innovative technology are mainly focused on sensing, ranging from the detection of biomolecules, disease markers, biological analytes, pass-through monitoring of biophenomena, up to medical applications (actuators, prosthetics, etc.). Indeed, the growing interest of the scientific community on bioelectronics is increasingly concerned with conceiving and realizing advanced tools and systems, which are expected to revolutionize the concept of personal care and healthcare, aiming at ease of use, low cost, high performance, and complete adaptability/integrability with living beings.

Contemporary work continues to be carried out on the research of innovative advancements and outperforming technologies for biomedical applications. One of the main aspects on which the efforts of researchers have focused concerns the discovery and improvement of materials with better chemical, electrical and mechanical properties, according to the specific biomedical applications to which they are intended. In this context, as proposed in this book, the superior properties of silicon carbide (SiC) have made it the material of choice for an increasing number of promising biomedical devices.

In chapter: Silicon Carbide Materials for Biomedical Applications an in-depth introduction to SiC was presented. Here a fast review of the critical material characteristics is presented in order to set the stage for using SiC for bio- and electrochemical sensing. It is worth summarizing briefly that the constituent silicon and carbon atoms composing the crystal structure of SiC are arranged in SiC_4 tetrahedrons forming a bilayer structure. SiC peculiarity consists in its ability to exhibit different crystal forms, that is, more than 250 special polymorphs [2]. In particular, SiC exhibits a particular two-dimensional polymorphism, called polytypism [3]. Polytypic compounds are characterized by stacking layers sharing the same structure for the lattice planes and differing for the relative positions of the adjacent stacked planes. SiC polytypes that result upon crystallization in binary structures made of identical layers with a cubic (C), hexagonal (H), and rhombohedral (R)-like arrangement of SiC bilayers are also characterized by the number of stacked layers constituting a repetitive stacking sequence in the direction perpendicular to the basal plane. Most of the possible polytypes are unstable, and within the previous scheme, the most frequent SiC polytypes for electronic device applications are the cubic 3C and both the hexagonal 4H [considered as the benchmark material within the family of wide bandgap (WBG) semiconductors, due to its widest bandgap] and 6H. The physical properties of SiC polytypes are almost identical, while the electrical ones depend specifically on the order in which the layers are arranged and on the number of layers in a unit cell. Some of the main physical and electronic parameters shown by 3C–SiC, 4H–SiC, and 6H–SiC are reported in Table 5.1.

Due to its robust properties, SiC is historically considered the most suitable material for the manufacturing of high-power, high-frequency electronic devices with superior performance with respect to Si-based ones.

Table 5.1 Physical and Electronic Parameters of 3C–, 4H–, and 6H–SiC Polytypes [4]

	3C–SiC	4H–SiC	6H–SiC
Dielectric constant, ε	9.72	9.66	9.66
Bandgap (eV)	2.39	3.26	3
Thermal conductivity (W/cm°C)	3–4	3–4	3–4
Saturation drift velocity, v_s (cm/s)	2.7×10^7	2×10^7	2×10^7
Electron mobility @ 300 K, μ_n (cm^2/V s)	≤ 1000	≤ 850	≤ 450
Hole mobility @ 300 K, μ_p (cm^2/V s)	≤ 40	≤ 120	≤ 100

Therefore, currently SiC is ideally considered as the best material for high-gain and high-speed solid-state devices. However, SiC shows several properties suitable for biologically related applications as discussed in chapter: Cytotoxicity of 3C–SiC Investigated Through Strict Adherence to ISO 10993, chapter: Study of the Hemocompatibility of 3C–SiC and *a*-SiC Films Using ISO 10993-4 and chapter: *In Vivo* Exploration of Robust Implantable Devices Constructed From Biocompatible 3C–SiC. Hereinafter, the most important ones are listed:

- *Biocompatibility*: the proved lack of cytotoxic effects when interfaced with several mammalian cell cultures makes SiC a better substrate in terms of both cell adhesion and proliferation [5].
- *Chemical inertness* in harsh environments: a dissolution rate below 30 nm per year [6,7] and an excellent corrosion resistance under normal biological conditions imply no adverse reaction and long-term stability when SiC-based devices are implanted in the body.
- Ideal *mechanical* properties for smart and long-term implants: Young's modulus higher than that of Si, high hardness (5.8 GPa), high elastic modulus (424 GPa), and low friction coefficient (0.17) prevent easy breakage of SiC when inserted in living tissues [5].
- *Electrical* properties for highly sensitive devices: SiC bandgap increases its sensing capabilities by reducing the number of electron–hole pairs formed from the thermal activation across the bandgap; high breakdown field ~2 MV/cm is two-fold higher than that of Si [8]; very high p- and n-type charge carrier mobility; and metallic conductivity achieved by heavily doping it with boron, aluminum, or nitrogen [9].
- *Optical* properties: high refractive index, transparency to visible light, and UV wavelength absorption; SiC nanoparticle photoluminescence makes it an ideal material for optical-based biosensing.
- Ease of *functionalization*: the wide reactivity of C and Si terminations on the SiC surface allow a large number of functionalization routes

(based, for instance, on self-assembled monolayers made of organosilanes or short-chain alkenes used as a first step for the immobilization of other biomolecules), in order to obtain a controlled and stable inorganic/(bio) organic interface [10].

- High *thermal conductivity*, similar to that of copper at room temperature [11].

Given the many outstanding properties that make SiC a particularly suitable material for biomedical applications, a wide number of devices based on SiC have been described so far in the literature, including cardiac stents, biomembranes, orthopedic implants, imaging agents, drug delivery systems, tissue engineering scaffolds, and biosensors [12]. Among these, biosensors able to detect biologically relevant molecules, or signals directly produced by living tissue, promise to have a central role in the development of technologies for point-of-care diagnostics and fast, reliable, sensitive, and portable sensors for healthcare applications. In general, biosensors are "analytical devices" incorporating a biological or biologically derived system and are intimately associated with or integrated within a physicochemical transducer, which in turn may be an optical, electrochemical, thermometric, piezoelectric, magnetic, or micromechanical tool. Their range of use covers medicine, biomedical research, drug discovery, environment, food monitoring, process industries, security, and defense. The scientific community in the biosensing field is paying close attention to SiC technology in the attempt to overcome the real and potential limitations showed by current techniques dedicated to the clinical diagnosis of many diseases or the detection of pathogenic agents and biomolecules. Interestingly, the majority of SiC-based biosensors reported so far in the literature exploit electrical detection mechanisms, mainly electrochemical ones (eg, differential pulse voltammetry and cyclic voltammetry), where SiC is mainly used as an electrode material or as a functionalized system for electrodes.

Recent achievements in bioelectronics show that SiC is one of the best alternatives to the fast flourishing class of the organic semiconductors (OS). OS are, in fact, gaining increasing interest with a number of working and quite promising biomedical applications reported already [13]. These materials, including poly(3-hexylthiophene) (P3HT), polyaniline (PAni), polypyrrole, and poly(3,4-ethylenedioxythiophene):poly(styrene sulfonic acid) (PEDOT:PSS), to cite few among a growing number, are characterized by a π-conjugated structure and have become a complementary alternative to conventional inorganic semiconductors such as silicon, GaAs, etc. [14,15]. Some of the advantages of these compounds are of a technological nature, the main one being the easy tailoring of their (multifunctional) properties by chemical design, incorporating functional groups while manipulating the physical conditions to meet the specific low-cost requirements of processing conditions. In fact, organic thin films are fabricated by inexpensive solution-based processing techniques (eg, spin coating and inkjet printing) or thermal evaporation at room temperature [15] even though quite sophisticated processes have also been proposed to fully control their final properties in the solid state [16,17]. Moreover, conformability, biocompatibility, and the peculiar conductive properties of many of these organic compounds make them very promising for innovative bioelectronic devices. Indeed, one of the most interesting properties of OS for biomedical applications is the porous structure available by some of them (PEDOT:PSS being one of the best examples). OS direct exposure to biological media demonstrates one of their major advantages consisting of mixed ionic–electronic conduction, which is their ability to change their conductive properties as a result of the exchange of ionic species with the liquid environments they are exposed to [18]. As a direct consequence PEDOT:PSS has been used for several applications, including displays [19], organic thin film transistors (OTFTs) [20], microelectrode arrays (MEAs) [21–23], memory devices [24,25], as well as chemical and biological sensors [14]. It is worthwhile briefly discussing the role played by PEDOT:PSS since it is by far the most used and promising OS material in neuronal interfaces and, even to a much less extent, in biosensing. OTFTs based on PEDOT:PSS belong to the class of organic electrochemical transistors (OECTs). Basically, OECTs are three-terminal devices, mainly made of PEDOT:PSS thin layers, where the transistor channel is defined as the region comprised between two planar electrodes, source and drain, while the gate electrode is separated from the transistor channel by an ionically conducting, electrically insulating electrolyte. The working principle of an OECT is based on the reversible electrochemical doping/dedoping of the transistor channel by injection of cationic species from the electrolyte upon application of a positive voltage to the gate electrode (ie, field-effect gating). OECTs based on PEDOT:PSS have been successfully used as biosensors for applications ranging from enzymatic and nucleotide sensing to cellular toxicology and for neurophysiology monitoring [26,27]. All of the applications of PEDOT:PSS-based OECTs reported that the electrochemical nature of the detection/working principle plays a key role, where electrochemistry comes into play not only in the reversible doping/dedoping of the transistor channel, but also in the specific reactions that can take place at the gate/electrolyte interface upon appropriate choice of the gate material and/or functionalization of the gate surface. To date, the main drawback shown by OECTs is the lack of an intrinsic selectivity even though several pathways via functionalization have been demonstrated. This is one aspect that enforces the potential of alternative sensors based on SiC, since SiC-based technology allows reliable strategies for specific selectivity versus a given analyte, due to the high versatility of SiC in terms of the numerous chemical functionalization pathways for this material. This is mainly due to the material possessing the chemistry of both Si and C depending on the termination layer at the surface.

Due to the relevant role of electrochemistry in the sensing mechanism of a majority of SiC-based biosensors, in this chapter particular care will be given to electrochemical devices based on SiC for biosensing applications. The aim is to contextualize these applications within the wide set of examples reported in the literature, where OS, and PEDOT:PSS in particular as a case material of reference, plays a dominant role. Hence, we will briefly describe the biomedical applications of SiC, from tissue engineering and bio-inspired ceramics to biosensors, providing a detailed overview of the current literature existing on SiC-based biosensors. A special focus on electrochemically operated devices, where the detection of analyte molecules takes place due to electrooxidation/reduction (redox) mechanisms, will be given and, whenever possible, a direct comparison between the performance of SiC-based and PEDOT:PSS-based biosensors will be made.

5.2 SiC AND BIOMEDICAL APPLICATIONS

In the previous section it was mentioned that SiC has emerged, since the beginning of the new millennium, as a very promising material for a wide range of biomedical applications, as demonstrated by several examples already available in the literature [28]. Although the growth trend of the SiC market is fast [29], most of the market will continue to be dedicated to power electronic applications (Schottky diodes and transistors). The market forecast is the expected outcome that, despite the strong interest toward SiC bioapplications, this highly promising technology is still at an early stage. Prior to extensively describing SiC-based biosensors in the next section in terms of the most recently demonstrated biomedical applications, a brief overview of the biomedical applications, complementary to that of other chapters in this book, is provided where SiC has already gained a significant role.

So far, among the main SiC biomedical applications there are those where mechanical and chemical properties of this material play the major role, such as coatings for coronary heart stents, biomorphic ceramics, and scaffolds for tissue engineering. The application of SiC as a passive stent coating was described in the early 2000s by several groups [30–34]. In particular, a marked reduction in fibrin and thrombus deposits around the SiC-coated implants was confirmed as well as negligible ion release into the surrounding biological environment [30]. Biomorphic SiC ceramics were used as a base material for dental and orthopedic (mainly bone) implant technology, taking advantage of their enhanced mechanical and biochemical properties. In particular, SiC was exploited as an interesting alternative to titanium and titanium alloy implants, in terms of both strength and density. Biomorphic SiC ceramics coated with bioactive glass were reported for bone substitution [35], while SiC as a ceramic coating material of titanium-based total hip replacement (THR) implants was also investigated with encouraging results [36]. SiC was also used as a scaffold material for tissue engineering, with the purpose of mimicking and/or replacing biological functions as a strategy for restoring the functionality of damaged tissues, such as bones, skin, cartilage, etc. SiC's interconnected porosity facilitates cell seeding and diffusion throughout the whole structure of both cells and nutrients. This allowed use of SiC composites as bone scaffolds by supporting human osteoblast attachment and growth [35,37–39]. In this respect, suitable surface treatments improve cell attachment. As an example, amine-functionalized SiC nanowires (SiCNWs) cross-linked to the extracellular matrix of a porcine tendon were described, with the purpose of optimizing the mechanical strength, the resistance to degradation, and the overall biocompatibility of the porcine tendon [40].

Nanoporosity of SiC was also exploited for the fabrication of freestanding semipermeable membranes used as filters in liquid environments toward biological molecules such as proteins [41]. The SiC-based membranes showed a selective permeability toward proteins according to their molecular weight, passing proteins of up to 29 kD and excluding larger molecules with a molecular weight of 44 kD (corresponding to a size of 4.7 and 5.0 nm, respectively). Interestingly, resistance to protein fouling of SiC biomembranes was also demonstrated [42].

Systems based on SiC are also used for the development of optically based devices. For instance, 3C–SiC nanoparticles are added to 3T3-L1 fibroblasts as bioimaging agents. Analysis with fluorescent microscopy allowed localizing the 3C–SiC nanoparticles inside the cells, verifying their cell nontoxicity after 1 week of incubation [43]. An *a*-Si/*a*-SiC photosensor has also been developed for DNA detection [44], while a 3C–SiC-based superlens for near-field scanning optical microscopy (NSOM) has been realized in order to promote biological imaging in fluidic environments [45]. Finally, an evanescent waveguide was described as a useful system for chemical sensing [46].

5.3 ELECTROCHEMICAL BIOSENSORS

SiC as active material in biosensors represents one of the most recent and exciting solutions for biomedical and point-of-care diagnostics technology. This class of devices has been investigated only in the last 5–6 years, making this new branch of research one of the most innovative applications of this outstanding material. Several biosensor structures where SiC plays a relevant role have been described so far, due to its outstanding chemical, electronic, and mechanical properties. Biosensors involving SiC as an active material in the biorecognition process are mainly electronically based. In this case, SiC is most often used as an electrode substrate or as a functionalizing element of common electrode materials, such as glassy carbon, both in the form of all-SiC nanostructures or nanocomposites in combination with Au, Pt, Pd, TiO_2 and NiO nanoparticles. In particular, the electronic properties of SiC allow the fabrication of devices where the biosensing events take place due to changes in surface impedance or to electrochemical reactions resulting from specific interactions between the SiC nanocomposite and the analyte under analysis. As described in other specifically devoted chapters of this book, SiC properties have been explored not only for devices conceived for detecting and quantifying biomolecules dissolved in biological fluids, but also in probes for *in vivo* monitoring of physiological signals.

5.3.1 SiC Biosensors

Biosensors exploiting SiC as an electrode or substrate material have been used for high-sensitivity detection of several biomolecules. For instance, Yang et al. originally reported the covalent bonding of DNA molecules on 3C–SiC films used as electrochemical electrodes [47]. The 3C–SiC surface was first functionalized with diazonium salts, and then DNA molecules were immobilized onto the so-modified surface. The DNA hybridization was detected by recording the cyclic voltammetry (CV) curves of the ss- and ds-DNA strands immobilized on the SiC-modified electrode surface (Fig. 5.1a) immersed in a PBS buffer solution. The DNA probe was labeled with a redox center of ferrocene portion. Two redox waves of ferrocene portions present in the CV curve of ss-DNA disappeared after hybridization

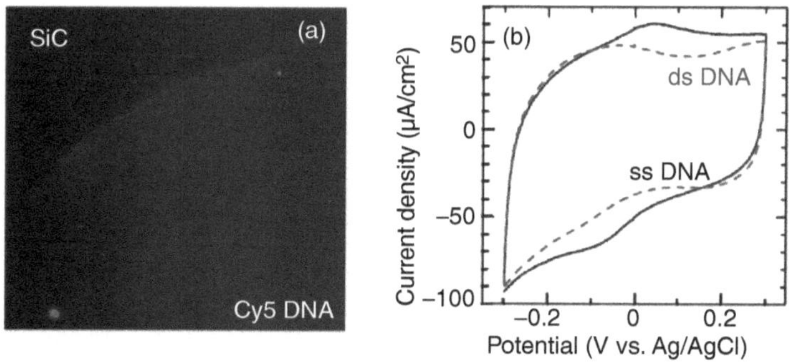

Figure 5.1 (a) Fluorescence micrograph of a SiC electrode functionalized with ds-DNA. (b) CV of ss-DNA (solid line) and ds-DNA (dashed line) modified SiC electrode showing the presence of redox waves according to the hybridization of the probe DNA. *(Reprinted with permission from [47]. Copyright (2011) American Chemical Society.)*

of ss-DNA with complementary DNA, providing a fingerprint of the DNA hybridization (Fig. 5.1b). DNA qualitative detection was also confirmed by a decrease of the redox current.

SiC was also used by Oliveros et al. to build a long-term device to predict silent myocardial infarction via the detection of myoglobin, a protein prevalent in the blood stream during an acute myocardial infarction [48]. To this end, the authors covered a SiC surface with self-assembled monolayers (SAMs) of aminopropyldiethoxymethylsilane (APDEMS) and (3-aminopropyl)-triethoxysilane (APTES), and successfully functionalized the SAM surfaces with the antimyoglobin antibody. When myoglobin interacts with the antibodies, a change in device impedance can be measured using an electrochemical cell. On this basis, they took advantage of electrochemical impedance spectroscopy (EIS) for myoglobin detection, studying in sequence: the antimyoglobin immobilization; the changes in flat band potential, changes of impedance in the space charge region of 3C–SiC after formation of an APTES SAM; and, afterwards, the impact of adding a fixed concentration of myoglobin (1 μg/mL) on top of the sensor surface. Although promising results are described for myoglobin detection, the binding of myoglobin on SiC-modified surfaces was not easy to identify, due to a certain variability of the data that they obtained.

Singh et al. studied a composite SiC-carbon fiber electrode for the detection of neuronal activity and biochemical sensing of electroactive neurotransmitters, such as dopamine hydrochloride and vitamin C [49]. They used CV measurements and EIS for the electrochemical characterization of the device, while detection of both neurotransmitters was performed exploiting the analogous redox activity of labile hydroxyl groups on their molecules, reaching a limit of detection below 10 μM. The authors also reported the *in vivo* recording of action potentials from an anesthetized rat's brain with very high signal-to-noise (S/N) ratio (20–25 compared to 4–5 obtained with commercial tungsten electrodes). The SiC-based electrodes showed a reversible electrochemical behavior revealed by their ability to transfer electrical charge with the charging/discharging of the electrical double layer. One of the most important findings of their work is that electrodes show a capacitive interaction, which is highly desirable for chemical and electrochemical stability of the electrodes.

Finally, electrodes made of SiC were used for *in vivo* measurement of biological signals also by Godignon's group, who has exploited the superior characteristics of SiC to develop temperature and impedance sensors for monitoring both myocardial ischemia and transplantation organ/tissue ischemia [50–52]. They fabricated needle-shaped, minimally invasive impedance probes, reported in Fig. 5.2a, based on SiC substrates with lower leakage current compared to that of Si-based impedance sensors, thus avoiding false impedance changes. The authors compared Si and

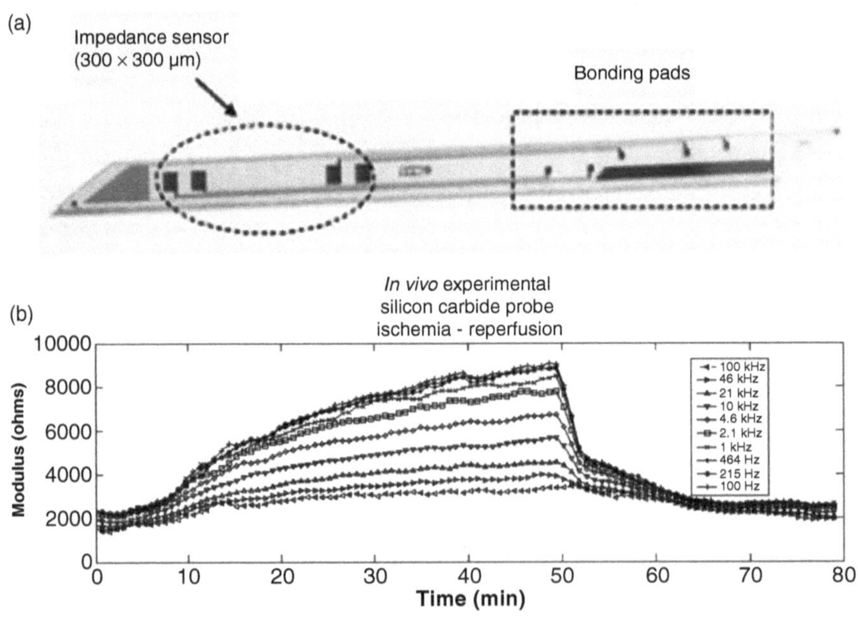

Figure 5.2 (a) Needle-shaped SiC probe for impedance measurements. (b) Impedance modulus from *in vivo* experiments for SiC-based impedance probes in rat kidney warm ischemia. *(Adapted with permission from Springer Science and Business Media [51].)*

SiC mechanical and electrical properties, demonstrating the superiority of SiC for impedance-based needle probes [52]. In fact, their results indicate that SiC outperforms Si in every aspect, showing: SiC needles with a mechanical fracture 2.5 times higher than for Si needles (1053 ± 200 mN vs. 452 ± 37 mN) allowing for critical loads; a four times higher modulus of rupture for SiC devices (774 MPa vs. 188 MPa); and a 10-fold increase in the frequency range for electrical measurements in SiC-based probes. A typical *in vivo* experiment performed by inserting SiC-based impedance probes in a rat kidney affected by warm ischemia, reported in Fig. 5.2b, shows an impressive operation of these tools for the impedance/modulus spectroscopic analysis. These results suggest that SiC should be preferably used over Si in all biomedical applications in which device breakage must be avoided or very precise electrical measurements are required.

Recently, SiC was also used as biocompatible substrate material for an implantable biosensor, whose scheme is reported in Fig. 5.3a, employing radio frequency (RF) for long-term, real-time, and continuous glucose monitoring as reported in chapter: SiC RF Antennas for *In Vivo* Glucose Monitoring and WiFi Applications [53]. The sensing mechanism is based on the shift of the resonant frequency as a function of glucose level (the higher the glucose level, the lower the resonance frequency), which is due to glucose-related changes in blood permittivity and conductivity. Interestingly, the device did not require direct contact with the liquid environment containing the analyte with the active material (SiC). Resonance frequency shifts corresponding to a 97- and 67-kHz shift per 1 mg/dL change in blood glucose were measured (Fig. 5.3b). This innovative SiC-based RF

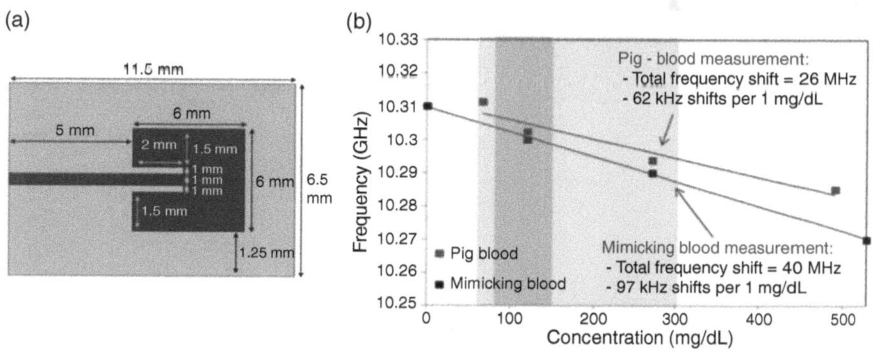

Figure 5.3 (a) Schematic of the SiC-based antenna glucose sensor. (b) Measured relative resonance frequency shift as a function of glucose concentration as presented in chapter: SiC RF Antennas for *In Vivo* Glucose Monitoring and WiFi Applications. *(Adapted with permission from [53].)*

antenna was shown to work efficiently across the glucose level range of medical interest, from the normal clinical condition to the monitoring of critical health conditions such as hypoglycemia (below 60–65 mg/dL) and hyperglycemia (above 240 mg/dL).

The properties of nanostructured SiC have been exploited in order to obtain selective functionalizing systems suitable for the detection of biomolecules. Uniform and highly dispersed gold–silicon carbide (Au@SiC) nanocomposites, prepared via ascorbic acid-assisted chemical reduction, were used by Yang et al. for label-free electrochemical immunosensing of human chorionic gonadotrophin (hCG) [54]. The device, using Au@SiC-modified glassy carbon electrodes (GCE) as shown in Fig. 5.4a, provides a good linear response range from 0.1 to 5 IU/L and 5 to 1000 IU/L with a detection limit of 0.042 IU/L (Fig. 5.4b). The immunosensor showed good selectivity and acceptable stability and reproducibility, resulting in a promising approach for the detection of hCG in clinical diagnostics. Yang et al. achieved the simultaneous electrochemical determination of hydroquinone (HQ) and bisphenol A (BPA) using a titanium dioxide–silicon carbide nanohybrid (TiO_2–SiC) [55]. A GCE was modified with the TiO_2–SiC nanohybrid and monodispersed ultrafine palladium nanoparticles were anchored

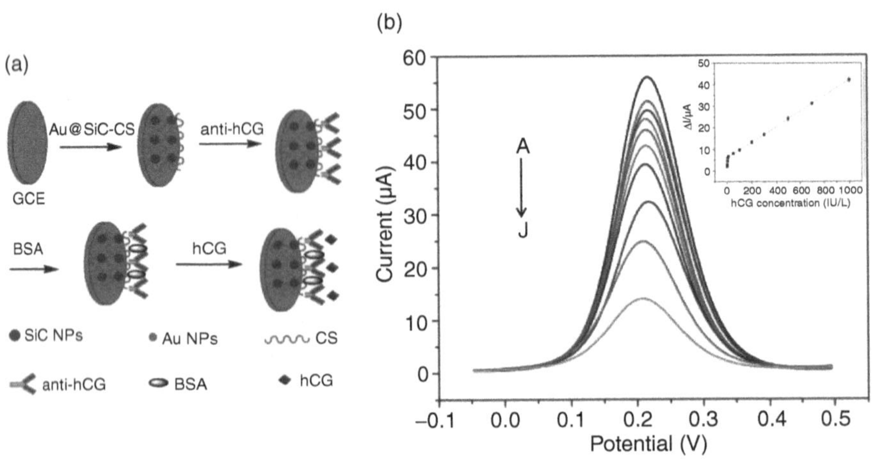

Figure 5.4 (a) Sketch of the fabrication process of the SiC-based hCG immunosensor. (b) Differential pulse voltammetry (DPV) curves of an immunosensor incubated with different concentrations of hCG standard solution (from A to J): 0.0, 1.0, 10.0, 50.0, 100.0, 200.0, 300.0, 500.0, 700.0, and 1000.0 IU/L in 0.1 M pH 7.4 PBS containing 2 mM $[Fe(CN)_6]^{3-/4-}$ and 0.1 M KCl. Pulse width, 0.05 s; amplitude, 0.05 V. The inset shows the calibration plot of the reduction current change (ΔI) versus concentration of hCG under optimal conditions. *(Adapted from [54]. Copyright (2014), with permission from Elsevier.)*

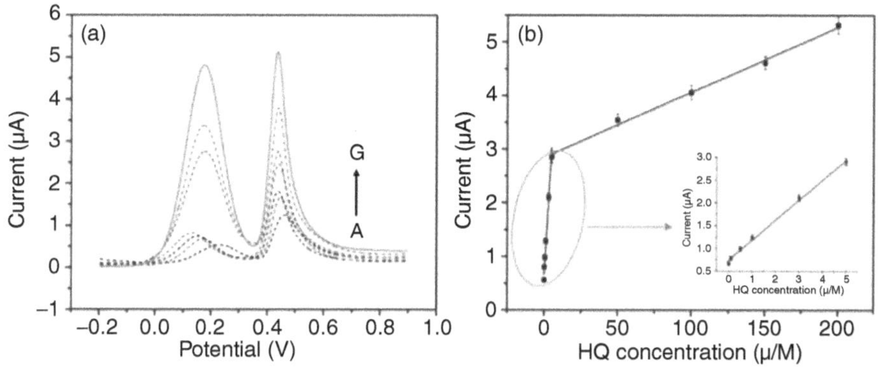

Figure 5.5 (a) DPVs obtained for the oxidation of mixture containing 200.0 mM HQ and 200.0 mM BPA at bare GCE (A), SiC/GCE (B), Pd/GCE (C), TiO_2–SiC/GCE (D), Pd@TiO_2/GCE (E), Pd@SiC/GCE (F), Pd@TiO_2–SiC (G) in 0.1 M, pH 6.0 PBS containing 10% ethanol. (b) Calibration plots of the oxidation current at Pd@TiO_2–SiC/GCE versus concentration of HQ. *(Adapted from [55]. Copyright (2014), with permission from Elsevier.)*

on the TiO_2–SiC surface. The Pd@TiO_2–SiC nanohybrid-modified GCE exhibited electrochemical performance superior to those of TiO_2–SiC and SiC in the simultaneous detection of HQ and BPA (Fig. 5.5a). The detection limits of the Pd@TiO_2–SiC nanohybrid electrode for HQ and BPA were 5.5 and 4.3 nM, respectively (Fig. 5.5b). The authors demonstrated the practical application of the modified electrode by the simultaneous detection of HQ and BPA in tap water and wastewater samples.

SiC nanoparticles were used for electrode modification as electrocatalysts for oxidation of insulin at physiological pH by Salimi et al. [56]. Their device proved to work properly in a linear dynamic range up to 600 pM, with sensitivity of 710 pA pM^{-1} cm^{-2} and a detection limit of 3.3 pM. In addition, the authors also show that the interference effect of cooxidized compounds can be diminished or eliminated by covering the surface of the modified electrode with a Nafion film. The oxidation of insulin at the GC electrode modified with SiC nanoparticles occurred at reduced overpotentials in comparison to a bare GC electrode. This work proved SiC nanoparticle-modified GC electrodes as detectors of insulin at the picomolar level when coupled with a flow system. Rafiee-Pour et al. describe a biosensor for micromolar-level detection of superoxide based on direct voltammetry of immobilized superoxide dismutase (SOD) onto the surface of a GC electrode modified with SiC nanoparticles [57]. The excellent biocatalytic activity of SOD toward superoxide dismutation was exploited and a pair of well-defined and nearly reversible redox peaks was reproducibly recorded.

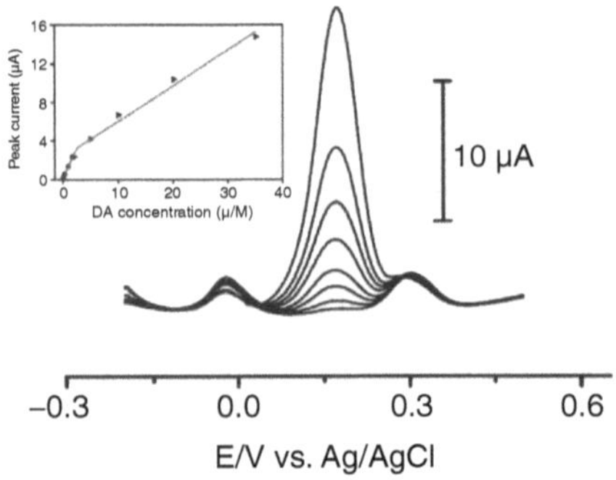

Figure 5.6 Square-wave voltammetric responses as a function of different concentrations of DA (0.2–35 mM) in the presence of 100 mM AA and 1.5 mM UA in 0.05 M phosphate buffer solution at a SiC coated GCE. In the inset the corresponding calibration curve is reported. *(Reproduced from [58], with permission of the Royal Society of Chemistry.)*

The authors reported a sensitivity and detection limit of 1.416 nA/μM, 1.66 μM, and 1.375 nA/μM, 2.1 μM for cathodically or anodically detection of superoxide, respectively. Dopamine (DA), a neurotransmitter of the central nervous system whose abnormal values can be related to neurodegenerative diseases (Alzheimer's, Parkinson's), has been detected with high sensitivity and selectivity by Wu et al. using GCE coated with SiC nanoparticles (SiC-GCE) [58]. CV measurements were performed to highlight the electrocatalytic activity of SiC-GCE with respect to bare GCE (vs. Ag/AgCl) toward DA, ascorbic acid (AA), and uric acid (UA) (Fig. 5.6). Well-defined and resolved CV responses were obtained for the oxidation of DA, AA, and UA with the SiC-GCE electrode, showing a similar electrochemical behavior for these three molecules, expressed by a negative voltage shift of the CV curve. In order to demonstrate the selectivity of the SiC-GCE electrodes for the DA detection, a batch with a mixture of the three analytes was prepared and CV curves acquired versus a reference Ag/AgCl electrode. Separated oxidation peaks could be observed, suggesting the possibility of discriminating between AA, UA, and DA, with a sensitivity toward DA of 16.9 A/M cm^2 and a detection limit of 50 nM. Another application of SiC nanoparticles-modified GCE (SiCNP/GCE) sensors for DNA bases was developed by Ghavami et al. The device was shown to be able to perform a selective and simultaneous determination of purine (guanine G and adenine A) and pyrimidine (thymine T and cytosine C) with a sensitivity down to the micromolar range [59]. The oxidation of individual and a mixture of DNA bases on SiCNP/GCE electrodes was investigated in a PBS solution by acquiring CV curves and differential pulse voltammetric

(DPV) responses and comparing them to those obtained using a bare GCE electrode. Independent oxidation peaks appeared for T, C, and for G and A bases when the SiCNP/GE electrode was used, while negligible oxidation peaks were found in the case of a bare GCE electrode. From the linear relationship between the DPV oxidation peak current of each DNA base and their concentration, detection limits of 0.015 μM for G and A, and 0.14 μM for T and C were calculated, respectively. The very easy electrochemical detection of DNA bases at a GCE electrode modified with SiC nanoparticles results in an easier and cheaper sensor technology than those generally employed for DNA bases detection.

An electrochemiluminescence (ECL) sensor for adrenaline, based on a GCE modified with a composite tyrosinase/SiC/chitosan film, was developed by Ye et al. (Fig. 5.7a) [60]. The working principle of this ECL sensor is based on the tyrosinase catalysis of adrenaline to produce hydrogen peroxide. In order to realize a step forward toward real applications, like in medical samples, the authors also studied the electrochemical response of several possible interferents, such as citric acid, noradrenaline, vitamin C, and DA. In the presence of these species (whose concentrations were 50 times higher than that of adrenaline) the response of the ECL sensor had no significant enhancement. In addition, several metal ions, such as Fe^{3+}, Ca^{2+}, Mg^{2+}, Pb^{2+}, Cu^{2+}, and Zn^{2+}, did not cause interference, demonstrating the very high selectivity of this ECL sensor toward adrenaline, with a detection limit of 0.5 nM and with an S/N ratio of 3 (Fig. 5.7b).

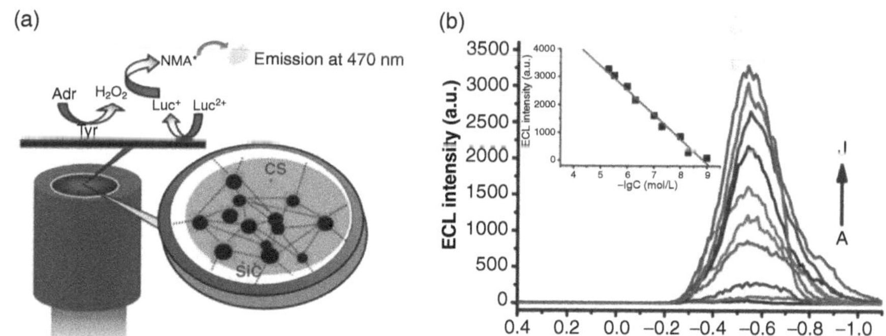

Figure 5.7 (a) Schematic representation of the ECL sensor for adrenaline assay. (b) The ECL spectra of the Tyr/CS/SiC-modified electrode obtained at different adrenaline concentrations. Inset: calibration curve with the ECL intensities versus the adrenaline concentrations. Tyr, tyrosinase; Luc, lucigenin; SiC, carborundum; Adr, adrenaline; NMA, *N*-methylacridone; CS, chitosan. *(Adapted from [60], with permission of the Royal Society of Chemistry.)*

Table 5.2 Comparison of Electrochemical Detection Performance of 600-NiO/SiC With Other Glucose Sensors

Electrode	Linear range (mM)	LOD (μM)	Sensitivity (mA/mM cm²)	References
GOx/PtNP/PAni/Pt	0.01−8	0.7	0.096	[12]
Cu—N—G	0.004−4.5	1.3	0.681	[35]
GM—NiO	0.001−0.1	1.6	0.918	[36]
3D graphene/Co_3O_4	<0.08	<0.025	3.39	[37]
Au/MNE	0.1−3000	0.5	2.978	[38]
600—NiO/SiC	0.004−7.5	0.32	2.037	[61]

References underlined follow the numeration in the original paper.
Reproduced with permission from [61].

Finally, SiC-functionalized electrodes were also used for sensitive glucose detection. In the study by P. Yang et al., NiO/SiC nanocomposites were prepared for the first time by the atomic layer deposition (ALD) technique and used as nonenzymatic electrocatalysts for glucose sensing [61]. NiO nanoparticles were grown on SiC particles by ALD with highly uniform particle size (4.2 nm), while SiC was used as the substrate due to its biocompatibility, low background current, and wide electrochemical window. A glucose detection limit of 0.32 μM was calculated for the NiO/SiC-based biosensor (S/N = 3), with a sensitivity of 2.037 mA/mM cm² and a linear response range up to 7.5 mM. Interestingly, the NiO/SiC-based biosensor showed high selectivity for glucose detection, since addition of electroactive interference species (ascorbic acid, uric acid, and DA) caused only small changes in the device signal compared to glucose addition. Table 5.2, reported from the original work, presents an interesting comparison with the performances of other glucose sensors (although not treated in this chapter).

5.4 SiC- AND PEDOT:PSS-BASED BIOSENSORS—A COMPLEMENTARY COMPETITION

PEDOT:PSS is one of the most popular semiconducting polymers, because of its high conductivity (1–1000 S/cm) and good stability over a wide range of pH values. Due to these main strengths it is already available on the market [62]. Given its widespread use and demonstrated performance it may well be considered a sort of benchmark for polymer-based electrochemical sensing. It consists of PEDOT, the semiconductor itself, and PSS, the most common counter ion added to PEDOT in order to circumvent its insolubility in water and as a dopant to further enhance its conductivity.

As described in the Introduction to this chapter, the mixed ionic–electronic conduction of PEDOT:PSS has given rise to a series of devices for applications where the presence of a liquid environment, and in particular water, is involved. In particular one of the main applications is biological sensing, intended for the detection of biological molecules, such as glucose, proteins, neurotransmitters, DNA bases, or signals related to living cells and whole tissue. Upon considering the scenario of biosensing applications of PEDOT:PSS described so far in literature, in this section we proceed with a comparison between performance, advantages/disadvantages, and perspectives of PEDOT:PSS-based devices with the corresponding ones based on SiC. In particular, we directly compare these two technologies by describing the sensing of specific biological analytes or signals (namely, glucose, DA, DNA, adrenaline, and action potentials), which have been reported using both PEDOT:PSS- and SiC-based sensors.

One of the first and probably the most relevant applications of PEDOT:PSS OECT sensors was the detection of glucose, which dates back to 2004 with the work of Malliaras's group [63]. Glucose was detected with a millimeter-sized device (Fig. 5.8a) using a sensing mechanism based on the detection of hydrogen peroxide (H_2O_2) generated as a by-product of the reaction between glucose and the enzyme glucose oxidase (GOx) added to the electrolyte (Fig. 5.8b). GOx is known to catalyze the reaction of glucose in the presence of oxygen to produce gluconolactone and H_2O_2. The catalytic ability of Pt, used as a gate material, for the oxidation of H_2O_2 was exploited to create an electron that is transferred to the gate. In order to preserve the charge neutrality in the electrolyte, a cation is injected into the

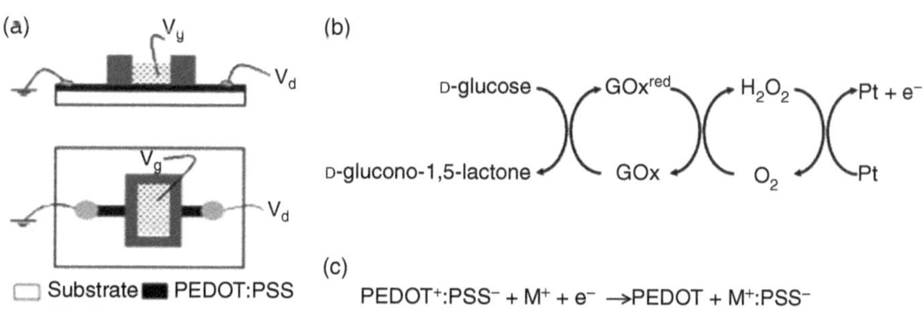

$$PEDOT^{+}{:}PSS^{-} + M^{+} + e^{-} \rightarrow PEDOT + M^{+}{:}PSS^{-}$$

Figure 5.8 (a) Scheme of the PEDOT:PSS device configuration used for glucose sensing (top panel: side view, bottom panel: top view). (b) Reaction cycle at the gate electrode involved in the detection of glucose using GOx. (c) Reaction taking place at the PEDOT:PSS channel. *(Adapted from [63] and [64] with permission of the Royal Society of Chemistry.)*

PEDOT:PSS film and compensates the PSS anion, thus reducing $PEDOT^+$ to the less conductive form $PEDOT^0$ (Fig. 5.8c). Since the concentration of H_2O_2 is directly correlated to that of glucose, H_2O_2 was used to detect and measure the amount of glucose present in the solution. A response was detected for concentrations as low as 100 μM of glucose, which is compatible with physiological levels of glucose in human body fluids (for instance, saliva and sweat). Interestingly, the authors foresee further improvements of the sensitivity by means of device optimization (amount of GOx present in the solution and thickness of the PEDOT:PSS film). More refined results in glucose sensing by OECTs were obtained by Bernards et al., who used millimeter-sized OECTs to also improve the understanding of the device physics underlying enzymatic sensing by OECTs [64]. Device operation was described in terms of charge transfer reactions with a Faradaic contribution that alters the potential at the gate/electrolyte interface, and that can be detected through changes in the source–drain current flowing into the PEDOT:PSS channel. This approach led to the detection of signals with concentrations as low as 1 μM of glucose, resulting in a device response logarithmically correlated to the glucose concentration. Comparable results were obtained by Shim et al. [65], who used a ferrocene mediator in order to enhance the electron transfer to the gate and achieved low-cost single-step devices entirely made of PEDOT:PSS. Ionic liquids, endowed with a wide electrochemical operation range, high ionic strength, low or negligible evaporation, and stabilization of the enzyme structure and functions, were used by Yang et al. [66] as an electrolyte in the OECT, resulting in sensing capabilities for glucose in the range from 10^{-7} to 10^{-2} M. Subsequently, further improvement in the biosensing performance of OECTs was achieved by means of surface functionalization of the active elements of the device. To this regard, Yang's group increased sensitivity and selectivity of OECTs for glucose sensing in complex liquid matrix by functionalizing the gate electrode with enzymes and nanomaterials [67,68]. On the one hand, Tang et al. [67] functionalized a Pt gate electrode with an enzyme (GOx) and Pt nanoparticles, dramatically improving the detection limit of the OECT for glucose sensing (5 nM) in comparison with analogous devices without nanomaterials. The astonishing performance reported was attributed to the excellent electrocatalytic properties of the Pt nanoparticles, together with a higher immobilization of the enzyme due to the higher surface area offered by the nanomaterials. On the other hand, Liao et al. [68] demonstrated that functionalization of a Pt gate electrode with an enzyme (GOx), graphene-based nanomaterials, and biocompatible polymers (chitosan or Nafion) that enhanced both the sensitivity and selectivity of the device, through an improved immobilization of the enzyme and a higher charge transfer at the gate/electrolyte interface. The authors reported a linear response in the concentration range from 10 to 1 μM of glucose, with a detection limit as low as 10 nM.

Some interesting considerations can hence be made from comparing work reported on glucose sensing based on SiC in [61] and on PEDOT:PSS in [63–68]. First, the use of nanosystems is a common feature in order to enhance the selectivity and detection limit of biosensors. Higher active area and unique catalysis properties directly arising from the nanometer size of systems such as nanoparticles, nanotubes, nanoflakes, etc., make them the elements of choice in all cases for device functionalization. Second, all glucose sensors developed so far, both based on PEDOT:PSS and SiC, show limits of detection that fall within the concentration range of physiological interest, resulting in promising point-of-care devices suitable for sensitive and fast measurement of biological signals. Dealing with the specific performance of SiC- and PEDOT:PSS-based glucose sensors, these two technologies showed both advantages and disadvantages. In favor of NiO/SiC nanosystems is that glucose sensing has been achieved in a nonenzymatic way, hence avoiding complicated procedures for enzyme immobilization, risks of poor long-term stability due to enzyme degradation, and dependence of the device response to pH and temperature during the measurements. Furthermore, SiC is an excellent substrate material that ensures biocompatibility, low background current, and a broad potential window. On the other hand, PEDOT:PSS glucose sensors demonstrated the ability to reach unprecedented detection limits as well as extreme selectivity upon suitable functionalization of the active elements of the device. Low cost and conformability complete the set of properties that make conducting polymers like PEDOT:PSS very good candidates for next generation, user-friendly, disposable and possibly wearable devices for personal care.

Electrophysiological recording of brain activity is under current investigation by many research groups worldwide because this could unveil brain functioning as well as help in detecting neuropathological conditions while paving the way to novel perspectives in prosthetics. To this end, several technologies are under study with the goal of extreme sensitivity (eventually at the resolution of single action potential) and least invasive for the patient. In this scenario, PEDOT:PSS has recently proved to be a suitable candidate for highly performing devices for recording action potentials, the well-defined peaks of potentials used by single neurons to communicate within neural networks. An array of PEDOT:PSS OECTs was successfully exploited by Yao et al. for *in vitro* monitoring of the action potentials of cardiac electrogenic muscle cells (HL-1) [69]. OECTs based on PEDOT:PSS and fabricated both on rigid and flexible substrates were a few tens of microns in size, and allowed the detection of local field potential changes caused by ion flows during action potential operation. The current modulation signals recorded by OECT were converted to voltages by using the transconductance curve and the OECT response showed regularly spaced spikes expressing the action potential of the cells. The transistor configuration represents a useful tool to measure action potentials because of the inherent amplification capacity of OECTs that makes it possible to tune the signal amplitude by varying the gate voltage applied and because of an excellent S/N ratio (routinely larger than 4 and up to 10) resulting from the high transconductance of these devices. The OECT showed performance comparable to those of previously reported inorganic transistor-based devices, but required less sophisticated fabrication processes, resulting in good candidates for low-cost implants. Very recently a neural interface array, called the NeuroGrid, was developed for *in vivo* recording of local field potentials and action potentials from superficial cortical neurons without penetrating the brain surface [70]. The NeuroGrid consists of an array of PEDOT:PSS electrodes with a size comparable with that of single neurons. The NeuroGrid showed stable performance in contact with living tissue up to 1 week and a high S/N ratio resulting from an efficient interface of the electrodes with the brain. Other advantages of the NeuroGrid are its lightweight and conformable structure and the neuron-sized density of electrodes that allow recording spikes from individual neurons. PEDOT:PSS and SiC probes demonstrated both advantages and disadvantages also in the case of action potential measurements. Indeed, although PEDOT:PSS devices offer considerably high S/N ratio, SiC probes demonstrated their ability to exceed their organic counterpart, mainly because of the high conductivity of SiC polytypes compared to that of PEDOT:PSS, the latter ranging between 1 and 1000 S/cm. Moreover, due to the very high chemical stability of SiC, especially in harsh environments such as in biological fluids, SiC-based implants, described in more detail in chapter: *In Vivo* Exploration of Robust Implantable Devices Constructed From Biocompatible 3C–SiC of this book, are expected to possess higher long-term stability than PEDOT:PSS implants. However, PEDOT:PSS probes showed several other advantages, such as their extreme conformability to brain tissue due to the flexible nature of organic semiconductors that leads to the possibility of depositing conducting polymers on a wide class of flexible substrates [71,72]. Conformability and flexibility allow for the realization of probes with inherent, less invasive properties, which results in significant benefit for patients under analysis.

In 2004 Krishnamoorty et al. reported, for the first time, a PEDOT-based device used as a label-free DNA sensor [73]. The sensor could detect complementary target DNA in PBS solution with a sensitivity down to 8×10^{-8} g/mL. The sensing mechanism is based on the immobilization of ss-DNA probes on PEDOT during polymerization: due to the formation of a DNA duplex, hybridization of DNA causes the modification of PEDOT structure chains that, in turn, induces a variation in conductivity. P. Lin et al. reported a label-free DNA sensor realized with an OECT based on PEDOT:PSS and integrated on a flexible substrate made of polyethylene terephthalate [74]. Flexibility is an important advantage for practical applications and the device showed stable performance (transfer curves) under different bending states. The DNA probes (ss-DNA) were immobilized on the surface of the gate electrode (Au gate) and hence the device was used as a sensor to detect the complementary DNA targets (ds-DNA). The transfer curves showed a gate voltage shift after hybridization of the DNA probes with DNA complementary targets, with a sensitivity down to 1 nM in static conditions. When DNA hybridization was enhanced by a pulsed electric field, the detection limit achieved values down to 10 pM. For the SiC electrode a pretreatment with a nitrophenyl linking layer was required to bind a DNA molecule to the SiC surface (for more on this topic see chapter: SiC Nanowire Based Transistors for Electrical DNA Detection where SiC nanowires are also used in label-free DNA detection). In the case of a PEDOT:PSS-based OECT sensor, the deposition of DNA was performed on the Au gate electrode, which induces a modulation of the surface potential at the Au gate. As a consequence PEDOT:PSS was able to transduce a current modulation change.

An OECT based on PEDOT:PSS as a transistor channel was employed for the first time for detecting DA in 2001 by Tang et al. [75]. The transistor employed a Pt wire electrode as the gate material immersed in a solution containing DA. The device was characterized before and after the addition of DA into the solution, and showed a detection limit to DA down to 50 nM. Interestingly this is the same sensitivity limit reported in the literature when a SiC-based working electrode was used [58]. The DA detection with OECT was assumed to be due to the electrooxidation of DA at the surface of the Pt electrode, with a sensing mechanism similar to that reported for the detection of H_2O_2 [64]. The sensitivity and selectivity to

DA detection with OECT was improved significantly by Liao et al. by employing a gate electrode of graphene coated with biocompatible polymers (Nafion or chitosan) on the surface [76]. With this method the detection limit reached 5 nM. SiC electrodes look very interesting when compared with a PEDOT:PSS-based OECT for the detection of DA, mainly because of their excellent electrocatalytic activity. Therefore they are well suited for developing electrochemical sensors. SiC in its pristine state can selectively detect DA with a detection limit of 0.05 μM, which is a higher limit with respect to the 5 nM value obtained with an PEDOT:PSS-based OECT with a graphene-modified electrode. In contrast the SiC electrode is capable of discriminating different analytes, such as inteferents with DA [58] without any chemical functionalization.

Adrenaline has been selectively detected in physio-fluids through an organic electrochemical transistor integrated on a textile substrate [77]. The detection of adrenaline is based on the electrooxidation of adrenaline at the surface of a Pt electrode, with the formation of adrenaline–quinone and adrenochrome. This is therefore very similar to the sensing of H_2O_2 [64], or DA [75] and eumelanine [78]. Although the OECT-based sensor does not show the best sensitivity (10^{-5} M), the work by Coppedè et al. reports the use of a real-complex physiological fluid, such as sweat, in which the detection is performed, thus improving the OECT performance with a step forward to very practical applications. The sensing of adrenaline performed by a SiC-modified electrode showed a much lower detection limit (0.5 nM) that falls in the concentration range of adrenaline, for example, in human plasma (0.2 nM). Therefore, it is a good candidate for practical uses, even if the film preparation is complex and the acquisition of ECL spectra is not so straightforward. In contrast, the PEDOT:PSS-based sensor shows higher detection limit, not suitable for practical application at this date, but yields the advantage of being easily fabricated on a flexible and e-textile integratable substrate. An additional advantage of the PEDOT:PSS sensor is that the fabrication of the device is versatile and easy, along with fast device response and straightforward data acquisition.

Performance of both SiC- and PEDOT:PSS-based sensing devices, together with the related references, are summarized in Table 5.3. Beyond the previous analysis of the features and demonstrated performance of SiC and PEDOT:PSS biosensors, it is also useful to imagine a combined use of PEDOT:PSS thin films and SiC-based electrodes that could represent a quite interesting perspective not yet explored. For example, we propose here that considering that the enhancement of OECTs' performance is

Table 5.3 Comparison Between SiC- and PEDOT:PSS-Based Biosensors

	SiC-based biosensors			PEDOT: PSS-based biosensors	
Analyte	**LOD**	**SiC role**	**References**	**LOD**	**References**
DNA	—	Electrode	[47]	10 pm	[73,74]
	0.015 μM	SiC nanoparticles	[59]		
Myoglobin	1 μg/mL	Electrode	[48]		
Dopamine	<10 μM	Electrode	[49]	5 nM	[75]
	50 nM	SiC nanoparticles	[58]		
Ischemia	—	Electrode	[50–52]		
Glucose	—	Substrate material	[53]	5 nm	[63–68]
	0.32 μM	NiO/SiC	[61]		
gCG	0.042 IU/L	Electrode	[54]		
HQ/BPA	5.5 nM/ 4.3 nM	TiO_2–SiC nanohybrid	[55]		
Insulin	3.3 pM	SiC nanoparticles	[56]		
Superoxide anion	1.66 μM	SiC nanoparticles	[57]		
Adrenaline	0.5 nM	Electrode	[60]	10 μM	[77]
Action potential		Electrode	[49]		[69,70]

related to electrochemical reactions at the gate electrode/electrolyte interface and given the electrocatalytic capability of bare and functionalized SiC structures, the use of this material as a gate electrode in OECTs could pave the way towards an intrinsic selectivity for PEDOT:PSS-based sensors. This, coupled with SiC specificity, which has been demonstrated to be effective in distinguishing different analytes in a solution [58], and using OECTs, may offer a very promising advanced sensor strategy that combines the strengths of both materials.

5.5 SiC-BASED FIELD EFFECT TRANSISTORS IN BIOSENSING: PERSPECTIVES AND CHALLENGES

FETs represent a fundamental playground in the development of Biosensing devices since they can ideally transduce physical–chemical and biological processes in several different ways. In addition they can control and amplify signals, thus exploiting the best opportunities of integrating the sensing function in more complex structures and systems. There are several types of devices operating in transistor mode that are implementable using polytypic forms of SiC. Typical SiC-based transistors, massively developed worldwide for power electronics applications, can be fabricated using various types of architectures. In most cases, architectures deal with the ability of implementing field effect doping, as it is in the cases of metal-oxide semiconductor field effect transistors (MOSFETs) operating at high temperatures, up to 1000 K [79], and metal-semiconductor field effect transistors (MESFETs) [80]. This is also the case for bipolar devices exploiting the diffusion of charge carriers across a junction, such as bipolar junction transistors (BJTs) [81], or insulated gate bipolar transistors (IGBTs) [82]. All of the earlier structures, fabricated by properly assembling SiC-based p–n junctions, are realized via local doping in the bulk of the SiC films. Hence, SiC technology is well suited to develop architectures exploiting field effect doping, except for the case in which the local doping technique is not employed. This happens for devices based on layered structures fabricated upon successive deposition of device layers, such as thin film transistors (TFTs). In TFTs it is more difficult to efficiently adopt SiC as an active layer, due to the high interfacial defect density at the SiO_2/SiC junction formed during the growth process, which is responsible for lower channel mobility [83]. In particular, the transport properties at the SiC interface obtained by growing cubic SiC on top of hexagonal SiC substrates [84] affect the device performance and limit its employment to large-area applications, as described by a two-dimensional electron gas (2DEG) in the case of heterojunction structures. Strategies aimed at obtaining films with a superior quality for the manufacturing of TFTs are required in order to overcome the factors that limit the interfacial transport. In fact, the heteropitaxial growth of SiC onto Si and SiO_2 substrates suffers for the mismatching induced by impurities, but also for the interfacial reactivity and stress effects induced in thin films by wafer bending [85].

Even if the TFT architecture is largely used for the manufacturing of unipolar devices based on amorphous and OS materials [86], the literature widely reports graphene TFTs made of multilayered epitaxial graphene (MEG) synthesized on SiC substrates [87]. Interestingly, MEG physical properties can be tuned by the substrate underneath them: for instance, a p-type dominant behavior has been obtained for MEG on the C-face while n-type charge carriers dominate the transport on Si-face MEG [88]. The combination of SiC with MEG is of great interest and we believe will bring in another very promising perspective from the point of view of sensing over a broad range of applications including biomedical. The combination of their properties and the control of the interfaces are, in fact, ideally suited for such applications.

The previously mentioned device structures pave the way to sensors where SiC plays an active role in the sensing mechanism. We will summarize the wide range of envisaged sensing applications by SiC-based FETs by following a classification based on the type of analyte and on the sensing features used. Specifically, SiC gas and vapor sensors operating in transistor mode have been successfully used to detect analytes of various nature and phase aggregation, so that organic molecules such as methanol in the concentration range 0.3–5% have been sensed by operating the device between 150 and 350°C [89], while gaseous systems, such as H_2, NH_3, CO, and O_2 [90], have been detected as well. Then, SiC-based field effect devices have been used for monitoring environmental chemical–physical properties, so that efficient pH detection with a sensitivity of 52.35 mV/pH, lower than that theoretically expected for electrolyte–insulator–semiconductor devices (of 59 mV/pH), has been demonstrated [91]. A key feature of SiC-based devices is their ability to operate under environmental conditions, such as high temperatures (up to 923 K) and harsh chemical and bioenvironments, where other materials easily fail or have much shorter lifetimes [92]. Sensing platforms have also been implemented where the role of SiC is not active, whereas even though acting as substrate, it is used to modify in a peculiar and controlled way the morphology of graphene sheets [93] and to induce the opening of a bandgap controlled by quantum confinement of carriers at the graphene–substrate interface. Clearly, it has been noted that the control of graphene properties on SiC is desirable for the promotion of the device specificity [94].

Despite the massive use of SiC for the manufacturing of electrodes suitable to carry out electrochemical- and biosensing, the literature lacks concrete examples dealing with applications of SiC-based transistors in the context of biosensing, conversely to what happens in the case of biocompatible organic materials [86]. Currently, beyond the SiC biocompatibility confirmed by an impressive lack of cytotoxicity [5], it is worth underlining the stability, which is practically unique, of this robust semiconductor under chemically harsh environments, with very low dissolution rates [6], and by far the higher mobility compared to that of OS, by at least three orders of magnitude [86]. Interestingly, recent findings demonstrate that SiC technology is also suitable for fabricating flexible, lightweight, and conformable thin films [95]. All of the previously mentioned features, together with the wide range of functionalization stages available for modifying the SiC surface by means of physical and chemical processes [10], make this material the perfect candidate for sensing and monitoring bioanalytes and for functionally interfacing biosystems in general. In addition, a concrete perspective can be envisaged in the manufacturing of SiC-based implantable transistors that can both grip large tissues and be used for improved electrical interfaces for nervous system tissues, as discussed in chapter: *In Vivo* Exploration of Robust Implantable Devices Constructed From Biocompatible 3C–SiC and chapter: Amorphous Silicon Carbide for Neural Interface Applications.

The discussion concerning the employment of WBG materials such as SiC, ZnO, and GaN as transducer elements in biosensing is still at a sort of preliminary stage, even if it can be dated back by one decade [96]. Although the transduction of ionic signals into electronic currents is intrinsically peculiar for organic semiconductors showing a mixed ionic/electronic conduction, WBG semiconductors require a surface biofunctionalization in order to execute an efficient transducing operation [96,97]. Transistors responsive to environmental changes modulated by a biological reaction are commonly referred to as ion-sensitive field effect transistors (ISFETs). These kinds of devices represent the natural interface between biosystems like cells (since cells fulfill their function by means of ionic fluxes) and complex biomolecules, where the presence and position of specific functional groups determine their reactivity in the context of the role that they play in physiological activities. ISFET operation is specifically based on the interaction of H^+ ions present in the liquid environment (acting as the gate electrode) and the surface of an insulating oxide layer, as accurately described by the site-binding model. This interaction induces an electric field acting on the current flowing through the channel electrodes, the source, and drain, the effect of which induces a shift of the device threshold voltage V_{TH}, which is the gate-to-source voltage required for the formation of the device conducting channel.

A stimulating development is reported by Lloyd-Spetz et al. that suggests the use of SiC for the manufacturing of ISFETs. They developed a multifunctional sensor system with an integrated transistor, resonator, and resistivity change measurements tool (Fig. 5.9) [98]. They show that the device, made of a SiC/ZnO bilayer with a ZnO surface silanized by self-assembling of APTES, works efficiently for human breath analysis aimed at sensing nitrogen oxide. The characterization of the MOS capacitor shown in this work is intended as a preliminary step towards the manufacturing of an ISFET biosensor with a porous gate metal as a semireference electrode. Except for this proof of concept device, the scarce literature produced so far is an indication of how low has been the consideration of SiC for fabricating ISFET structures for biological applications. Hence, the main question

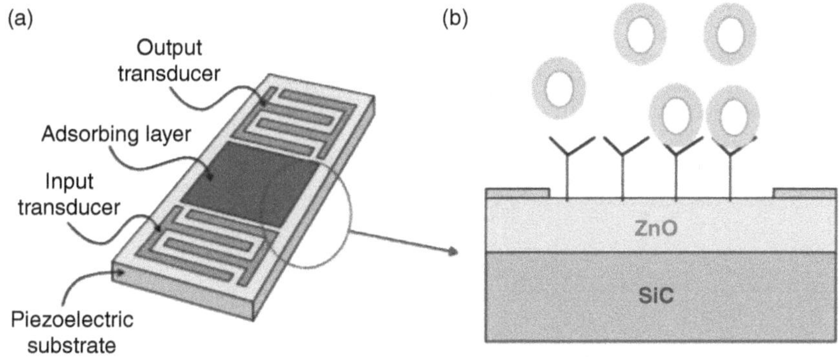

Figure 5.9 ***Top view (a) and side view (b) of a multifunctional sensor device with an integrated transistor, resonator, and resistivity change measurements.*** *(Reproduced with permission from [98].)*

concerns whether SiC polytypes can substitute actively organic and ordinary semiconductors, and also the substrates, for the manufacturing of ISFETs working efficiently as biosensors. We believe that the great potential in this direction is widely unexplored and misjudged for reasons that are partially accidental.

We believe that a possible right strategy could be pursued by taking into account the charge-modulated field-effect transistor (CMFET), a device proposed by Barbaro et al. with the aim of detecting small changes of the net electric charge related to a wide set of biomolecular processes via field effect modulation. In this specific case the device, conceived with the aim of performing label-free DNA hybridization detection and realized in standard CMOS technology, ensures a concrete applicability in terms of large-scale integration [99]. The device structure consists of a floating gate transistor, a MOS [99] rather than an OTFT [100], biased through a control capacitor, and an active probe area constituted by an exposed portion of a floating gate. In particular, the sensing area can be functionalized ad hoc with chemical groups acting as an ultrathin, insulating spacer with the aim of favoring the anchoring of the surface charges, Q_S, that one would sense the interaction with detectable analytes (Fig. 5.10).

CMFETs, unlike the case of ISFETs, work without any external reference electrode. The working principle of CMFETs is based on a redistribution of the electric charge in the floating gate, specifically a separation of charges driven by the electric field that is generated by the charges immobilized in the probe area. In this way, the capacitive coupling generated between the floating and control gates is able to modulate the current in

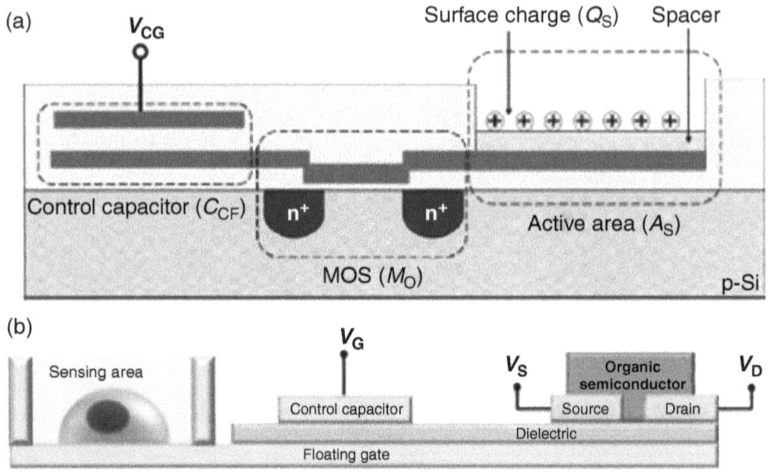

Figure 5.10 ***Basic structure of the charge-modulated field effect transistor based on (a) a floating gate MOS transistor [98] and (b) a standard OTFT [100].*** *(Adapted from [98]. Copyright (2006), with permission from Elsevier. Authors acknowledge the original author and source of publication [100] for providing with the figure.)*

the transistor channel. Namely, the redistribution depends on the sign of the anchored charges, so it induces a current decrease and/or increase that depends on the sign of charges generated by the probed molecules onto the exposed floating gate surface. In a floating gate transistor biased by a control capacitor, the threshold represents an effective threshold voltage, V_{THF}. Its variation ΔV_{THF} depends on the presence of charges anchored onto the floating gate, rather than on the bias stress effect of the FET device, and it can be expressed as a function of the net charge Q_S and C_{TOT} (C_{TOT} is the overall capacitance given by the sum of the capacitances between the floating gate and other electrodes) as follows:

$$\Delta V_{THF} \approx -\frac{Q_S}{C_{TOT}}$$

At a first glance, FET architectures based on MOS capacitors are ideally suitable for realizing charge-modulated transistors by means of conventional manufacturing protocols used for SiC-based MOSFET fabrication. Since the device operation is quite independent from the nature and properties of the channel materials, no particular drawbacks can be envisaged in this case and the fast response and high-gain typical of SiC-MOSFETs should be ideally suited for detecting very small charges, down to one electron charge per μm^2 [99].

TFT structures based on the MEG/SiC interface usually exploit a top-gated configuration with a dielectric layer made of hafnium oxide, deposited on graphene by ALD [101], while bottom-gated TFTs based on buried SiC gates, fabricated by including, for instance, nitrogen atoms in a SiC crystal, are less commonly used [102]. However, the latter structures are completely compatible with the scheme proposed in Fig. 5.10b. In addition, considering that SiC, as reported in Table 5.1, shows a high dielectric constant capable of a strong capacitive coupling between the metallic floating gate and the control gate, bottom-gated TFT structures can equally be well considered for manufacturing well-performing competitive ion-sensitive devices. Finally, taking into account the high electronic performance of SiC and considering that SiC functionalization relies on a wide set of strategies and functionalizing materials, it emerges that SiC/dielectric oxide interfaces can be eventually exploited to replicate the CMFET structure in Fig. 5.10b. In this case SiC substrates can efficiently play the role of a floating gate, functionalized ad hoc in order to immobilize specific analytes, thus efficiently generating a surface charge on it.

5.6 CONCLUSIONS

Even though the recognized versatility of SiC has pointed towards innumerable applications relying upon its physical, chemical, and electronic properties, only in the last two decades has it been considered for the design and implementation of electronic devices specifically designed for applications in biosensing and bioelectronics. On the other hand, biocompatibility, chemical inertness in harsh environments, ideal mechanical properties, and ease of functionalization are some of the peculiar features marking this material as being well suited for biomedical and biosensing applications. In this chapter, after a brief description of biomedical applications based on the use of SiC as coatings for coronary heart stents, biomorphic ceramics for prosthetics and scaffolds for tissue engineering, we have discussed some very promising applications in the biosensing field where the biorecognition process is electronic and/or electrochemical while the mechanism revealing the presence of analytes mainly relies on electrochemical reactions (label free). What emerges is a widely underestimated panorama where SiC biosensors have been demonstrated to detect efficiently and with high sensitivity electroactive neurotransmitters, such as DA, or several biomolecules, such as DNA, adrenaline, glucose, and myoglobin. This restriction is driven by the necessity of framing SiC-based biosensors in a more general overview regarding the leading biosensors, which currently are based, in most cases, on organic semiconductors. In fact biosensors based on PEDOT:PSS lead the field by far, and often operate in the presence of electrochemical reactions. This is specifically why we have compared SiC-based biosensors with those based on PEDOT:PSS. The comparison shows what we could well define as virtuous complementary competition. On the one hand, for some given analytes PEDOT:PSS-based biosensors still outperform SiC-based ones in terms of lower detection limits. On the other hand, SiC biosensors show some advantages with respect to PEDOT:PSS-based devices, since SiC electrodes, for instance, are capable of discriminating different analytes, even without any functionalization, and SiC-based probes show a very high S/N ratio. Even though the flexibility and direct ion to electron amplifying transduction typical of organic sensors makes them unique for several biomedical applications, we believe, on the basis of our discussion here, that there will be much more intensive use of SiC in multifunctional biosensor operations where the monitoring and control of biophenomena needs robust materials showing simultaneously a wide set of properties, from chemical–physical robustness to electrical and mechanical performance. Finally considering that, in our opinion, the application of SiC-based transistors in biosensing has not been adequately explored, with a view to enlarging the opportunities of biosensing devices fully exploiting SiC, we conclude in the last section of this chapter by proposing a novel architecture for high-speed, implantable detectors that is introduced and discussed in the light of ion-sensitive transistors. It is our firm belief that by combining the superior characteristics of SiC and PEDOT:PSS-based devices the community can realize truly innovative and widely used biosensors for today's personal healthcare market.

REFERENCES

[1] Willner I. Biomaterials for sensors, fuel cells, and circuitry. Science 2002;298:2407–8.

[2] Cheung R. Silicon carbide microelectromechanical systems for harsh environments. London: Imperial College Press; 2006.

[3] Schneer CJ. Polymorphism in one dimension. Acta Crystallogr 1955;8:279–85.

[4] Lebedev AA. Heterojunctions and superlattices based on silicon carbide. Semicond Sci Technol 2006;21:R17–34.

[5] Coletti C, Jaroszeski MJ, Pallaoro A, Hoff AM, Iannotta S, Saddow SE. Biocompatibility and wettability of crystalline SiC and Si surfaces. Proc 29th Annu Int Conf IEEE Eng Med Biol Soc 2007;5850–3.

[6] Bolz A. Applications of thin-film technology in biomedical engineering. New York, NY: Marcel Dekker; 1995.

[7] Harder C, Rzany A, Schaldach M. Coating of vascular stents with antithrombogenic amorphous silicon carbide. Prog Biomed Res 1999;4:71–7.

[8] Harris GL. Basic physical properties. Properties of silicon carbide. United Kingdom: IET; 1995.
[9] Kimoto T, Cooper JA. Fundamentals of silicon carbide technology: growth, characterization, devices and applications. Singapore Pte Ltd.; John Wiley & Sons 2014.
[10] Yakimova R, Petoral RM, Yazdi GR, Vahlberg C, Lloyd Spetz A, Uvdal K. Surface functionalization and biomedical applications based on SiC. J Phys D Appl Phys 2007; 40:6435–42.
[11] Wright NG, Horsfall AB. SiC sensors: a review. J Phys D Appl Phys 2007;40:6345–54.
[12] Saddow SE, editor. Silicon carbide biotechnology: a biocompatible semiconductor for advanced biomedical devices and applications. 1st ed. UK: Elsevier Ltd; 2011.
[13] Malliaras GG. Organic bioelectronics: a new era for organic electronics. Biochim Biophys Acta – Gen Subj 2013;1830:4286–7.
[14] Liao C, Yan F. Organic semiconductors in organic thin-film transistor-based chemical and biological sensors. Polym Rev 2013;53:352–406.
[15] Kumar B, Kaushik BK, Negi YS. Organic thin film transistors: structures, models, materials, fabrication, and applications: a review. Polym Rev 2014;54:33–111.
[16] Iannotta S, Toccoli T, Biasioli F, Boschetti A, Ferrari M. Highly ordered films of quaterthiophene grown by seeded supersonic beams. Appl Phys Lett 2000;76:1845–7.
[17] Coppedè N, Toccoli T, Pallaoro A, Siviero F, Walzer K, Castriota M, Cazzanelli E, Iannotta S. Polymorphism and phase control in titanyl phthalocyanine thin films grown by supersonic molecular beam deposition. J Phys Chem A 2013;111:12550–8.
[18] Rivnay J, Owens RM, Malliaras GG. The rise of organic bioelectronics. Chem Mater 2014;26:679–85.
[19] Andersson P, Forchheimer R, Tehrani P, Berggren MBM. Printable all-organic electrochromic active-matrix displays. Adv Funct Mater 2007;17:3074–82.
[20] Lin P, Yan F. Organic thin-film transistors for chemical and biological sensing. Adv Mater 2012;24:34–51.
[21] Cui X, Martin DC. Electrochemical deposition and characterization of poly (3,4-ethylenedioxythiophene) on neural microelectrode arrays. Sens Actuators B Chem 2003;89:92–102.
[22] Blaua A, Murrb A, Wolffc S, Sernagord E, Medinia P, Iurillia G, Zieglerb C, Benfenati F. Flexible, all-polymer microelectrode arrays for the capture of cardiac and neuronal signals. Biomaterials 2011;32:1778–86.
[23] Sessolo M, Khodagholy D, Rivnay J, Maddalena F, Gleyzes M, Steidl E, Buisson B, Malliaras GG. Easy-to-fabricate conducting polymer microelectrode arrays. Adv Mater 2013;25:2135–9.
[24] Scott JC, Bozano LD. Nonvolatile memory elements based on organic materials. Adv Mater 2007;19:1452–63.
[25] Romeo A, Dimonte A, Tarabella G, Angelo PD, Erokhin V, Iannotta S. A bio-inspired memory device based on interfacing *Physarum polycephalum* with an organic semiconductor. APL Mater 2015;3:014909.
[26] Tarabella G, Mahvash Mohammadi F, Coppedè N, Barbero F, Iannotta S, Santato C, Cicoira F. New opportunities for organic electronics and bioelectronics: ions in action. Chem Sci 2013;4:1395.
[27] Strakosas X, Bongo M, Owens RM. The organic electrochemical transistor for biological applications. J Appl Polym Sci 2015;41735:1–14.
[28] Mahmoodi M, Ghazanfari L. Fundamentals of biomedical applications of biomorphic SiC. Properties and applications of silicon carbide. Rijeka, Croatia: INTECH Open Access Publisher; 1999. p. 297–345.
[29] MarketsandMarkets, Silicon Carbide (SiC) in Semiconductor Market by Technology, Product, and Application (Automotive, Defense, Computers, Consumer Electronics, ICT, Industrial, Medical, Power, Railways, and Solar) by Geography – Forecast and Analysis to 2013–2020; 2014.

[30] Rzany A, Harder C, Schaldach M. Silicon carbide as an anti-thrombogenic stent coating: an example of a science-based development strategy. Prog Biomed Res 2000;5:168–78.
[31] Scheller B, Hennen B, Severin-Kneib S, Zbek C, Schieffer H, Markwirth T. Long-term follow-up of a randomized study of primary stenting versus angioplasty in acute myocardial infarction. Am J Med 2001;110:1–6.
[32] Dinne I, Kalnins U, Kumsars I, Erglis A, Jegere S. Clinical outcomes of silicon carbide coated stents in patients with coronaryartery disease. Med Sci Monit 2002;8:16–20.
[33] Dzyadevych SV, Soldatkin AP, El'skaya AV, Martelet C, Jaffrezic-Renault N. Enzyme biosensors based on ion-selective field-effect transistors. Anal Chim Acta 2006;568:248–58.
[34] Hehrlein C. aSiC: H-coated stents stent passivation with silicon carbide as a possible alternative to drug-eluting stents – a comprehensive review of pre-clinical and clinical results. Interv Cardiol 2009;4:60–3.
[35] Gonzalez P, Serra J, Liste S, Chiussi S, Leon B, Perez-Amor M, Martinez-Fernandez J, de Arellano-Lopez AR, Varela-Feria FM. New biomorphic SiC ceramics coated with bioactive glass for biomedical applications. Biomaterials 2003;24:4827–32.
[36] Ikeuchi K, Kusaka J, Yoshida H. Tribology for all-ceramic joint prostheses. J Ceram Process Res 2000;1:53–6.
[37] Thian ES, Huang J, Best SM, Barber ZH, Bonfield W. A new way of incorporating silicon in hydroxyapatite (Si-HA) as thin films. J Mater Sci Mater Med 2005;16:411–5.
[38] Rokusek D, Davitt C, Bandyopadhyay A, Bose S, Hosick HL. Interaction of human osteoblasts with bioinert and bioactive ceramic substrates. J Biomed Mater Res Part A 2005;75:588–94.
[39] Saki M, Kazemzadeh Narbat M, Samadikuchaksaraei A, Ghafouri HB, Gorjipour F. Biocompatibility study of a hydroxyapatitealumina and silicon carbide composite scaffold for bone tissue engineering. Yakhteh Med J 2009;11:55–60.
[40] Deeken CR, Fox DB, Bachman SL, Ramshaw BJ, Grant SA. Characterization of bionanocomposite scaffolds comprised of amine-functionalized gold nanoparticles and silicon carbide nanowires crosslinked to an acellular porcine tendon. J Biomed Mater Res B Appl Biomater 2011;97B:334–44.
[41] Rosenbloom AJ, Sipe DM, Shishkin Y, Ke Y, Devaty RP, Choyke WJ. Nanoporous SiC: a candidate semi-permeable material for biomedical applications. Biomed Microdevices 2004;6:261–7.
[42] Bai S, Ke Y, Shishkin Y, Shigiltchoff O, Devaty RP, Choyke WJ, Strauch D, Stojetz B, Dorner B, Hobgood D, Serrano J, Cardona M, Nagasawa H, Kimoto T, Porter LM. Four current examples of characterization of silicon. Mat Res Soc Symp Proc 2003;742:K3.1.1.
[43] Botsoa J, Lysenko V, Géloën A, Marty O, Bluet JM, Guillot G. Application of 3C-SiC quantum dots for living cell imaging. Appl Phys Lett 2008;92:173902.
[44] Caputo D, de Cesare G, Nascetti A, Scipinotti R. Two-color sensor for biomolecule detection. Sensor Lett 2008;6:542–7.
[45] Taubner T, Korobkin D, Urzhumov Y, Shvets G, Hillenbrand R. Near-field microscopy through a sic superlens. Science 2006;313:2006.
[46] Pandraud G, French PJ, Sarro PM. Fabrication and characteristics of a PECVD SiC evanescent wave optical sensor. Sens Actuators A Phys 2008;142:61–6.
[47] Yang N, Zhuang H, Nebel CE. Nanocrystalline 3C-SiC electrode for biosensing applications. Anal Chem 2011;83:5827–30.
[48] Oliveros A. Myoglobin detection on SiC: immunosensor development for myocardial infarction. PhD, University of South Florida; 2013.
[49] Singh S, Buchanan RC. SiC-C fiber electrode for biological sensing. Mater Sci Eng C 2007;27:551–7.
[50] Godignon P. SiC materials and technologies for sensors development. Mater Sci Forum 2005;483–485:1009–14.

[51] Gomez R, Ivorra R, Godignon JP, Millan J, Erill I, Solà A, Hotter G, Luis P. A SiC microdevice for the minimally invasive monitoring of ischemia in living tissues. Biomed Microdevices 2006;8:43–9.
[52] Gabriel G, Erill I, Caro J, Gomez R, Riera D, Villa R, Godignon P. Manufacturing and full characterization of silicon carbide-based multi-sensor micro-probes for biomedical applications. Microelectr J 2007;38:406–15.
[53] Afroz S, Thomas SW, Mumcu G, Saddow SE. Implantable SiC based RF antenna biosensor for continuous glucose monitoring. IEEE Sensors Conf 2013;1–4.
[54] Yang L, Zhao H, Fan S, Deng S, Lv Q, Lin J, Li C. Label-free electrochemical immunosensor based on gold–silicon carbide nanocomposites for sensitive detection of human chorionic gonadotrophin. Biosens Bioelectr 2014;57:199–206.
[55] Yang L, Zhao H, Fan S, Li B, Li C. A highly sensitive electrochemical sensor for simultaneous determination of hydroquinone and bisphenol A based on the ultra fine Pd nanoparticle@TiO2 functionalized SiC. Anal Chim Acta 2014;852:28–36.
[56] Salimi A, Mohamadi L, Hallaj R, Soltanian S. Electrooxidation of insulin at silicon carbide nanoparticles modified glassy carbon electrode. Electrochem Commun 2009;11:1116–9.
[57] Rafiee-Pour H-A, Noorbakhsh A, Salimi A, Ghourchian H. Sensitive superoxide biosensor based on silicon carbide nanoparticles. Electroanalysis 2010;22:1599–606.
[58] Wu W-C, Chang H-W, Tsai Y-C. Electrocatalytic detection of dopamine in the presence of ascorbic acid and uric acid at silicon carbide coated electrodes. Chem Commun 2011;47:6458–60.
[59] Ghavami R, Salimi A, Navaee A. SiC nanoparticles-modified glassy carbon electrodes for simultaneous determination of purine and pyrimidine DNA bases. Biosens Bioelectr 2011;26:3864–9.
[60] Ye H, Xu H, Xu X, Zheng C, Li X, Wang L, Liu X, Chen G. An electrochemiluminescence sensor for adrenaline assay based on the tyrosinase/SiC/chitosan modified electrode. Chem Commun 2013;49:7070–2.
[61] Yang P, Tong X, Wang G, Gao Z, Guo X, Qin Y. NiO/SiC nanocomposite prepared by atomic layer deposition used as a novel electrocatalyst for nonenzymatic glucose sensing. ACS Appl Mater Interfaces 2015;7:4772–7.
[62] Elschner A, Kirchmeyer S, Lövenich W, Merker U, Reuter K. PEDOT, principles and applications of an intrinsically conductive polymer. Boca Raton, FL: CRC Press; 2010.
[63] Zhu Z, Mabeck JT, Zhu C, Cady NC, Batt A, Malliaras GG. A simple poly(3,4-ethylene dioxythiophene)/poly(styrene sulphonic acid) transistor for glucose sensing at neutral pH. Chem Commun 2004;13:1556–7.
[64] Bernards DA, Macaya DJ, Nikolou MJ, DeFranco A, Takamatsu S, Malliaras GG. Enzymatic sensing with organic electrochemical transistors. J Mater Chem 2008;18:116.
[65] Shim NY, Bernards DA, Macaya DJ, DeFranco JA, Nikolou M, Owens RM, Malliaras GG. All-plastic electrochemical transistor for glucose sensing using a ferrocene mediator. Sensors 2009;9:9896–902.
[66] Yang SY, Cicoira F, Byrne R, Benito-Lopez F, Diamond D, Owens RM, Malliaras GG. Electrochemical transistors with ionic liquids for enzymatic sensing. Chem Comm 2010;46:7972–4.
[67] Tang H, Yan F, Lin P, Xu J, Chan HLW. Highly sensitive glucose biosensors based on organic electrochemical transistors using platinum gate electrodes modified with enzyme and nanomaterials. Adv Funct Mater 2011;21:2264–72.
[68] Liao C, Zhang M, Niu L, Zheng Z, Yan F. Highly selective and sensitive glucose sensors based on organic electrochemical transistors with graphene-modified gate electrodes. J Mater Chem B 2013;1:3820.
[69] Yao C, Li Q, Guo J, Yan F, Hsing I-M. Rigid and flexible organic electrochemical transistor arrays for monitoring action potentials from electrogenic cells. Adv Healthc Mater 2014;4:528–33.

[70] Khodagholy D, Gelinas JN, Thesen T, Doyle W, Devinsky O, Malliaras GG, Buzsáki G. NeuroGrid: recording action potentials from the surface of the brain. Nat Neurosci 2015;18:310–5.
[71] Liao C, Zhang M, Yao MY, Hua T, Li L, Yan F. Flexible organic electronics in biology materials and devices. Adv Mater 2014;1–35.
[72] Iannotta S, D'Angelo P, Romeo A, Tarabella G. Scalable and flexible bioelectronics and its applications to medicine. In: Caironi M, editor. Large area flexible electronics. Weinheim, Germany: Wiley-VCH; 2015.
[73] Krishnamoorthy K, Gokhale RS, Contractor AQ, Kumar A. Novel label-free DNA sensors based on poly(3,4-ethylenedioxythiophene). Chem Commun 2004;7:820–1.
[74] Lin P, Luo X, Hsing I, Yan F. Organic electrochemical transistors integrated in flexible microfluidic systems and used for label-free DNA sensing. Adv Mater 2011;23(35):4035–40.
[75] Tang H, Lin P, Chan HLW, Yan F. Highly sensitive dopamine biosensors based on organic electrochemical transistors. Biosens Bioelectr 2011;26:4559–63.
[76] Liao C, Mak C, Zhang M, Chan HLW, Yan F. Flexible organic electrochemical transistors for highly selective enzyme biosensors and used for saliva testing. Adv Mater 2014;27(4):676–81.
[77] Coppedè N, Tarabella G, Villani M, Calestani D, Iannotta S, Zappettini A. Human stress monitoring through an organic cotton-fiber biosensor. J Mater Chem B 2014;2:5620–6.
[78] Tarabella G, Pezzella A, Romeo A, D'Angelo P, Coppedè N, Calicchio M, D'Ischia M, Mosca R, Iannotta S. Irreversible evolution of eumelanin redox states detected by an organic electrochemical transistor: en route to bioelectronics and biosensing. J Mater Chem B 2013;1:3843–9.
[79] Palmour JW, Kong HS, Davis RF. High-temperature depletion-mode metal-oxide-semiconductor field-effect transistors in beta-SiC thin films. Appl Phys Lett 1987;51:2028.
[80] Hilton KP, Uren MJ, Hayes DG, Wilding PJ, Johnson HK, Guest JJ, Smith BH. High power microwave SiC MESFET technology. In: 1999 Symposium high performance electron devices for microwave and optoelectronics applications EDMO (Cat. No. 99TH8401), IEEE; 1999. p. 71–74.
[81] Wang J, Williams BW. The 4H-SiC npn power bipolar junction transistor. Semicond Sci Technol 1999;14:1088–97.
[82] Katakami S, Fujisawa H, Takenaka K, Ishimori H, Takasu S, Okamoto M, Arai M, Yonezawa Y, Fukuda K. Fabrication of a P-channel SiC-IGBT with high channel mobility. Mater Sci Forum 2013;740-742:958–61.
[83] Ouisse T, Bano E. Electronic properties of the SiC-SiO/sub 2/interface and related systems. In: 1997 international semiconductor conference 20th Ed. CAS '97 Proc., IEEE; 1997. p. 101–110.
[84] Chandrashekhar MVS, Thomas CI, Lu J, Spencer MG. Observation of a two dimensional electron gas formed in a polarization doped C-face 3C/4H SiC heteropolytype junction. Appl Phys Lett 2007;91:033503–133503.
[85] Ferro G. Overview of 3C-SiC crystalline growth. Mater Sci Forum 2010;645:49–54.
[86] Sun Y, Liu Y, Zhu D. Advances in organic field-effect transistors. J Mater Chem 2005;15:53.
[87] de Heer WA, Berger C, Wu X, Sprinkle M, Hu Y, Ruan M, Stroscio JA, First PN, Haddon R, Piot B, Faugeras C, Potemski M, Moon J-S. Epitaxial graphene electronic structure and transport. J Phys D Appl Phys 2010;43:374007.
[88] Spencer M, Singh A, Uddin MA, Nomani MWK, Tompa G, Sbrockey N, Tolson J, Shields V, Hwang J, Koley G. Graphene on different substrates for sensing applications. In: 2012 twelfth IEEE international conference on nanotechnology. IEEE; 2012. p. 1–5.
[89] Kanungo J, Anderson M, Darmastuti Z, Basu S, Käll P-O, Ojamäe L, Spetz AL. Development of SiC-FET methanol sensor. Sens Actuators B Chem 2011;160:72–8.

[90] Andersson M, Pearce R, Lloyd Spetz A. New generation SiC based field effect transistor gas sensors. Sens Actuators B Chem 2013;179:95–106.
[91] Lin Y-T, Huang C-S, Chow L, Lan J-M, Yang C-M, Chang L-B, Lai C-S. Light-immune pH sensor with SiC-based electrolyte–insulator–semiconductor structure. Appl Phys Express 2013;6:127002.
[92] Soong C-W, Garverick SL, Fu X-A, Patil AC, Mehregany M. A fully monolithic 6H-SiC JFET-based transimpedance amplifier for high-temperature capacitive sensing. IEEE Trans Electron Devices 2013;60:4146–51.
[93] Yakimova R, Iakimov T, Yazdi GR, Bouhafs C, Eriksson J, Zakharov A, Boosalis A, Schubert M, Darakchieva V. Morphological and electronic properties of epitaxial graphene on SiC. Phys B Condens Matter 2014;439:54–9.
[94] Zhou SY, Gweon G-H, Fedorov AV, First PN, de Heer WA, Lee D-H, Guinea F, Castro Neto AH, Lanzara A. Substrate-induced bandgap opening in epitaxial graphene. Nat Mater 2007;6:770–5.
[95] Chen J, Liu W, Yang T, Li B, Su J, Hou X, Chou K-C. A facile synthesis of a three-dimensional flexible 3C-SiC sponge and its wettability. Cryst Growth Des 2014;14:4624–30.
[96] Yakimova R, Steinhoff G, Petoral RM, Vahlberg C, Khranovskyy V, Yazdi GR, Uvdal K, Lloyd Spetz A. Novel material concepts of transducers for chemical and biosensors. Biosens Bioelectr 2007;22:2780–5.
[97] Rohrbaugh N, Bryan I, Bryan Z, Arellano C, Collazo R, Ivanisevic A. AlGaN/GaN field effect transistors functionalized with recognition peptides. Appl Phys Lett 2014;105:134103.
[98] Lloyd Spetz A, Pearce R, Hedin L, Khranovskyy V, Söderlind F, Käll P-O, Yakimova R, Uvdal K. New transducer material concepts for biosensors and surface functionalization. In: Schmid U, editor. SPIE European microtechnologies new millennium, International Society for Optics and Photonics; 2009. p. 736206–736208.
[99] Barbaro M, Bonfiglio A, Raffo L. A charge-modulated FET for detection of biomolecular processes: conception, modeling, and simulation. IEEE Trans Electr Devices 2006;53:158–66.
[100] Spanu A, Lai S, Cosseddu P, Tedesco M, Martinoia S, Bonfiglio A. An organic transistor-based system for reference-less electrophysiological monitoring of excitable cells. Sci Rep 2015;5:8807.
[101] Kim M, Hwang J, Lepak LA, Lee J, Spencer MG, Tiwari S. Improvement of carrier mobility of top-gated SiC epitaxial graphene transistors using a PVA dielectric buffer layer. Nanotechnology 2012;23:335202.
[102] Jouault B, Camara N, Jabakhanji B, Caboni A, Consejo C, Godignon P, Maude DK, Camassel J. Bottom-gated epitaxial graphene suitable for half-integer quantum metrology? Condens Matter; 2011. arXiv:1106.5923.

CHAPTER 6

SiC RF Antennas for *In Vivo* Glucose Monitoring and WiFi Applications

Fabiola Araujo Cespedes, Sylvia Thomas, Shamima Afroz and Stephen E. Saddow
Department of Electrical Engineering, University of South Florida, Tampa, FL, United States of America

Contents

6.1 INTRODUCTION

The American Diabetes Association estimates that nearly 10% of the US population has diabetes and that, by 2050, one in three Americans will have diabetes. Diabetes mellitus is a metabolic disease in which the body is unable to produce or properly use insulin, leading to elevated glucose levels in the blood, known as hyperglycemia. A person with frequent or extended episodes of hyperglycemia can suffer from complications in the nervous system, blood vessels, and other organs, as well as heart disease, kidney disease, strokes, vision loss, and amputation [1,2,3]. Therefore, maintaining a healthy glucose level is essential in a person's life.

Unfortunately a long-term *in vivo* glucose sensor has yet to enter the market with typical maximum time of use on the order of 4–7 days. This chapter discusses ongoing research at the University of South Florida in the development of a bio- and hemocompatible glucose sensor based on the use of a radio frequency (RF) antenna constructed using 4H–SiC. *In vivo* tissue studies have also been conducted and will be presented in chapter: *In Vivo* Exploration of Robust Implantable Devices Constructed From Biocompatible 3C–SiC where a 30-day experiment in pig muscle showed no immune system response to a PEG coated the sensor as well as variants in sensor materials (3C–SiC, amorphous SiC [*a*-SiC]), thus further motivating this research.

6.2 BLOOD-GLUCOSE MONITORING METHODS

A person can monitor their glucose levels by the conventional self-monitoring glucose (SMG) method or through a continuous glucose monitoring (CGM) means. Clinical studies have proven that SMG levels helps facilitate treatment decisions in insulin and noninsulin use patients with diabetes [2]. The Central and Eastern Europe Diabetes Technologies and Therapeutics stated that SMG is an "essential tool that should be accessible to all patients with diabetes." The organization also provided a report of numerous benefits and recommendations [4]. Self-monitoring helps to improve an individual's overall health and prompts one to make proper health decisions based on their glucose levels [2,3,5]. For SMG, the patient draws a drop of blood into a capillary by taking a prick from the fingertip to measure the blood-glucose levels at that instant in time. Since, in some cases, patients are recommended to check their blood glucose at least 8–10 times a day, the repeated procedure of finger pricking can be painful [6] and cause calluses over time, thus increasing pain and measurement accuracy errors [7]. Additionally, since the results show the instantaneous blood-glucose levels, patients with extreme conditions could suffer episodes of dangerous levels of glucose at times when the blood glucose is not being monitored [6]. Furthermore, this method is extremely difficult for patients suffering from type 1 diabetes who are often young children, many of who are not yet able to utilize SMG themselves and require a healthcare professional or parent/teacher to maintain healthy insulin levels throughout the day [8]. Clearly, a reliable CGM approach is needed to ease the pain and suffering of millions of patients. While CGM systems do exist on the market, they require weekly sensor wire replacement at a typical cost of $90 per sensor. Fig. 6.1 shows the steps required to change out the sensor weekly, which clearly is an involved process that proves difficult for many patients, especially the critically ill and adolescent.

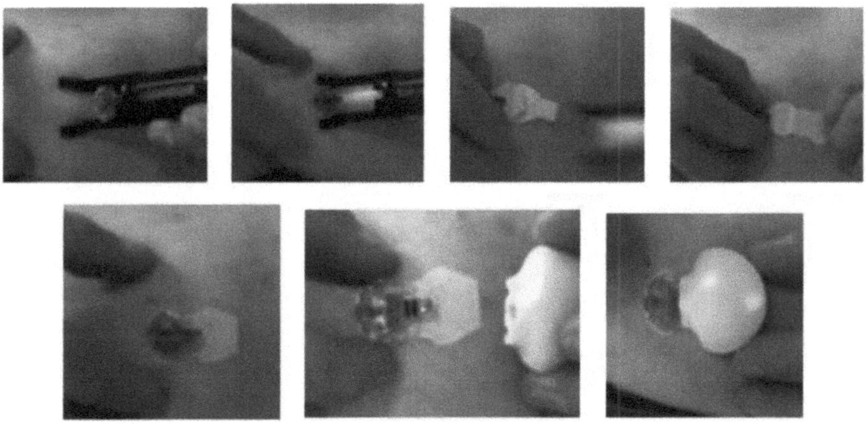

Figure 6.1 ***Photographic sequence showing the insertion of a CGM sensor in the belly of a diabetic patient.*** The needle sensor must be replaced every 3–7 days and requires calibration with a conventional SMG system [8].

All currently approved US Food and Drug Administration (FDA) CGM require a disposable needle-like insertion into the body, which lasts only up to a week [3,4,6] (see Fig. 6.1). In addition, CGM also requires calibration four times a day with the finger-pricking blood sample technique [3,4] since the measurement is not done directly on the blood glucose, but rather measures the glucose of the interstitial fluid (ISF). These CGM systems, such as those marketed by Medtronic, DexCom STS, and Abbott FreeStyle Navigator, can result in elevated costs, not only due to the device itself, but also due to the cost of the disposable sensor needles, adding a monthly cost of around $300 [6].

Alternative proposals for CGM include noninvasive and/or long-term implantable glucose-sensing approaches. No noninvasive alternative has found success in the market or has been approved by the FDA as additional development is needed to overcome numerous reliability issues and limitations. Long-term implantable glucose-sensing research has emerged as a promising alternative, but the developing designs are predicted to last only a maximum of 18 months *in vivo* [9]. While this is a major step forward, a longer-term solution would be of great benefit worldwide and we believe that a silicon carbide (SiC)-based CGM is one possible candidate for this challenging application.

6.2.1 Self-Monitoring Glucose

The search for an SMG solution started in the early 1970s. In 1971, Anton Clemens from the United States filed a patent for a blood-glucose monitor for point-of-care use in diabetic patients [10]. A glucose-monitoring device, called an Ames Reflectance Meter, automatically assessed enzyme-based reagent strips by analyzing changes in color with glucose. Prior to this, blood glucose had been estimated by reading the color change from a chart by eye. The Ames Reflectance Meter was succeeded by the Ames Eyetone, but it was restricted to clinical areas, such as hospital wards and physicians' offices. In 1978, the first portable devices for the self-monitoring of blood glucose became available. Self-monitoring of glucose was acceptable to patients; it also demonstrated that the technology could help patients with glycemic control. More importantly, self-monitoring also reduced the frequency of hypoglycemia, thus greatly extending patient quality of life.

Several alternatives for SMG have been researched over the last several years, striving to overcome the pain and hassle of sensor reinsertion every 3–7 days and finger-pricking calibration four times a day. Unfortunately, these new alternatives are faced with numerous limitations and accuracy problems.

One approach to addressing these issues is noninvasive technology, but no noninvasive technology has successfully functioned in the market [11]. For example, the GlucoWatch that came out on the market was discontinued after 2007 due to demonstrating poor results and the serious defect of burning patients [11,12], while other developments, such as the Glucoband, never made it into the market [11,13]. Several others are still under development such as Biosensors [14], I-Sugar X Meditech [15], Easy Check [16], and EyeSense [17]. Still others are not yet FDA approved, such as the GlucoTrack [18] and SCOUT [19]. Table 6.1 summarizes a list of noninvasive devices/companies for glucose sensing, with their status as of July 2014. The table does not include discontinued devices, such as the GlucoWatch, Diasensor, and Pendra [20], or products that have disappeared from the market, like the Cnoga Medical Glucometer, Glucoband, or C8MediSensors.

Most noninvasive glucose-monitoring systems face the challenge of being susceptible to external interference from other factors such as body temperature, perspiration, skin moisture, changes in skin thickness, and body movement [15]. For instance, infrared spectroscopy, including near infrared (NIR) spectroscopy and far infrared (FIR) spectroscopy, depend on optical transmission and reflection measurements, which are subject to interference from external factors that affect the reflection measurement. For this reason, NIR requires frequent recalibration. In FIR, the emitted energy that is absorbed by glucose and measured is so small that this method has not yet

Table 6.1 Summary of noninvasive devices/companies status[a]

Device/company	Technology	Status	Sources
Biosensors	SEMP technique (bioimpedance spectroscopy)	Under development [14]	http://www.biosensors-tech.com/
I-Sugar X Meditech	Fluorescent technology	Under development [15]	http://www.freedom-meditech.com/
Easy Check, Positive ID EyeSense	Chemical sensing in exhaled breath fluorescent technology	Under development [16]	http://www.positiveidcorp.com/products_easycheck.html
EyeSense	Fluorescent technology	Under development. Launch postponed to 2015 [17]	http://en.eyesense.com/
GlucoTrack	Thermal, ultrasound, and electromagnetic technologies	Approved 2013 in Europe. Not yet approved by FDA [18]	http://www.integrity-app.com/
Grove Instruments	NIR spectroscopy	In clinical trial [19,21]	http://groveinstruments.com/
OrSense Ltd.	Occlusion technology	Not concluded [22]	http://www.orsense.com/Glucose
SCOUT DS, Veralight Inc.	Fluorescent spectroscopy	Not FDA approved [19]	http://www.veralight.com/products.html

[a] Verified list from source [20], excluding no longer available devices.

been proven to be accurate [7,12]. In other methods, such as Raman spectroscopy, the measurement of light scattering that is caused by generated oscillations such as laser oscillation in the ocular fluid is subject to interference from other molecules [8,20]. In thermal spectroscopy, the IR radiation that is emitted from the body is also affected by other factors external to glucose concentration [8]. Another example is the technology based on measuring the ISF that is secreted from the skin to measure the glucose levels, which presents a time lag deficiency [23]. Overall, noninvasive technologies lack accuracy due to being susceptible to external factors such as transpiration, temperature, positioning, and/or displaying time lag problems of up to 20 min, making the technology not yet reliable [20]. This has led to discontinued devices such as the Glucowatch, Diasensor, and Pendra [14,20] or has postponed the estimated launching date for others, such as the case for the EyeSense.

Another approach toward SMG is a fully implantable glucose-monitoring system. These medical devices face other types of challenges such as *in vivo* inflammatory reaction and foreign body reaction, posing risk for the patients and hence the need for biocompatibility tests on any implantable device [23]. No current biosensor can operate for longer than a few days *in vivo* [24]. In reaction to nonbiocompatible foreign materials, the body forms an encapsulation around the object that is generally 100 μm thick, which not only acts as a diffusion barrier but is also electrically insulating. Therefore, long-term implantations are subject to gradual loss of sensor functionality and stability due to fibrosis encapsulation and tissue changes in the proximity of the sensor [24–26]. Therefore, implantable devices, including antennas, are coated with biocompatible materials, but in the aspect of communicating externally from the body, this adds to the effects of sensor encapsulation [27].

In addition to the issue of biocompatibility, the life cycle of the implanted device is relevant to factors such as the sensing mechanism inside the body and the energy required to function properly. For instance, the device in development by GlySens is expected to be ready by 2017 with a life-cycle expectancy of only 1 year after implantation [9]. There is currently a wide research gap pertaining to implantable devices, where no known development has been proposed on devices that would function longer than 18 months. This is likely due to one or more of the previously mentioned challenges faced by implantable devices for a truly "long-term" implantable solution.

6.2.2 Continuous Glucose Monitoring

For an ideal glucose sensor, accurate response to changing glucose concentrations should be fast. Under physiological conditions, it should have a long operational lifetime, but most of all it must be acceptable to the patient. Therefore, it would ideally provide real-time continuous information regarding patient glucose level and should not require any user calibration. CGM provides real-time data to the patients and alarms the times of hypoglycemia or rapid glucose change by providing data about the direction, magnitude, duration, frequency, and potential causes of fluctuations in blood-glucose levels. Many technologies including noninvasive, minimally invasive, or invasive are being developed to realize novel continuous glucose sensors and they may depend on ISF, transdermal technologies, or direct blood sampling.

Different noninvasive glucose sensors, such as optical transducers and different spectroscopic methods (infrared absorption spectroscopy; mid infrared or near infrared, etc.) face multiple problems, including light scattering, as well as specific absorption bands for accurate blood-glucose sensing [28]. Another approach is to use such techniques as impedance spectroscopy and electromagnetic coupling, which are based on impedance measurement of the dielectric parameters of blood [29,30]. This impedance spectroscopy/electromagnetic coupling technique has been used in different noninvasive glucose-sensing applications where the resonant frequency, as well as the level of the return loss of the sensors, was shown to change with different glucose levels [31,32]. On the other hand, invasive glucose sensors such as subcutaneous needle-type sensors use enzyme electrodes. Enzymes are used which catalyze reduction–oxidation (redox) reactions; therefore in doing so electrons are accepted or donated. Hydrogen peroxide is generated from glucose oxidase (GOx) and a platinum electrode is used to measure this amperometrically, which is directly proportional to the glucose concentration. A subcutaneous needle-type amperometric enzyme electrode based on GOx is used by Medtronic (Medtronic Diabetese, Northridge, California) for a continuous glucose-monitoring system (CGMS). It has a 4-min response time with an average of 63 h of recording time. Unfortunately, significant sensor drift requires four calibrations per day. This sensor drift is postulated to be partly due to the alteration of local blood flow as a result of protein and cell coating of the sensor, variable tissue oxygen tension, tissue interference, and wound response to the sensor. After explanting the sensor, the sensitivity remains low but after repeated washing it improves; therefore it is reversible [33,34].

Some newer devices by DexCom STS (Dexcom, San Diego, California) and Abbott FreeStyle Navigator (Abbott Diabetes, Alameda, California) also use a subcutaneous enzyme-based electrode. FreeStyle Navigator has a recording time of 5 days whereas the DexCom is a wireless system using low-power radio and has a recoding time of 72 h. Therefore, these sensors all have similar life time limitations.

Microdialysis systems, which use a fine, hollow microdialysis fiber placed subcutaneously, are another type of subcutaneous sensor. ISF glucose freely diffuses into the fiber where the fiber is perfused with isotonic fluid from an *ex vivo* reservoir. Since the sensor is *ex vivo* it avoids the problems associated with fouling by proteins and cells. This sensor has less drift than CGMS, but still requires one calibration per day. To assess clinical performance, microdialysis and subcutaneous needle-type sensors have been compared to assess clinical performance; the microdialysis probe displayed greater accuracy with an additional lag of 7 min than the needle-type sensor. Therefore, microdialysis systems have an inherent physical lag time for the dialysate to be pumped to the sensor, which uses more power. The device is larger because it requires perfusate solution, which makes this device very uncomfortable to patients. In addition the physical and chemical properties of the membrane may change over time, as will the tissue characteristics such as volume, temperature, hydration, and pressure. In summary, for CGM, commercial sensors have several issues/limitations associated with their systems—the glucose sensor needs replacement frequently; the RF unit and other external devices are bulky adding discomfort for the patient already wearing an insulin pump; and the RF unit is continuously powered requiring frequent battery replacement. In addition the FDA still recommends a confirmation test using conventional blood sample from the finger; and most importantly the sensors have a noninstantaneous response time.

The key to the success of all implantable glucose sensors is the need to ensure reliable performance when they are exposed to living tissue and fluids. The term biofouling has gained common usage as a description of the build-up of proteins and fibroblasts that accompany the body's immune reaction to the sensor. The body attempts to ward off the foreign material by forming an encapsulant, which is typically 100 μm or more thick. This layer is problematic for any device that seeks to sample serum or blood since it acts as a diffusion barrier, not to mention an electrically insulating layer. Encasing the implant with a biocompatible material is necessary to prevent the implant from harming the surrounding tissue and to reducing the effects of the capsular contracture [35]. A better solution would be a sensor made from a fully biocompatible material that would allow for longer periods of functionality in the body and overcomes the current sensor lifetime limitations. We believe that a body fluid contacting sensor made from biocompatible SiC is one possible solution, which we will now discuss.

6.3 SiC FOR RF BIOTECHNOLOGY

SiC has long been used in advanced radio frequency (RF) systems due to its excellent carrier transport and high-power properties [36]. Many of today's point-of-use healthcare systems are hybrid implantable systems that combine RF (think WiFi) and biosensor technologies. The biosensors usually rely on relevant physiological parameters for continuous monitoring and an integrated antenna is employed in order to send the received data to an external receiver [37]. Therefore, implantable antennas have attracted considerable attention in recent years as a potential solution for communicating with these implantable biosensors. These hybrid systems combining RF (antenna/wireless communication) and biosensor technologies are the key to developing the next generation of continuous monitoring systems including implantable pacemakers and defibrillators, glucose monitors, insulin pumps, hearing aids, healthcare facility communication, medical and emergency equipment tracking, and remote patient monitoring just to name a few.

For decades silicon (Si) has been the material of choice for designing and fabricating numerous devices due to its excellent mechanical and electronic properties, particularly for sensing applications. While Si may continue to be suitable for short-term microelectromechanical systems and nanoelectromechanical systems applications, a biocompatible and hemocompatible device material for long-term biomedical applications is highly desirable. Many biomedical devices, such as biochemical sensors, biologically interfaced neural networks, and smart biomedical implants, require materials that have inherent sensing capability. In addition, it is necessary that sensing be performed using not only mechanical means, but also through electrical means as well. To date, only a few crystalline semiconductors with efficient electrical and mechanical properties have been investigated, with Si and diamond (carbon [C]) drawing most of the attention. There has been significant activity in the area of ceramic materials such as titanium dioxide (TiO_2) and zinc oxide (ZnO), but these are not easy materials to realize electronic functionality with. Unfortunately, Si and its native oxide silicon dioxide (SiO_2) have been shown to display different degrees of cytotoxicity, mostly due to its instability in aqueous solutions with subsequent formation of silica and silicates, which are known for their harmful effects on cells as was discussed earlier in the second and third chapters of this book. In addition TiO_2, which can become a semiconductor upon ion implantation, does not display sufficiently satisfying electronic properties that would justify its adoption for long-term sensing applications. Therefore, there is a global initiative to advance implantable medical device materials beyond Si technology as discussed in the first edition of this book [38]

Silicon carbide is a semiconducting material that appears ideally suited to fill this existing gap in materials for biomedical device development. It has also been found that it is a semiconductor material that is chemically inert to acids, alkalis and salts, does not expand in liquid environments, displays biocompatibility, longevity, and great sensing potentiality, and can be processed with the existing methods developed for the silicon industry [36,39]. SiC is physically strong and chemically inert, which hypothetically makes it ideal for interaction with the harsh body environment. *a*-SiC and polycrystalline SiC have been used as coating materials for many implantable biomedical devices as mentioned in chapter: Silicon Carbide Materials for Biomedical Applications. These materials have also been used for clinical studies, such as bone prosthetics and heart stents, which confirmed the biocompatibility of these forms of SiC [40–42]. *a*-SiC was also used as a thin-film coating for the Utah introcortical neural prosthetic constructed from Si [43]. Like Si, SiC is a semiconducting material, which enables the creation of complex electrical devices upon doping incorporation. On the other hand, unlike many semiconductors, SiC may be found/synthesized in many crystal forms, known as polytypes, having more than 170 different observed crystal structures, but with a common stoichiometry (i.e., all forms contain bilayers of Si and C) [36]. 4H–SiC and 6H–SiC, two of the most common hexagonal polytypes of SiC, have been commercially available as substrates and are routinely fabricated into commercially available electronic devices [43,44]. These materials have shown good biocompatibility, nontoxicity, and low biofouling when used as electrode myocardial biosensors [45,46]. More recently our group has shown that the only purely cubic form, 3C–SiC, is even better for most biomedical applications and is also hemocompatible (see chapter: Study of the Hemocompatibility of 3C–SiC and *a*-SiC Films Using ISO 10993-4 and Ref. [36]).

Previously, we discussed how one of the major problems that affects the lives of patients with a permanent disease, such as chronic diabeties, is obtaining daily blood samples multiple times to monitor their glucose levels.

Such intermittent testing is laborious, often painful, and requires significant diligence to adequately monitor true glycemic patterns. The development of implantable biosensors has paved the road for developing CGMSs [47–50]. Despite the intense research on implantable biosensors, present day biosensors cannot operate *in vivo* for more than 10 days before issues of biofouling impact the sensing capabilities. In addition, as the demand for interoperable communications between sensors and data transfer into and out of the body evolves with point-of-care service, there is a clear need to design and fabricate *in vivo* antennas. When fabricated using a fully biocompatible and hemocompatible material, such as 3C–SiC, these antenna sensors can curtail adverse tissue reaction, which can endanger the life of a patient [51,52].

6.4 SiC RF ANTENNA DEVELOPMENT FOR CGM

A CGM sensor employing RF signals has been developed using the biocompatible material silicon carbide [53]. The fabricated implantable RF antenna utilizes SiC as the material of choice due to its great potential benefit as a robust bio- and hemocompatible material for long-term *in vivo* systems. Distinguished by its innovation to utilize SiC as the device material, this system removes the need to encase this biomedical device with biocompatible materials and addresses the short lifetime issue of current sensors.

Unlike biosensors that require direct contact with ISFs to trigger chemical reactions to operate, this biocompatible SiC sensor does not require a direct interface to bodily fluids. The sensing mechanism is based upon a shift in the antenna resonant frequency as a function of change in glucose levels, which electrically manifests itself as a change in blood permittivity and conductivity.

To test the sensor as a function of glucose level, measurements using synthetic body fluid, which is electrically equivalent to blood plasma, and pig blood were performed *in vitro*. Changes in sensor performance to varying glucose levels were measured and a shift in resonant frequency to lower values was observed with increasing glucose level. *In vitro* sensor performance from an antenna operating at 10 GHz demonstrated that the sensor showed a dose-dependent response to glucose concentration from 120 to 530 mg/dL. A shift of 40 and 26 MHz for blood-mimicking liquid and pig blood were observed corresponding to a 97 and 67 kHz shift per 1 mg/dL change in blood glucose, respectively [53].

For *in vivo* applications the antenna sensor needs to operate inside the body environment, and it has been determined that the best operational location of this biocompatible SiC biosensor is within fatty tissue in close proximity to blood vessels. *In vivo* tissue reactions to SiC-based continuous glucose sensors were evaluated by subcutaneous implantation in pigs. A total of five different antenna sensors (glucose sensor) were coimplanted: (1) bare SiC-based titanium/gold (Ti/Au) antenna, (2) an "all-SiC" antenna composed of 3C–SiC with the Si backside coated with *a*-SiC, (3) an SiO_2-coated Ti/Au antenna (SiO_2 serves as the bonding layer of polyethylene glycol (PEG) to improve biocompatibility), (4) covalently bonded PEG on a Ti/Au antenna surface to reduce nonspecific protein adsorption, and finally (5) an *a*-SiC-coated Ti/Au antenna sensor. A silicon sample was used as a positive control and a 3C–SiC sample was used as a negative control. After 4 weeks, the implanted sensors and surrounding tissue were removed and processed for light microscopy and subsequent histological tissue evaluation. These results are discussed in chapter: *In Vivo* Exploration of Robust Implantable Devices Constructed From Biocompatible 3C–SiC where the histopathological tissue analysis of the SiC antennas showed no signs of any inflammatory response [54].

6.4.1 Implantable SiC RF Antenna Glucose Sensor

In recent years, considerable progress has been made in developing implantable biosensors that can continuously monitor different healthcare issues such as glucose levels of a diabetic patient [55,56]. In particular, the use of radio frequency identification (RFIDs) and systems to perform biotelemetry have been reported and it is now possible to implant wireless sensors inside the human body to monitor various bodily functions and transmit health-related information outside of the human body for remote monitoring [57–59]. Real-time detection and monitoring of blood glycemia with high precision, sensitivity, selectivity, and speed is required for biomedical applications and clinical monitoring. Other systems have been designed using implantable antennas for external communication for continuous monitoring systems [47–49,59]. SiC has been investigated as a sensor material and analyzed using the techniques of impedance spectroscopy and electromagnetic coupling, which are based on impedance measurement of the dielectric parameters of blood [60,61]. This impedance spectroscopy/electromagnetic coupling technique has been used in different noninvasive glucose-sensing applications where the resonant frequency, as well as the level of the return loss of the sensors, was shown to change with different glucose levels [32,62–64].

In the case of electromagnetic coupling-based sensors, a SiC antenna sensor does not require a direct interface with the interstitial fluid, thus offering a simpler monitoring technique as compared to conventional sensors. The antenna responds by exhibiting a shift in its resonance frequency as the varying glucose concentration changes the effective complex permittivity of the blood [62]. Different than previous antenna-based glucose-sensing configurations [32], the investigated operational position of the SiC biosensor was within a fatty tissue region that is in close proximity to a blood vessel. For demonstrating the sensing mechanism and characterizing the sensitivity, different glucose level measurements were performed *in vitro* by using a two-layered fat and blood layer model as shown in Fig. 6.2. Two types of blood media were used: blood-mimicking liquid and pig blood. In both instances the fatty layer was synthesized with fat-mimicking liquid. Using this *in vitro* model, a glucose level variation from 120 to 530 mg/dL was found to provide 40 and 26 MHz shifts in the maximum return loss of the sensor for blood-mimicking liquid and pig blood, respectively. The details of the design of the antenna sensor are presented here.

A microstrip patch antenna was designed to further demonstrate the feasibility of a biocompatible SiC *in vivo* biosensor for CGM. A conventional rectangular patch antenna was utilized due to its simplicity, compatibility with monolithic microwave integrated circuit designs, and versatility in terms of resonant frequency, polarization, radiation pattern, and impedance. A design frequency of 10 GHz was used for an antenna layout of $8.5 \times 11.5\ mm^2$ on a 0.37-mm thick semiinsulating biocompatible 4H–SiC substrate ($\varepsilon_r = 10$), known to exhibit very low microwave loss (Fig. 6.2a). After an RCA clean the semi-insulating 4H–SiC substrate was metallized with titanium/gold using E-beam evaporation and then patterned with standard photolithography: 20 nm of Ti was evaporated followed by

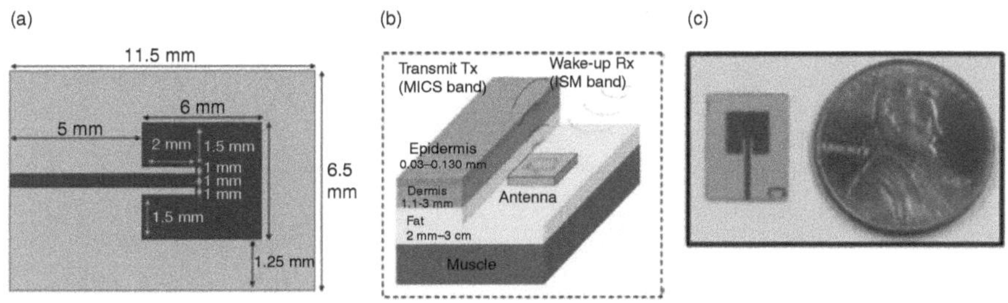

Figure 6.2 ***10 GHz SiC RF sensor.*** (a) 4H–SiC microstrip patch antenna dimensions, (b) schematic of sensor placement within the body, and (c) photograph of the 4H–SiC-based antenna realized with a Ti/Au metal conductor patch [24].

500 nm of Au. The geometrical design parameters were determined using the ANSYS High Frequency Structure Simulator (HFSS™), and then the inset feed length was adjusted to achieve optimal impedance matching. The simulated radiation efficiency in free space was 65% and simulated broadside realized gain was 3.94 dB.

When placed near a blood vessel, such an antenna can potentially sense the variation in blood permittivity, which is strongly dependent on glucose level [65]. The permittivity varies with blood-glucose altering antenna parameters, such as input impedance and reflection coefficient, S_{11} [66]. For diabetic patients, the concentration of other minerals present in the blood, such as calcium, chloride, potassium, and magnesium, has been shown not to vary as significantly as glucose [67–69]. Therefore, changes in permittivity are utilized to estimate the patient's plasma glucose levels.

The operational position of this implantable sensor was selected within a fatty tissue region that is in close proximity to a blood vessel. *In vitro* testing was performed using fatty tissue and blood-mimicking liquids modeled to represent different glucose levels within the blood, as shown in Fig. 6.3. The fatty tissue-mimicking liquid was prepared from oil (10.2%) and dibutyl glycol ether, 89.8%, and permittivity and conductivity measured by a dielectric probe kit (HP85070A) controlled by a vector network analyzer (VNA HP 8719D) [32]. Fig. 6.3 shows the measured real part of the permittivity for fat-mimicking liquid as $\varepsilon_r \approx 5$ and this is in accordance with

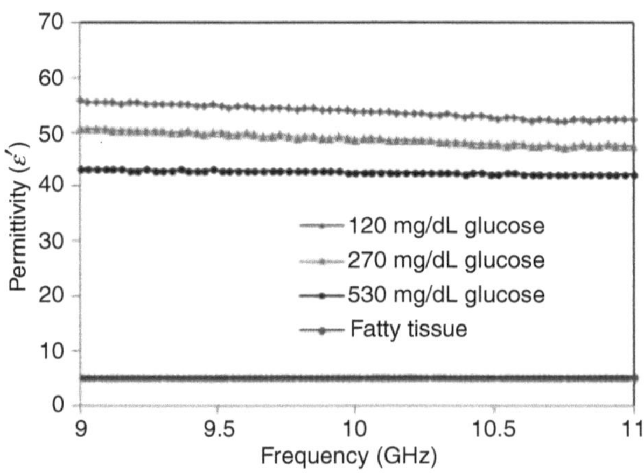

Figure 6.3 ***Measured relative permittivity (real part) of fat and blood-mimicking liquids versus RF frequency.*** The blood-mimicking liquid is altered by different glucose levels to simulate glucose levels found within the blood. Fatty tissue response is shown for reference [24].

the values reported in the literature [70,71]. The corresponding conductivity was calculated from the measured imaginary part of the permittivity as $\sigma \approx 0.51$ S/m.

Different weights of D-glucose were added to the blood-mimicking liquid in order to represent various glucose levels. The accuracy of the glucose levels was verified using a conventional glucometer (TRUEtrack™ blood-glucose monitoring system). Fig. 6.3 also shows the measured real part of the permittivity from the blood-mimicking liquids as a function of glucose concentration. As expected, the permittivity varies with glucose levels. In addition, the corresponding conductivities were calculated from the measured imaginary part of the permittivity as 18, 15, and 13 S/m for the blood-mimicking liquids exhibiting 120, 270, and 530 mg/dL glucose concentrations, respectively.

6.4.2 Sensor Characterization

The antenna sensor was tested via reflection coefficient measurements using an HP 8719D VNA. Fig. 6.4 depicts the experimental setup that was utilized to model the antenna placed inside a fatty layer in close proximity to blood. The thicknesses of the three blood-mimicking liquids exhibiting different glucose concentrations were utilized to observe the variation in resonance frequency of the antenna. The baseline reading (ie, the resonance frequency at which the antennas input reflection coefficient S_{11} is a minimum) was taken by using a blood-mimicking liquid without any glucose concentration. Specifically, as shown in Fig. 6.5, the measured shift in the resonance frequency was toward lower frequency with increasing glucose concentration, with a maximum shift of 40 MHz observed for the largest

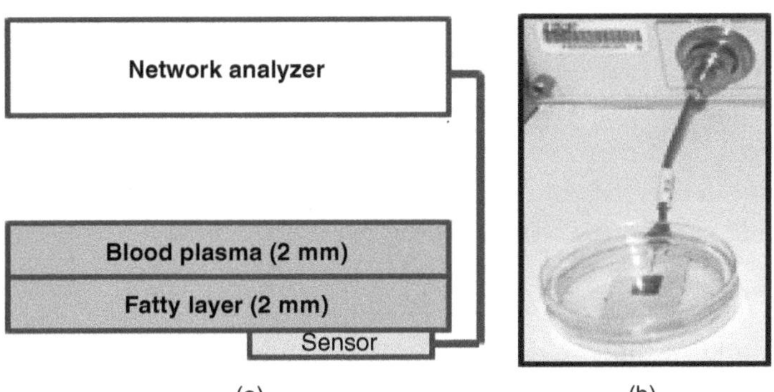

Figure 6.4 ***Setup for measuring the blood plasma glucose level.*** (a) Schematic diagram and (b) photograph of the experiment.

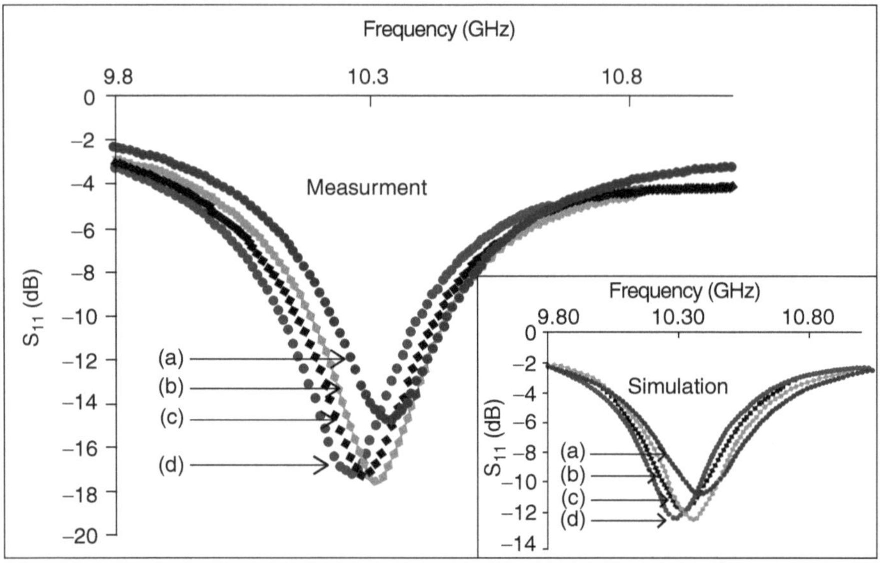

Figure 6.5 Measured and simulated (inset) S_{11} of the SiC patch antenna as a function of blood-mimicking liquid glucose concentration for (a) 0 mg/dL (no glucose added), (b) 120 mg/dL, (c) 270 mg/dL, and (d) 530 mg/dL. Note resonance shift to lower frequency with increasing glucose level.

glucose concentration of 530 mg/dL. This implies 97 kHz shift per 1 mg/dL change in blood glucose. The simulated S_{11} response of the antenna using the measured permittivity values of the fat and blood-mimicking liquids was found to be in good agreement with the experimental results.

After getting promising glucose-sensing results using blood-mimicking liquids, the sensor's performance was tested using pig blood. Ethylenediaminetetraacetic acid was used to avoid fast blood coagulation, and the blood was stored at 4°C for 3 days. Initially the pig blood had a glucose level of 67 mg/dL. Three additional samples were prepared with glucose levels of 120, 270, and 490 mg/dL by adding D-glucose to the blood. Similar to the glucose-sensing experiment using fat and blood-mimicking liquids, glucose sensing was demonstrated using pig blood. The observed shift in frequency was 62 kHz per mg/dL of glucose for pig blood.

In Fig. 6.6 it has been shown that the sensor performance was linear with the frequency shift being a direct function of glucose concentration. For the blood-mimicking liquid, the total frequency shift (Δf) was 40 MHz for a glucose level of 120–530 mg/dL. It has also been shown that there is a 97 kHz resonance frequency shift per 1 mg/dL of glucose. Similarly for pig blood, the total frequency shift (Δf) was 26 MHz over 67–490 mg/dL

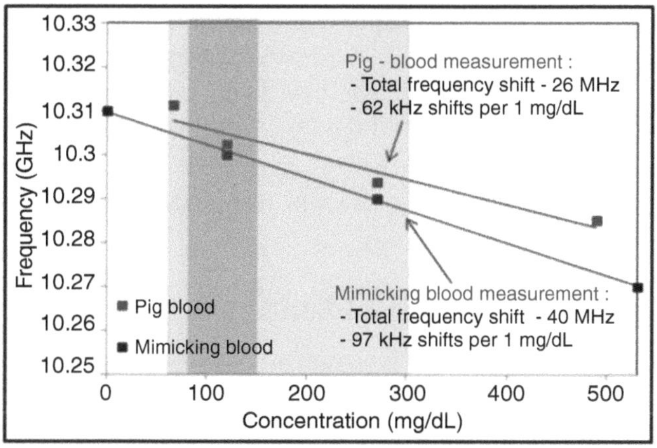

Figure 6.6 ***Measured relative resonance frequency shift dependence on glucose concentration.*** **The shaded regions are provided for reference - dark shading is the glucose range for a typical healthy patient while light shading shows low and high glucose levels typical of diabetic patients. Note fairly linear response versus glucose concentration.**

of glucose concentration. It also showed a resonance frequency shift of 62 kHz per 1 mg/dL of glucose in actual pig blood.

From these results, it was observed that the antenna response to blood-mimicking liquid and pig blood had similar frequency shifting trends—the resonance frequency decreases with an increase of glucose levels. For a nondiabetic healthy person, the American Diabetes Association recommends a fasting plasma glucose level of 70–130 mg/dL, which is shown in Fig. 6.6 (dark-gray shaded area). Below 60–65 mg/dL is known as hypoglycemia and above 240 mg/dL is known as hyperglycemia (shown in light-gray shaded areas). Note that the SiC-based sensor covers the normal to critical care application with a reasonable slope in the frequency versus glucose slope at 10 GHz. These results demonstrate the potential use of biocompatible SiC substrates for long-term, real-time CGM of diabetic patients where the resonance frequency decreases with an increase of glucose levels with linear sensor performance. Due to limitations in the medical bands allocated for such a device, the sensor was redesigned to operate in the industrial, scientific, and medical (ISM) band (2.45 GHz) and this work, while still ongoing, has been quite promising and is discussed next.

6.5 SENSOR PLATFORM DEVELOPMENT FOR THE ISM BAND

In the last section we saw the very encouraging results that demonstrated that a 4H–SiC RF antenna could sense changes in blood-glucose level *in vitro* at 10 GHz. This pioneering work was critical to the ultimate development of a CGM system, and in this section current R&D ongoing at the University of South Florida is presented. First the antenna was redesigned so that it could operate in the ISM band, which is one of the medical bands allocated for biomedical devices. Then research being performed to sense blood-glucose changes using a similar mechanism as in RFIDs is presented, whereby preliminary research shows that this sensing mode is indeed possible for this application.

6.5.1 SiC Antenna Design: ISM Band

The Federal Communication Commission established the Medical Implant Communication Services, operating in the frequency range from 402 to 405 MHz, due to this frequency being internationally available for this purpose. More importantly these frequencies are suitable for propagation within the human body [66]. This frequency range is viable to work alongside low-power circuits, and is in a relatively low-noise portion of the spectrum, while also allowing propagation in human tissue [62,67], and it is also accepted by the European Radiocommunications Committee [68].

Nevertheless, this frequency was not considered for two reasons. On the one hand, the lower the frequency, the bigger the required antenna patch and minimizing the size of the antenna is an important factor, both for cost and comfort issues. More importantly the lower the frequency, the less sensibility in shift of resonant frequency as the glucose changes. Other frequency bands are also reserved for medical purposes, such as the ISM band at 2.45 GHz and the wireless medical telemetry service band between 608 and 614 MHz, 1395 and 1400 MHz, and 1427 and 1429 MHz [56]. The ISM band offered a compromise between size and sensitivity, making this frequency band the ideal choice for this application.

As a preliminary trial, the antenna was designed using Roger's Duroid 6010™ as the substrate with a thickness of 50 mil. Fig. 6.7a shows the antenna patch geometry, with the corresponding dimensions listed in Fig. 6.7b. A photograph of the antenna is showed in Fig. 6.8. This was simulated using ANSYS HFSS resulting in the simulation beam profile shown in Fig. 6.9 and the S_{11} response shown in Fig. 6.10. These same antennas were used to conduct the preliminary trials as will be described in the following section.

The fabricated antenna is shown in Fig. 6.8 based on this geometry. The simulated radiation pattern and S_{11} response are shown in Fig. 6.9a and b, respectively.

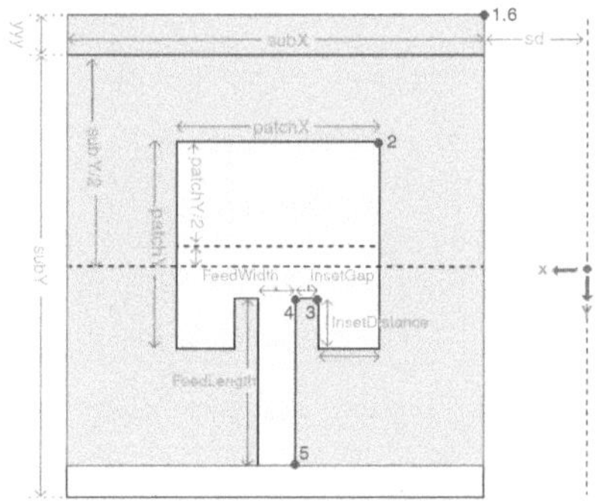

Name	Value	Unit	Evaluated value
subX	3.5	cm	3.5 cm
subY	4.4	cm	4.4 cm
subH	50	mil	50 mil
patchX	2.64	cm	2.64 cm
patchY	1.95	cm	1.95 cm
FeedWidth	0.119	cm	0.119 cm
InsetGap	0.06	cm	0.06 cm
InsetDistance	0.704	cm	0.704 cm
FeedLength	1.992	cm	1.992 cm
yy	0.1	cm	0.1 cm
yyy	subY/2-patchY/2+yy+insetdistance-feedlength		0.037 cm

Figure 6.7 ISM band patch antenna showing (a) antenna geometry and (b) antenna dimensions.

Figure 6.8 ***Photograph of ISM band microstrip patch antenna.*** Antenna substrate Roger's Duroid 6010, 50 mil thick. Dimensions shown in Fig. 6.7.

6.5.2 RFID Sensing Platform for CGM Using SiC RF Antenna

Ultimately the design of an implantable SiC RF antenna is desired for a long-term CGM sensor platform. A sensor platform approach for glucose sensing will have the capability to remotely detect variations in glucose levels as follows. A signal is sent from an external to the body transmitter (EBTx), which transmits a variable frequency signal toward the implanted sensor (SiC patch antenna). Then an external to the body receiver (EBRx) receives the reflected signal, as displayed in Fig. 6.10. Changes in the glucose levels that surround

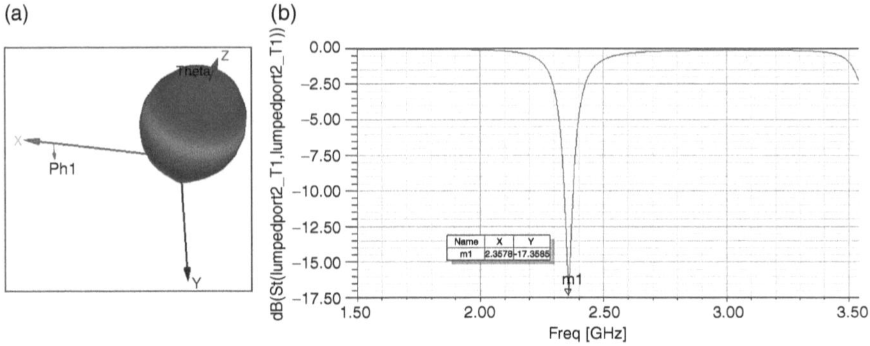

Figure 6.9 ISM band microstrip patch antenna simulated (a) radiation pattern and (b) S_{11} response. Simulation performed using ANSYS HFSS. Note that a symmetrical beam profile, as expected, is simulated and the resonant frequency of 2.45 GHz achieved with a return loss of ~17.5 dB in free space.

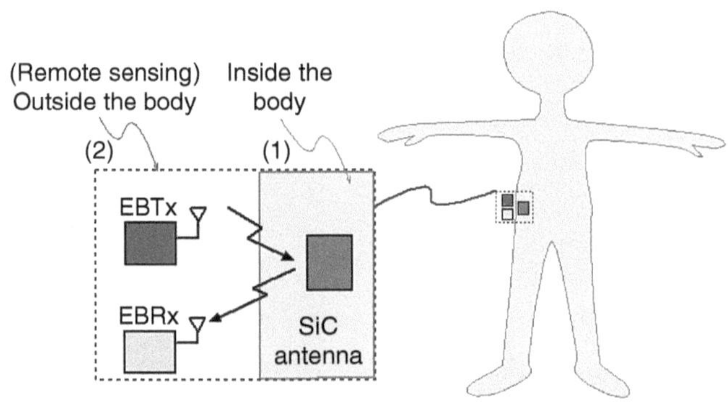

Figure 6.10 ***RFID-based CGM concept.*** (1) A biocompatible implantable SiC antenna sensor. (2) Remote detection through the use of a swept frequency signal from the EBTx, which is reflected back to the EBRx by (1). Changes in sensor resonant frequency will then be measured by changes in received power as the transmitter frequency is swept over a frequency range equal to the anticipated change in glucose levels.

the implanted antenna will result in a variation of its resonant frequency as was demonstrated at 10 GHz. The power received at the EBRx will change directly in relationship to the variations of the resonant frequency of the implanted antenna, and thus amplitude changes will be an indirect, but accurate, measurement of the patient's glucose level. The sensor platform is designed to sweep the EBTx signal frequency in order to detect the variation of power over the range of frequencies used to identify the glucose level of the patient.

In the first phase of the trials, three antennas were constructed. Two of the antennas were connected in close proximity to each other, facing a third antenna as shown in Fig. 6.11.

Figure 6.11 ***RFID detection RF front end showing the antenna location for the preliminary experimental trial.*** Blood-mimicking liquid or pig blood will then be placed on the wooden platform above the antenna IMP (not shown) so that changes in synthetic glucose level can be measured. The antenna to IMP height is 26 mm. Note EBTx and EBRx antennas facing down with the integral ground plane visible.

This setup, as shown in Fig. 6.12, allowed a Petri dish to be inserted at a fixed distance of 9 mm between the EBTx and EBRx antennas and the IMP antenna (ie, the implanted antenna). The Petri dish, as shown in Fig. 6.12, was filled with animal fetal blood albumens (FBA) combined in each test with set ratios of calculated D-glucose. The proportion of glucose

Figure 6.12 ***Petri dish inserted between EBTx/EBRx antennas and IMP antenna.*** Spacing X mm. Petri dish contains FBA with varying concentrations of D-glucose and the RFID sensing mechanism was verified using this setup.

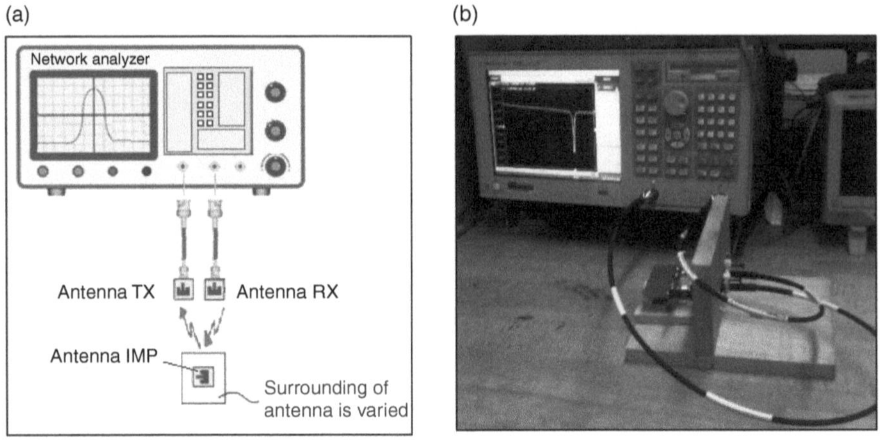

Figure 6.13 RFID sensor concept showing (a) antenna connections to the network analyzer used to obtain S_{11} and S_{21} as a function of glucose concentration and (b) photograph of RFID sensor under test.

in the FBA was both calculated and measured using the commercial glucose meter TRUEtrack.

During the first set of trials, the EBRx and EBTx antennas were connected directly to the network analyzer in order to record the variations of the parameters S_{11} and S_{21} of the antennas as a function of the glucose levels of the FBA in the Petri dish. Fig. 6.13a shows a schematic representation of this setup, while Fig. 6.13b shows a photograph during an S_{11} measurement.

The second phase of the RFID sensor trials consisted of connecting the antenna to the AD8302. The AD8302 is a fully integrated system for measuring gain/loss and phase in numerous receive, transmit, and instrumentation applications. The AC-coupled input signals can range from −60 to 0 dBm in a 50-Ω system, from low frequencies up to 2.7 GHz. The outputs provide an accurate measurement of either gain or loss over a ±30-dB range scaled to 30 mV/dB, and with phase over a 0–180-degree range scaled to 10 mV/degree. By taking the difference of their outputs, a measurement of the magnitude ratio, or gain, between the two input signals is available. These signals may even be at different frequencies, allowing the measurement of conversion gain or loss [72]. The AD8302 board was connected to the RFID antennas as shown in Fig. 6.14.

Recording of the sensor output vector fluxgate magnetometer (VMAG) was made using a digital oscilloscope. The oscilloscope was set to 500 mV per division. The first test was done by recording the output without connecting the receiving antenna on the board and then by adding the receiving

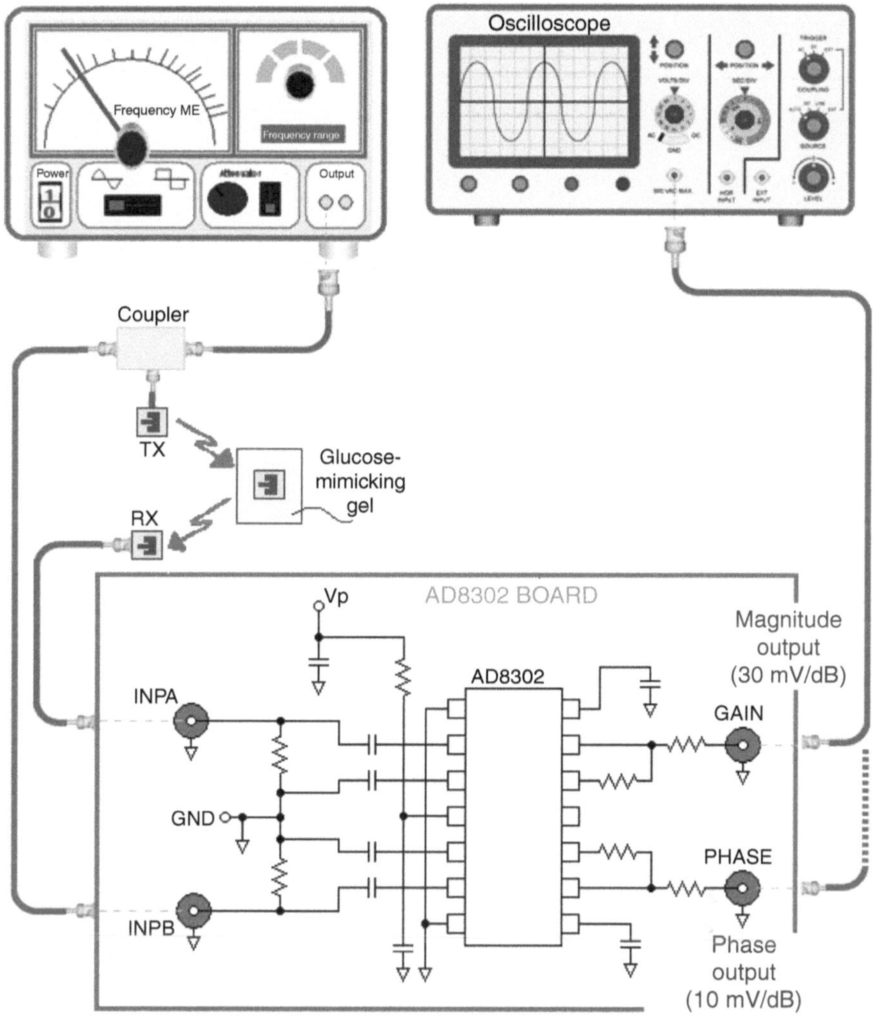

Figure 6.14 ***Schematic representation of connections of RFID sensor antennas to the AD8302 board.*** Note output gain (VMAG in text) and phase are shown on the right.

antenna. A value greater than 1 V of difference was observed. The test was continued by comparing the results of no reflecting antenna (in open space) and a reflecting antenna. Adding the reflecting antenna resulted in a received signal amplitude increase of ~125 mV in VMAG, which validated the RFID RF setup. Finally, the test was repeated with a blood-glucose concentration of 100 and 580 mg/dL, and an easily measured variation in output voltage from the AD8302 board of approximately 3 mV was observed.

6.6 SUMMARY AND FUTURE WORK

This work shows that SiC RF technology has the potential to be used as a glucose sensor to monitor patient blood-glucose levels in real time. This claim is predicated on the *in vitro* (chapter: Cytotoxicity of 3C–SiC Investigated Through Strict Adherence to ISO 10993) and *in vivo* (chapter: *In Vivo* Exploration of Robust Implantable Devices Constructed From Biocompatible 3C–SiC) data that shows that SiC is biocompatible, as was reported in the first edition of this book. In addition, these results can be transferred to various healthcare systems in the future, especially as a WiFi node for the emerging biosensor/biosensor system market. This technology can provide an increase in the understanding of biosensing/detecting and continuous diagnostic feedback and can deliver a platform for system integration of implantable WiFi devices using SiC RF antennas. The implanted antenna can remain in the body for years, and the external antennas can be externally, outside the body, connected to an electronic circuitry to display the glucose levels real time, without the need for continuous blood-glucose measurements via finger pricks, which is known to be both painful and unreliable. At the University of South Florida we are continuing this work and the next phase is to build the RFID sensor system, in the ISM band, using 4H–SiC antennas. After this is completed full *in vivo* sensor trials will be conducted in female farm pigs and blood-glucose levels continuously monitored for a period of 30 days to demonstrate our technology and prepare for more extensive clinical trials.

REFERENCES

[1] Fast facts: data and statistics about diabetes. American Diabetes Association; 2013.
[2] Executive summary: standards of medical care in diabetes – 2013. Diabetes Care (American Diabetes Association) 36; 2013.
[3] Holt P. Blood glucose monitoring in diabetes. Nurs Stand 2014;28(27):52–8.
[4] Czupryniak L, Barkai L, Bolgarska S, Bronisz A, Broz J, Cypryk K, Honka M, Janez A, Krnic M, Lalic N, Martinka E, Rahelic D, Roman G, Tankova T, Varkonyi T, Wolnik B, Zherdova N. Self-monitoring of blood glucose in diabetes: from evidence to clinical reality in Central and Eastern Europe – recommendations from the International Central-Eastern European Expert Group. Diabetes Technol Ther 2014;16(7):460–75.
[5] Alzaid A, Schlaeger C, Hinzmann R. 6th annual symposium on self-monitoring of blood glucose (SMBG) applications and beyond, April 25–27, 2013, Riga, Latvia. Diabetes Technol Ther 2013;15(12):1033–52.
[6] Myers W. Managing diabetes with continuous glucose monitoring. In: Bass P, editor. Everyday Health Media, LLC; 2013.
[7] Non-invasive monitors and glucose sensors: what's new. The Cleveland Clinic Foundation; 2005.

[8] Puttkammer B. Continuous glucose monitor (CGM) – setup & insert walk through. Available from: http://youtu.be/ReZX7rLi5D8; 2014
[9] GlySens. (Still) developing implantable CGM. Diabetes Mine; 2014.
[10] Clemens AH. Reflectance Meter, US patent No. 3,604,815; 1971.
[11] Hoskins M. Apple creating an iWatch that could monitor glucose? Diabetes Mine; 2013.
[12] Freer B. Feasibility of a non-invasive wireless blood glucose monitor; 2011.
[13] Freer B, Venkataraman J. Feasibility study for non-invasive blood glucose monitoring. Proceedings of the IEEE Antennas and Propagation Society International Symposium (APSURSI '10). IEEE, Toronto, Canada, July 2010. p. 1–4.
[14] Biosensors. Available from: http://www.biosensors-tech.com/; 2014.
[15] I-Sugar X Meditech. Available from: http://www.freedom-meditech.com/; 2014.
[16] Easy Check, Positive ID EyeSense. Available from: http://www.positiveidcorp.com/products_easycheck.html; 2014.
[17] EyeSense. Available from: http://en.eyesense.com/; 2014.
[18] Glucotrack – in the news: Integrity Applications, Inc. enters into manufacturing agreement with Wistron Corp. Available from: http://www.integrity-app.com/inthenews/integrity-applications-inc-enters-into-manufacturing-agreement-with-wistron-corp-2/; 2014.
[19] SCOUT DS, Veralight Inc. Available from: http://www.veralight.com/products.html; 2014.
[20] So C, Choi K, Wong T, Chung J. Recent advances in noninvasive glucose monitoring. Medical devices: evidence and research. Auckland, New Zealand: Dove Medical Press Limited; 2012.
[21] Grove Instruments. http://groveinstruments.com/; 2014.
[22] OrSense. Available from: http://www.orsense.com/; 2014.
[23] Oliver N, Toumazou C, Cass A, Johnston D. Glucose sensors: a review of current and emerging technology. Diabet Med 2009;26:197–210.
[24] Afroz A. A biocompatible SiC RF antenna for *in-vivo* sensing applications. PhD, University of South Florida; 2013.
[25] Onuki Y, Bhardwaj U, Pharm M, Papadimitrakopoulos F, Burgess D. A review of the biocompatibility of implantable devices: current challenges to overcome foreign body response. J Diabetes Sci Technol 2008;2(6):1003–15.
[26] Koschwanez H, Reichert W. *In vitro, in vivo* and post explantation testing of glucose-detecting biosensors: current methods and recommendations. Biomaterials 2007;28(25):3687–703.
[27] Skrivervik A, Merli F. Design strategies for implantable antennas. Laboratoire d'Electromagnetisme et d'Acoustique, Ecole Polytechnique Federale de Lausanne, Lausanne, Switzerland; 2011.
[28] Oliver NS, Toumazou C, Cass AEG, Johnston DG. Glucose sensors: a review of current and emerging technology. Diabet Med 2009;26(3):197–210.
[29] Tatara T, Tsuzaki K. Derivation of extracellular fluid volume fraction and equivalent dielectric constant of the cell membrane from dielectric properties of the human body. J Med Biol Eng Comput 2000;38:377–83.
[30] Hayashi Y, Oshige I, Katsumoto Y, Omori S, Yasuda A, Asami K. Temporal variation of dielectric properties of preserved blood. Phys Med Biol 2008;53:295–304.
[31] Kim S, Melikyan H, Kim J, Babajanyan A, Lee J-Ha, Enkhtur L, Friedman B, Lee K. Noninvasive *in vitro* measurement of pig-blood D-glucose by using a microwave cavity sensor. Diabetes Res Clin Pract 2012;96:379–84.
[32] Venkataraman J, Freer B. Feasibility of non-invasive blood glucose monitoring. 2011 IEEE international symposium on antennas and propagation and USNC/URSI; 2011. p. 603–606.

[33] Pickup JC, Claremont DJ, Shaw GW. Responses and calibration of amperometric glucose sensors implanted in the subcutaneous tissue of man. Acta Diabetol 1993;130:143–8.
[34] Thome-Duret V, Gangnerau MN, Zhang Y, Wilson GS, Reach G. Modification of the sensitivity of the glucose sensor implanted into subcutaneous tissue. Diabetes Metab 1996;22:174–8.
[35] Black RD. Recent advances in translational work on implantable sensors. IEEE Sens J 2011;11(12).
[36] Saddow SE, Agrawal A, editors. Advances in silicon carbide processing and applications. Norwood, MA: Artech House; 2004. p. 3171–82.
[37] Beach RD, Conlan RW, Godwin MC, Moussy F. Towards a miniature implantable *in vivo* telemetry monitoring system dynamically configurable as a potentiostat or galvanostat for two- and three-electrode biosensors. IEEE Trans Instrum Meas 2005;54(1):61–72.
[38] Saddow SE, editor. Silicon carbide biotechnology: a biocompatible semiconductor for advanced biomedical devices and applications. UK: Elsevier Ltd; 2011. p. 61–72.
[39] Cicero G, Catellani A. Towards SiC surface functionalization: an ab initio study. J Chem Phys 2005;122:214716.
[40] Li X, Wang X, Bondokov R, Morris J, An YH, Sudarshan TS. Micro/nanoscale mechanical and tribological characterization of SiC for orthopedic applications. J Biomed Mater Res B Appl Biomater 2005;72(2):353–61.
[41] Kalnins U, Erglis A, Dinne I, Kumsars I, Jegere S. Clinical outcomes of silicon carbide coated stents in patient with coronary disease. Med Sci Monit 2002;2:16–20.
[42] Santavirta S, Takagi M, Nordsletten L, Anttila A, Lappalainen R, Konttinen YT. Biocompatibility of silicon carbide in colony formation test *in vitro*: a promising new ceramic THR implant coating material. Arch Orthop Trauma Surg 1998;118(1–2): 89–91.
[43] Hsua J-M, Tathireddy P, Rieth L, Normann AR, Solzbacher F. Characterization of a-SiC(x):H thin films as an encapsulation material for integrated silicon based neural interface devices. Thin Solid Films 2007;516(1):34–41.
[44] Naji A, Harmand M. Cytocompatibility of two coating materials, amorphous alumina and silicon carbide, using human differentiated cell cultures. Biomaterials 1991;12(7):690–4.
[45] Yakimova R, Petoral RM Jr, Yazdi RM, Vahlberg GR, Spetz CL, Uvdal AK. Surface functionalization and biomedical applications based on SiC. J Phys D Appl Phys 2007;40:6435–42.
[46] Zorman CA, Mehregany M. Materials for microelectromechanical systems. In: Gad-el-Hak M, editor. The MEMS handbook: introduction, fundamentals. Boca Raton, FL: CRC Press; 2001. p. 15–23.
[47] Wilkins E. Towards implantable glucose sensors: a review. J Biomed Eng 1989;11: 353–61.
[48] Wilkins E. A rechargeable glucose electrode long term *in vitro* performances. Biomed Instrum Technol 1993;27:325–33.
[49] Pickup JC. Glucose sensors and closed-loop insulin delivery. Biotechnology of insulin therapy. Oxford: Blackwell; 1991. 126-153.
[50] Pickup JC. *In vivo* glucose monitoring: sense and sensorbility. Diabetes Care 1993;16:535–9.
[51] Coletti C, Jaroszeski MJ, Pallaoro A, Hoff AM, Iannotta S, Saddow SE. Biocompatibility and wettability of crystalline SiC and Si surfaces. Proceedings of the twenty-ninth international conference of the IEEE EMBS. Lyon, France; Aug. 23–26, 2007. p. 5849–5852.
[52] Kumari TV, Vasudev U, Kumar A, Menon B. Cell surface interaction in the study of biocompatibility. Trends Biomat Artif Organs 2002;15:37–41.
[53] Afroz S, Thomas SW, Mumcu G, Saddow SE. Implantable SiC based RF antenna biosensor for continuous glucose monitoring; 2013.

[54] Thomas SW, Afroz S, Saddow SE, Muffly K, Hoyte L. *In-vivo* evaluation of subcutaneous histological tissue reaction to SiC based continuous glucose sensor. J Histotechnol. Print ISSN: 0147-8885 Online ISSN: 2046-0236 (In Preparation, 2015).
[55] Wilson GS, Zhang Y, Reach G, Moatti-Sirat D, Poitout V, Thévenot DR, Lemonnier F, Klein JC. Progress toward the development of an implantable sensor for glucose. Clin Chem 1992;38(9):1613–7.
[56] Koschwanez HE, Reichert WM. *In vitro, in vivo* and post explantation testing of glucose-detecting biosensors: current methods and recommendations. Biomaterials 2007;28(25):3687–703.
[57] Nguyen KD, Chen IM, Luo ZQ, Yeo SH, Duh HBL. A wearable sensing system for tracking and monitoring of functional arm movement. IEEE/ASME Trans Mechatron 2011;16(2):213–20.
[58] Alahakone AU, Senanayake S. A real-time system with assistive feedback for postural control in rehabilitation. IEEE/ASME Trans Mechatron 2010;15(2):226–33.
[59] Karacolak T, Hood AZ, Topsakal E. Design of a dual-band implantable antenna and development of skin mimicking gels for continuous glucose monitoring. IEEE Trans Microw Theory Tech 2008;56(4):1001–8.
[60] Tatara T, Tsuzaki K. Derivation of extracellular fluid volume fraction and equivalent dielectric constant of the cell membrane from dielectric properties of the human body. J Med Biol Eng Comput 2000;38:377–83.
[61] Hayashi Y, Oshige I, Katsumoto Y, Omori S, Yasuda A, Asami K. Temporal variation of dielectric properties of preserved blood. Phys Med Biol 2008;53:295–304.
[62] Kim S, Melikyan H, Kim J, Babajanyan A, Lee J-Ha, Enkhtur L, Friedman B, Lee K. Noninvasive *in vitro* measurement of pig-blood D-glucose by using a microwave cavity sensor. Diabetes Res Clin Pract 2012;96:379–84.
[63] Kim J, Babajanyan A, Hovsepyan A, Lee K, Friedman B. Microwave dielectric resonator biosensor for aqueous glucose solution. Rev Sci Instrum 2008;79:086107.
[64] Lee K, Babajanyan A, Kim Ch, Kim S, Friedman B. Glucose aqueous solution sensing by a near-field microwave microprobe. Sens Actuators A 2008;148:28–32.
[65] Topsakal E, Karacolak T, Moreland EC. Glucose-dependent dielectric properties of blood plasma. URSI, IEEE 2011.
[66] Balanis CA. Antenna theory, analysis and design. John Wiley & Sons 2016.
[67] Schwan HP. Electrode polarization impedance and measurements in biological materials. Ann NY Acad Sci 1968;148:191–209.
[68] Hagness SC, Taflove A, Bridges JE. Wideband ultra-low reverberation antenna for biological sensing. Electron Lett 1997;33:1594–5.
[69] Hayashi Y, Livshits L, Caduff A, Feldman Y. Dielectric spectroscopy study of specific glucose influence on human erythrocyte membranes. J Phys D 2003;36:369–74.
[70] Soontornpipit P, Furse CM, Chung YC. Design of implantable microstrip antennas for communication with medical implants. IEEE Trans Microw Theory Tech 2004;52(8):1944–51.
[71] Gabriel C, Gabriel S, Courthout E. The dielectric properties of biological tissues: I. Literature survey. Phys Med Biol 1996;41:2231–49.
[72] AD8302. RF/IF Gain and Phase Detector. Analog Devices, Inc. Available from: www.analog.com; 2002.

CHAPTER 7

In Vivo Exploration of Robust Implantable Devices Constructed From Biocompatible 3C–SiC

Christopher L. Frewin*, Sylvia Thomas, C. Coletti† and Stephen E. Saddow****
*The Department of Biomedical Engineering, Erik Johnsson School of Engineering and Computer Science, University of Texas, Dallas, TX, United States of America
**Department of Electrical Engineering, University of South Florida, Tampa, FL, United States of America
†Center for Nanotechnology Innovation@NEST, Pisa, Italy

Contents

7.1 INTRODUCTION

Using manufactured devices to restore biological functionality is not a new concept, and in fact has been in practice since the beginnings of recorded history. There is evidence of humans fabricating and implanting artificial teeth constructed from metals, bamboo, and ivory as far back as 4000 BC in ancient China and Egypt [1,2]. Catheters constructed from hollow reeds, wood, and various precious metals have been used to treat issues within the urinary tract and have been found in Egypt, Greece, China, and India dating from approximately the same time periods as the dental implants. Internal implants, like replacements for bone, namely, hip ball joint replacements, are more recent, and were first implanted in the 1800s [3,4]. The common thread between all of these biomedical devices is that they are intended to replace a purely mechanical physiological function, like mastication, waste evacuation, or physical support and locomotion. Another common characteristic is that these devices were usually constructed from single materials, which was due directly to the simple, mechanical nature of the processes that the devices were designed to replace. It should also be noted that to this date, even though these devices are simple in construction, many of the materials used within these devices have issues existing within the body environment without initiating chronic inflammatory responses [5,6].

In the late 1700s, inventors were investigating a new type of interaction, one which would bring about a revolution which would forever increase the complexity of implantable devices. Benjamin Franklin used electricity to stimulate paralyzed limbs on a number of undisclosed patients in 1751 [7]. 1791 saw Luigi Galvani as one of the first scientists to characterize the electrophysiological muscle response. These experiments pioneered the possibility of using electricity as a therapeutic treatment for a myriad of disabilities, and in turn demonstrated the need for better understanding of the physiological aspects of stimulation as well as a better control when exciting the stimulation target. In 1855, Guillaume Duchenne discovered that electrical stimulation of neuromuscular activity with alternating current (AC) produced superior results when compared to direct current (DC). Further efforts by Alan Lloyd Hodgkin and Andrew Huxley led to the use of complex electrical circuits, mainly involving operational amplifiers, to quantify and model the basic electrical activity of the nervous system, the action potential, in 1952 [8]. This Nobel Prize winning discovery not only opened incredible avenues into the workings of electrically active cells, but it also demonstrated that complex electronics could be used to facilitate excellent control over physiological signals. However, until the advent of semiconductor electronics, the size and stability of complex electronics, which were vacuum tube driven, did not provide the practicality for use in the human physiological environment. However, semiconductors did not solve all of the physiological interaction issues. Just as with mechanical biomedical implant devices, the semiconductor materials required to achieve reliable complex electronics have been tied to issues within the physiological environment [10,11,13].

As discussed in chapter: Cytotoxicity of 3C–SiC Investigated Through Strict Adherence to ISO 10993, device failure mechanisms include both biotic and abiotic processes. The biotic mechanism is believed to originate from the chronic foreign body response following initial implantation and associated downstream processes [9–13]. The materials which have been previously used in the construction for electrical devices normally consist of the device support, electronics, conductors, and insulators. Typically complex electronics and device supports are constructed from silicon (Si) and its derivatives (SiO_2, Si_3N_4). Polymers like polyimide and parylene are used for insulators, and noble metals like platinum and its derivatives (Pt, Pt black, and Pt/Ir alloys), or ceramic semiconductors like iridium oxide, titanium oxide, and titanium nitride have been used as conductors [12,14,15].

There is a growing recognition that the physical and chemical properties of the material itself [16] can influence the biotic response and ultimately the device performance longevity. For example, polyimide-insulated tungsten microwire arrays chronically implanted in rats demonstrated insulation damage due to excessive oxidative stress via Fenton chemistry leading to deterioration in the recording sites [17]. This study suggests the importance of the physical and chemical resilience of the implant construction materials. In addition to the oxidative stress, polyimide has additional issues like water absorption, which leads to swelling strain [18,19]. Later formulations of polyimide improved water absorption rates, but chemical interactions with the environment increase local acidity [20]. Delamination of multilayer structures that constitute these implantable technologies likely contributes to device failure. For example, mechanically applied stress in parylene C generates microcracks (Fig. 7.1) and additional material failures arise from blistering/pore formation [21], which exposes the underlying materials to the external environment [22].

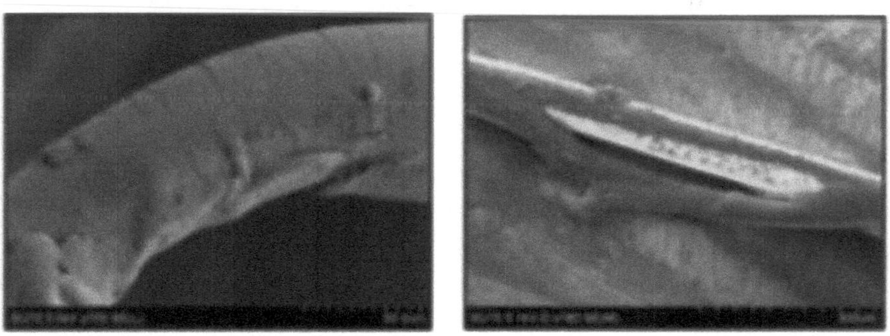

(a) (b)

Figure 7.1 ***Scanning electron microscope (SEM) micrographs of parylene C-coated tungsten wires.*** (a) With the application of tensile stress, multiple microcracks are revealed across the surface of the insulation. (b) Axial splitting and delamination of parylene-C due to weak bonding of the material to the tungsten wire [22].

Reliability of electrical devices within the body environment has been a major issue. Si electrical devices are extremely susceptible to contamination from metallic ions, such as sodium (Na), potassium (K), and iron (Fe), and extracellular fluid contains copious amounts of Na^+, K^+, Cl^-, Ca^{+2}, and Mg^{+2} ions, which will interact with the electrical device. This problem has been well known in the semiconductor industry, and is one of the driving forces behind the development of the cleanroom for processing electronic devices [23,24]. Metallic ions diffuse through the insulator, which is usually silicon dioxide, leading to fixed, embedded charges at the devices' interface, thus degrading device reliability and functionality [24–28]. Additionally, Si and SiO_2 have indicated that they are not necessarily biocompatible, and may generate cytotoxic and genotoxic effects within cells [12,29–44]. Replacing Si with another semiconductor may be the answer, but most of the semiconductors are composed of toxic materials (ie, gallium, arsenic, indium, antimony, etc.).

Silicon carbide is a compound semiconductor which is physically robust, chemically inert, and has shown evidence of biocompatibility and hemocompatibility [45]. As a semiconductor, it has the same inherent properties that make it ideal for complex electronics, and it can be used to construct any device that silicon can be used to create. In this chapter, we will demonstrate the chemical resilience of SiC within a physiological environment, especially against Si, the leading electronic material. We will also evaluate the *in vivo* reactions to SiC using murine and porcine animal models.

7.2 CORROSION AND CHEMICAL RESILIENCE

Quantifying the chemical stability and corrosion resistance of a biomedical device is an important factor in assessing overall device biocompatibility and future reliability. ISO 10993-13 through ISO 10993-19 deals with the chemical characterization of materials exposed to physiological conditions [46]. ISO 10993-14 provides methodology guidelines for the identification and quantification of degradation products from ceramics, to which semiconductor materials belong, in simulated physiological solutions [47]. ISO 10993-14 consists of two parts: the first part is to test the material in normal physiological solutions with a pH of 7.4 ± 0.2. The second part is immersion in an extreme solution of pH 3.0 ± 0.2, which simulates the pH of the human body under extreme acidic conditions, normally due to inflammatory reactions. Additionally, temperature can be used to produce accelerated aging conditions and has been found to follow a time dependence of $2^{\Delta T/10}$, where ΔT is the difference between the set temperature and 37°C [48,49]. Corrosion tests in simulated and extreme solution were performed on Si(100), cubic-silicon carbide (3C–SiC), amorphous silicon carbide (*a*-SiC) (four variant formulations as described in chapter: Study of the Hemocompatibility of 3C–SiC and *a*-SiC Films Using ISO 10993-4), and SiO_2. Characterization and data analysis for the material surface changes were characterized using the SEM, atomic force microscope (AFM), and optical profilometer before and after soaking testing of the samples.

7.2.1 Sample Preparation, Cleaning Procedures, and Analysis Techniques

Silicon wafers (University Wafer, Inc.) with the crystallographic orientation of (100) and thickness of 500 μm were selected as both test materials and as the substrate for all SiC variants as well as the SiO_2 samples. Thin films of 3C–SiC were grown at the University of South Florida by M. Reyes on the Si(100) substrates using the USF SiC Group's previously reported hot-wall chemical vapor deposition (CVD) process at the growth rate of ~5 μm/h at 1350°C [50,51]. Four different variants of *a*-SiC stacks were prepared by J. Register of the USF SiC Group using the USF Nanotechnology Research & Education Center Plasmatherm 700, plasma-enhanced chemical vapor deposition (PECVD) tool as described in chapter: Study of the Hemocompatibility of 3C–SiC and *a*-SiC Films Using ISO 10993-4 [52]. The *a*-SiC film was deposited on all substrates simultaneously in order to allow for direct comparison of the underlying material impact on the measured response. The 4-in. wafers were cut into coupons, 8 × 10 mm in size, using a diamond-blade dicing saw, with 68 coupons obtained from each wafer.

Our cleaning procedures are described as follows. Before all material depositions, the Si wafers were prepared using the two-stage RCA cleaning method [53]. In standard clean 1 (SC1) the wafers were immersed in a 1:1:5 solution of NH_4OH (ammonium hydroxide, 27%):H_2O_2 (hydrogen peroxide, 30%):H_2O at 80°C for 10 min to remove any organic contamination. SC1 treatment normally results in the formation of a thin layer of oxide on the surface of the wafers, which is removed by immersion in 5% hydrofluoric (HF) acid at 25°C for 1 min. The second stage of standard clean 2 (SC2) has the wafers immersed in 1:1:6 HCl:H_2O_2:H_2O at 80°C for 10 min to remove any trace metallic contamination, followed by a dip in 5% HF acid at 25°C for 1 min. The wafers were then rinsed with deionized (DI) water for 5 min followed by nitrogen gas drying [53].

Solvent clean starts with individual samples placed in a glass test tube with 5 mL of acetone (99% volume). The tubes were then sonicated for 10 min at 25°C with the intention of removing oily, organic contamination and debris. The acetone was replaced by 5 mL of isopropanol alcohol (IPA, 96% volume) and the samples returned to sonication for an additional 10 min at 25°C. The IPA was removed from the tubes and replaced by DI water with a final sonication for 10 min at 25°C. Piranha solution (1:2 H_2SO_4:H_2O_2) cleaning was prepared at a temperature of approximately 80–85°C. Samples were then immersed in piranha for 10 min. After piranha cleaning the samples were rinsed with DI water for 10 min. At the end of any cleaning procedure, samples were retrieved, dried using pure nitrogen gas (99% N_2), and kept in cleaned sample holders.

A Hitachi S800 SEM was used to detect qualitative evidence of material surface degradation at magnifications of 100×, 500×, 1000×, and 5000×. The SEM was used in field emission mode at 25 kV acceleration voltage using the secondary electron detection mode with a sample surface-tilting angle of 45 degrees to enhance surface feature observation. Quantitative evidence was provided through a Park Systems XE-100 AFM, which allowed nanometer-scale surface feature observation. AFM was performed in tapping mode with scan areas of 5 × 5, 10 × 10, and 45 × 45 μm at a scan rate of 0.2 Hz. Features such as surface roughness and feature size were gathered through the Park Systems XEI software.

Fig. 7.2 displays (a) 10 × 10 and (b) 5 × 5 μm AFM micrographs obtained from one of the 3C–SiC samples after the cleaning procedure. The sample is considered relatively flat, with an average surface roughness of ~7.2 ± 0.5 nm root mean square (rms). AFM data elaboration was performed using the XEI software (Park Systems) on the 5 × 5 μm micrograph and is shown in chapter: Study of the Hemocompatibility of 3C–SiC and *a*-SiC Films Using ISO 10993-4, Fig. 3.1. The measurements show that the terrace width on one of the 3C–SiC steps was ~41 nm, and the step height was about 4.57 Å. The lattice height for 3C–SiC is 4.36 Å, therefore the observed step height represents a single crystalline terrace and is thus indicative of high-quality crystal growth [54].

The commercial Si surfaces used in this study were atomically flat, which was accomplished through chemical–mechanical planarization, which uses caustic chemicals and fine nanoparticles to remove imperfections. Fig. 7.2 shows an AFM scan which indicates that the Si surface morphology was

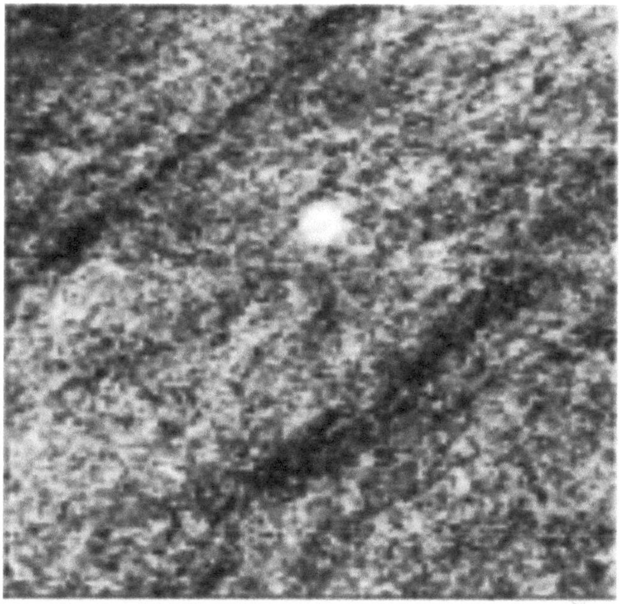

Figure 7.2 ***AFM micrograph of 5 ×5-μm scans of a Si sample.*** AFM measurement was in tapping mode with a 0.2-Hz scan rate. The micrographs are normalized to a Z height interval of (+10,−10) nm. The r_q value was 137 nm rms.

marked only by a few nanometer deep polishing scratches, and it possesses a surface rms roughness, r_q, within a 5 × 5 μm scan size of 137 pm rms. Anisotropic etching, or chemical etching, of Si(100) produces a distinctive pyramid shape due to faster etching rate of the (100) plane as compared to the (111) plane. Changes in the silicon due to any chemical reaction should be detectable through either AFM or SEM microscopy.

7.2.2 Corrosion Testing in Simulated Solution

In a previous investigation by Frewin et al. [44], it was observed that Si showed pitting and morphological surface degradation after the culture of H4 human neuroglioma and PC12 rat pheochromocytoma immortalized cell culture; however, at that time the exact source of this damage was not understood [44,55,56]. Goodwin et al. in 1997 and Pocock et al. in 2001 indicated in their work that the culture of hippocampal microglial cells could release nitric oxide (NO) [57,58]. Wink et al. in 1998 provided evidence about the role of NO in the biochemistry of neurological systems [59,60]. The release of NO, hydrogen peroxide, and other caustic enzymes due to the action of inflammatory cells can result in the formation of nitrous reactive species, which are known in the semiconductor processing industry as corrosive agents for silicon and its derivatives [61–64]. ISO 10993-14 suggests a set of corrosion tests prior to any *in vivo* application of the implantable materials and devices [47,65]. It was our desire to devise a method which incorporates the ISO 10993-14 standard with which to identify the source of any surface modifications seen after cell culture.

Although the *in vitro* culture environment is simplified as compared with *in vivo* physiology, there are multiple variables, which obfuscate the source of the corrosion which we observed on Si samples after cell culture. First, silicon reacts with water to form silicic acid through the process $SiO_2 + 2\ H_2O \rightarrow H_4SiO_4$. It was decided to examine not only the effects of the cells on the surface but to observe the effects of aqueous solutions as well. As we have observed that cells can attach to the surface of silicon, we examined the effects of serum-enhanced media, which could also be affecting its surface [43,44,66]. The experiment was broken down into three stages. First, we compared the surface morphology of semiconductors cultured directly with H4 human neuroglioma (H4 ATCC® HTB-148™, American Type Culture Collection) cells. The culture methods and protocol for the H4 cell line are described in Ref. [44]. Additionally, we examined samples soaked within complete Dulbecco's modified eagle medium or within artificial cerebrospinal fluid (ACSF) [67]. ACSF was chosen as it provided an *in vitro* environment similar to the human brain, the environment closest to that which the H4 cells were derived. The ACSF is formulated by adding concentrations of salts in DI water: 119 mM NaCl, 26.2 mM $NaHCO_3$, 2.5 mM KCl, 1 mM NaH_2PO_4, 2.5 mM $CaCl_2$, and 1.3 mM $MgCl_2$ 10 mM glucose. The ACSF solution was filtered through a 0.22 μm particulate filter and balanced to 7.4 pH through NaOH/HCl titration prior to the experiment.

Si(100) substrates diced into an 8 × 10 mm die were soaked within one of the two solutions or through incubation of cells directly on the surfaces for a total of 96 h at 37°C. After the 96-h immersion, the samples were individually rinsed under a flow of DI water for 10 min to remove ionic remnants. The samples were then cleaned using our solvent cleaning method (Section 7.2.1), which would target weakly bonded lipids and other polymeric materials, following which the surfaces were imaged using optical, SEM, and AFM microscopy. This same pattern of cleaning followed by microscopy was followed for the samples after using piranha and HF solution immersion.

7.2.3 Results and Discussion

Figs. 7.3 and 7.4 show AFM and SEM micrographs selected from the Si samples placed within the three media treatments and measured after 96 h. Table 7.1 summarizes the average surface roughness obtained from all of the Si surfaces. Clearly, the surface has an increased roughness due to increased surface morphological alterations detected on all of the surfaces regardless of the media treatment applied. However, the Si surfaces exposed to only the media and the ACSF show the smallest, which decrease in value as more aggressive cleaning procedures are applied, with the final HF treatment returning surface roughness to picometer roughness, which was ~130 pm rms before the applied treatment. The removal of these particulates indicates that they are most likely composed of both organic particulates and solidified salts, which were deposited on the surface of the materials during the soaking period. Although the SEM and AFM

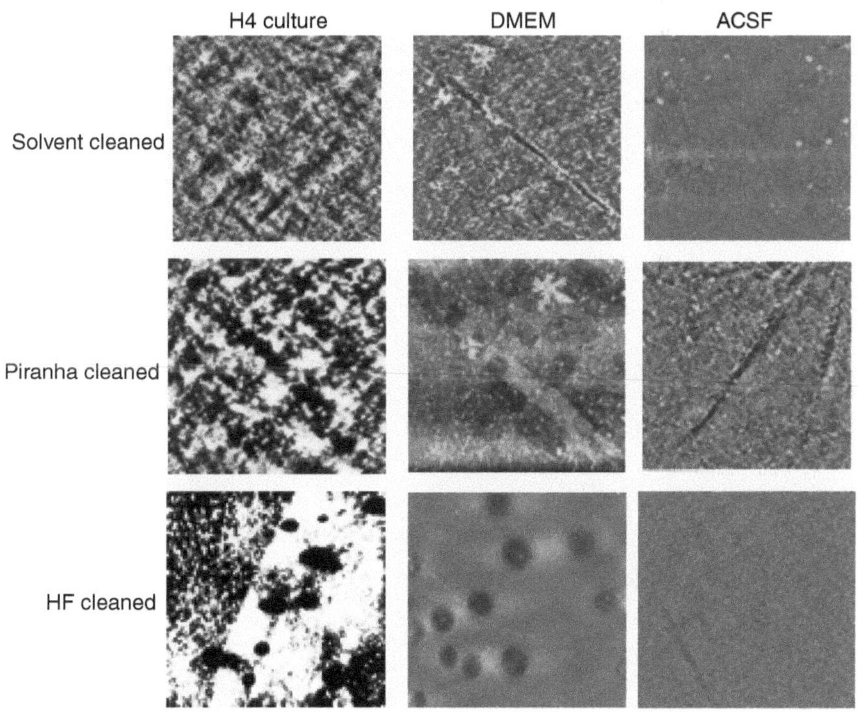

Figure 7.3 ***45 × 45 μm area AFM micrographs of Si samples immersed in cell media + H4 cells (left column), cell media (center column), and ACSF (right column) for 96 h at 37°C.*** Cleaned with solvents, solvents + piranha, and solvents + piranha + HF, respectively. The micrographs are normalized to a Z height interval of +10,–10 nm [68].

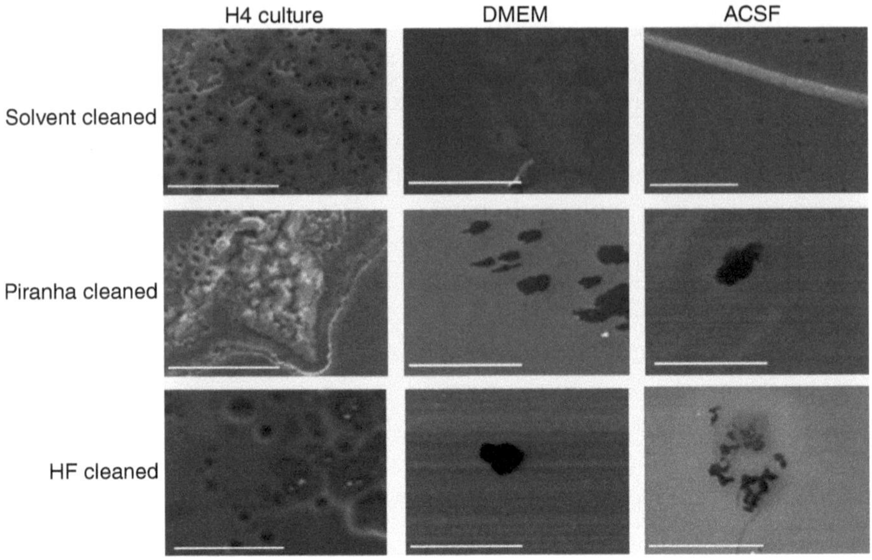

Figure 7.4 ***SEM micrographs of Si samples immersed in cell media + H4 cells, cell media and ACSF for 96 h at 37°C.*** Cleaned with solvents, the acceleration voltage was 25 kV, working distance was 5 mm, and sample surface tilting angle was 0 degrees; gold palladium coating was not used for these scans. Magnification 1000×. Scale = 50 μm in all the micrographs [68].

Table 7.1 AFM data extracted from *in vitro* tests

Cleaning method	Data	Cell media + H4 cells	Cell media	ACSF
Solvent cleaned	Surface roughness, r_q (rms)	4.859 nm	4.392 nm	2.550 nm
	Peak to valley	69.276 nm	78.378 nm	102.217 nm
Solvent + piranha cleaned	Surface roughness, r_q (rms)	10.315 nm	3.377 nm	2.669 nm
	Peak to valley	146.975 nm	75.099 nm	56.719 nm
Solvent + pirhana +HF cleaned	Surface roughness, r_q (rms)	30.305 nm	789.180 pm	808.417 pm
	Peak to valley	348.478 nm	14.865 nm	7.438 nm

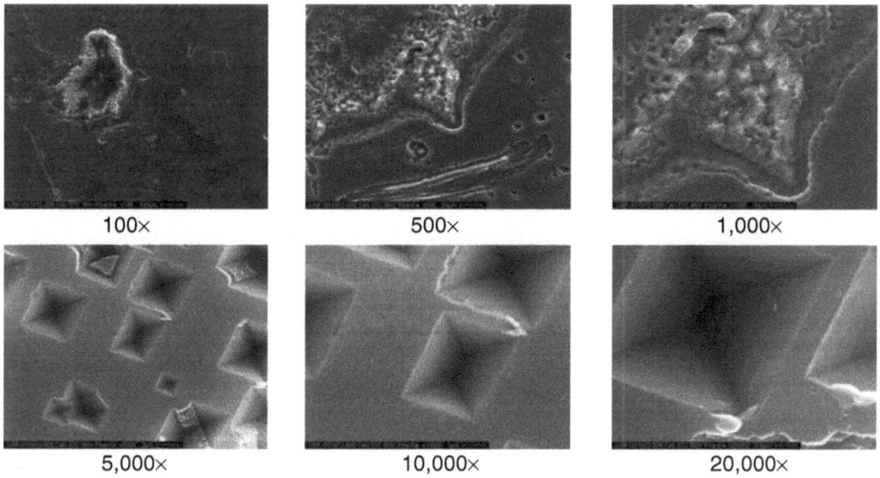

Figure 7.5 ***SEM micrographs displaying the surface of Si after* in vitro *culture with H4 human neuroglioma cells and subsequent cleaning with solvents, piranha solution, and HF.*** The revealed Si surface is covered with pyramidal etch pits, which is characteristic of anisotropic etching for Si(100). These etch pits were not seen in any previous cleaning procedure, which indicates that the surface may have been oxidized by the cell line [68].

micrographs did not display the characteristic presence of pyramidal etch pits, there remains obvious surface modifications, which we hypothesize could be due to the hydration of Si discussed earlier. It is also apparent that these small morphological changes were not the same as observed in Frewin et al. [44].

Alternatively, the morphology of the Si samples was severely altered after the removal of attached H4 cells. The surface roughness also increases with each subsequent cleaning method, which is the opposite observation with the Si exposed to complete media and ACSF, with the largest increase in feature size noticed once again after HF cleaning. Si has an extremely low etch rate when exposed to HF acid, but the acid can etch SiO_2 with very high etch rates depending on oxide quality [69]. The HF etch revealed that the surface of Si was covered with the characteristic pyramidal etch pits, which was especially apparent using SEM microscopy (Fig. 7.5).

Quantitative AFM analysis verified the presence of multiple pyramid-shaped pits on the surface of this type of material, with an average pit depth of 57 ± 8 nm and width of 4600 ± 70 nm. The deepest pit that was observed had a depth of 138 nm. Due to the absence of these features on the media and ACSF soaked samples, the presence of cells may be the source of the surface degradation. Inflammatory cells have been shown to release many caustic chemicals during frustrated phagocytosis, which is characterized by an attempt of the cell to remove an invading material, which it cannot envelop [53]. Furthermore, active neural inflammatory cells, like microglia, produce large amounts of NO, a free radical which is used in signaling and cellular defense, and hydrogen peroxide, an oxidizing agent [61]. NO in aqueous environments has displayed a proclivity to produce numerous subspecies, one of which is nitric acid, which is a known anisotropic oxidizing etchant of Si and forms SiO_2 [50,51,54]. As H4 neuroglioma are derived from cancerous glial cells, they may well be a source of this chemical excretion, and therefore locally oxidize the surface of the Si. However, future experiments would have to use alternative measurement techniques, like mass spectroscopy or Fourier transform infrared spectroscopy, to quantify the chemical species emitted by H4 cells and the chemical composition of the Si surface. Furthermore, other cell lines would need to be evaluated to verify that this particular effect is only characteristic to the H4 glial cell line, or if it is also observed with other cell types. Needless to say, Si has proven that it is not resilient within the physiological environment, and using it as a material for biological implant devices may lead to numerous issues during its lifetime.

7.2.4 Accelerated Aging Study

Exposure of SiO_2, 3C–SiC, and *a*-SiC to the aforementioned chemical treatments for 96 h at 37°C did not produce any notable surface modification. It may be that these surfaces are more resistant to the components in the treatments than bare Si, and 96 h may be too short an exposure time. It has been seen with polymeric materials tested within buffer solutions that temperature is linked to material degradation and can be used to simulate acceleration in aging [48,49]. Accelerated aging was found to follow the time dependence of $2^{\Delta T/10}$, where ΔT is the difference between the elevated temperature and 37°C. 1× phosphate buffer saline (PBS) solution was prepared by dissolving five tablets of PBS (Invitrogen Technologies) in grade 2 DI water. The pH was adjusted to 7.4 ± 0.1 by NaOH–HCl titration and passed through a 0.22-μm filter. Six samples of each material were placed in Pyrex® screw cap test tubes (Sigma-Aldrich) with 1.6 mL of 10 mM PBS to yield an electrolyte-to-surface area ratio of 1 mL/cm^2, and the cap was closed loosely to avoid any internal pressure increase inside the tubes. The pH of the solution was monitored using a PC-controlled SPER scientific pH meter. The pH was maintained at 7.4 ± 0.1 and the tests performed at 37° and 60°C for 30 days.

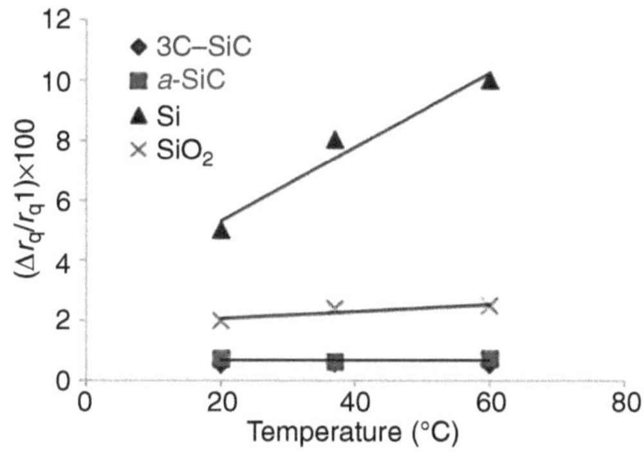

Figure 7.6 ***Surface roughness change of a-SiC, 3C–SiC, SiO$_2$, and Si(100).*** Note that the only significant change in surface roughness was observed for Si surfaces.

Quantitative rms surface roughness data were collected using AFM before and after the accelerated PBS corrosion test for *a*-SiC, 3C–SiC, SiO_2, and Si as displayed in Fig. 7.6. The average surface roughness remained consistent for 3C–SiC and *a*-SiC after a time period of 30 days at both 37 and 60°C. Surface degradation was observed for both Si and SiO_2, which increased in severity with increased temperature, signifying a time acceleration of 4.9 months. SiO_2 experienced a very slight surface degradation, whereas Si experienced significant surface degradation at 1 month at 37°C and even more at the 60°C accelerated test temperature. This observation follows the small surface modifications seen in Section 7.2.5 at 96 h in ACSF, which has a similar molar content of the ionic concentrations found in PBS.

7.2.5 Corrosion Study Conclusions

Evaluation of the material integrity for implantable biomedical devices has been one of the tools used in modern device testing to determine device material stability before implantation *in vivo*. We have used methodologies found in ISO 10993-14 and ISO 10993-15 to evaluate various semiconductor materials in simulated physiological environments as well as with exposure to cells [47,65]. Corrosion tests were performed in PBS, pH = 7.4 ± 0.2, for 30 days at physiological temperatures of 37°C as well as the accelerated regime of 60°C [47]. In all the experiments, 3C–SiC and *a*-SiC displayed no surface degradation, verifying their excellent chemical resilience. SiO_2 has nearly no degradation, but this material has many other issues, namely, it has been shown to be cytotoxic and absorbs metallic ions, which leads to degradation in its electrical functionality [27,30,31,70]. Si shows the greatest change in surface morphology, both in the physiological solutions as well as after interacting directly with cells. From a device standpoint, this change could lead to reliability issues in the long term. Silicon carbide is a much better option as it not only withstands the chemical environment, but it can be made into the same types of electrical devices as silicon.

7.3 *IN VIVO* PERFORMANCE

We have demonstrated that silicon carbide may be an excellent material for long-term use within the body. In chapter: Cytotoxicity of 3C–SiC Investigated Through Strict Adherence to ISO 10993, we showed that the cubic polytype, 3C–SiC, produced a cytotoxic reaction similar to culture-treated polycarbonate plastic using ISO 10993-5 standards. In chapter: Study of the Hemocompatibility of 3C–SiC and *a*-SiC Films Using ISO 10993-4, we showed that *a*-SiC and 3C–SiC not only did not produce an appreciable level of cytotoxic response, but they have demonstrated levels of static and dynamic hemocompatibility much better than Si and SiO_2 through ISO 10993-4 protocols. In Section 7.2, we used ISO 10993-14 and -15 methodologies to demonstrate that 3C–SiC and *a*-SiC were not susceptible to corrosion in the physiological environments. This accumulation of excellent *in vitro* evidence verifies that 3C–SiC and *a*-SiC may be materials that can be used to construct complex, electrically active biomedical devices. However, these previous examples were all performed within well-controlled laboratory simulated environments. The body is much more complex and variable, and thus the next stage is to expose the materials to an active physiological environment. In this section we will discuss experiments involving testing 3C–SiC and *a*-SiC *in vivo* within animal models, first within the brain of wild-type mice and second subcutaneously in female farm pigs.

7.3.1 Effects of 3C–SiC to Neural Tissue

Millions of people around the world have physical disabilities, which have arisen from damage to the central nervous system (CNS). In addition, thousands more have experienced the debilitating loss of a limb, which is unfortunately a growing trend due to wartime violence. The brain–machine interface (BMI) shows great promise to be able to provide therapeutics for these types of injuries [71–73]. One BMI type incorporates invasive implantable neural interfaces (INI) as its connection to the body, and has shown an increased level of utility in that they have the ability to function closed loop through bidirectional signaling between the device and specific neural populations [14,72,73]. Over the years the one major issue with these types of devices preventing their increased acceptance has been their questionable long-term reliability, which manifests itself as signal degradation to complete signal loss over time [10,74].

One of the major factors in the reliability issue has been linked to materials used in INI construction [9,11,12,75]. The saline-filled CNS environment, which is full of inflammatory microglia cells, poses great difficulties that need to be addressed for the INI. Harsh chemicals like NO, peroxides, oxidases, and numerous free radicals, as well as a wide range of mechanical forces, add to the difficulties that an INI device must contend with [60,76–78]. These secretions not only can damage the material of a device, which was investigated and demonstrated in Section 7.2, but they can affect the cells and tissue in the surrounding area. Fig. 7.7a shows a representation of the inflammatory response within the brain to neural implants. Materials which are not accepted within the brain will become surrounded and attached by activated inflammatory microglia glia cells and then fully encapsulated by astrocytes, which has the effect of increasing local electrical impedance and decreasing the ability of the device to send and receive signals. This effect can become chronic, leading to neural degeneration (necrosis). Fig. 7.7b provides an example of the necrosis around deep brain implants, which is especially apparent at the INI tips. A successful INI device therefore would most likely require superior chemical resilience and mechanical reliability in addition to the proper electrical properties and biocompatibility.

We have been investigating 3C–SiC as a possible material for INI devices [44,79]. 3C–SiC, the only cubic lattice of over the 250 lattice variations of SiC, has many advantages inherent to all SiC materials, such as chemical inertness, physical robustness, and, as a semiconductor, can be engineered into complex electrical devices [80]. The advantage of 3C–SiC is that it can be grown as a thin film on inexpensive silicon substrates and therefore is potentially much cheaper to manufacture than bulk SiC devices. The Si substrate also offers additional processing options not available when using bulk SiC substrates [80]. In this present study we detail the processing and subsequent *in vivo* implantation of passive 3C–SiC shanks into a murine model, demonstrating the outstanding performance of 3C–SiC using Si as a comparison material.

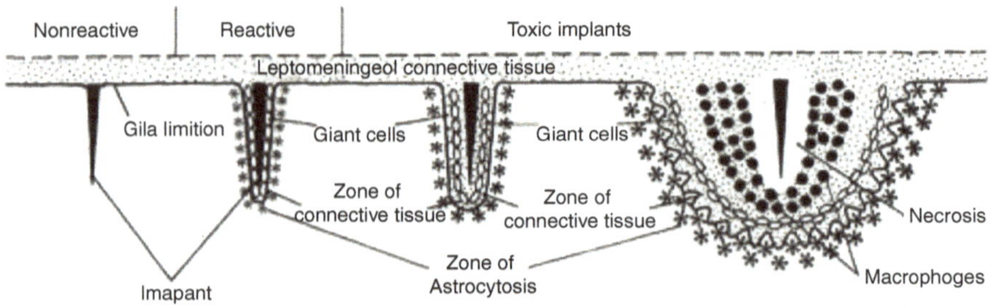

(a)

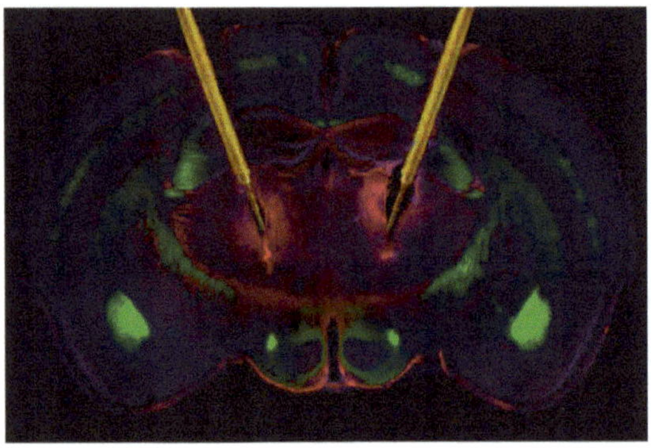

(b)

Figure 7.7 (a) A diagram demonstrating the neural immune response to cortical implant from [9]. The stages are marked by the activation and response of microglia, the gradual activation, and encapsulation of the device by astrocytes. Chronic response leads to the generation of giant, interconnected macrophages and tissue necrosis. (b) Fluorescent micrograph of deep brain stimulation electrodes implanted in the brain of a mouse. The chronic degeneration is marked by the presence of astrocytes stained against glial fibrillary acidic protein (GFAP) in red. Necrosis is especially evident around the active tips. *(Image from Ref. [81].)*

7.3.1.1 Manufacture of the 3C–SiC and Si Probes

In accordance with ISO 10993 standards, this evaluation of 3C–SiC is a continuation of our previous studies (chapters: Cytotoxicity of 3C–SiC Investigated Through Strict Adherence to ISO 10993; Study of the Hemocompatibility of 3C–SiC and *a*-SiC Films Using ISO 10993-4), which indicated 3C–SiC high biocompatibility and hemocompatibility through *in vitro* testing [82,83]. In this preliminary study, the tissue response around passive shanks fashioned from 3C–SiC epitaxial material was compared against a positive reaction control of Si. The implant structure for this test consisted of dual isosceles triangle shanks with a 250-μm base, a 7-mm length, an apex angle of 5 degrees, and a thickness of 15–20 μm. The base of the shanks wereattached to a 5-mm long, hemispherical tab, 1 mm in radius. The Si and 3C–SiC tabs were glued together with surgical cyanoacrylate glue as shown in Fig. 7.8a. The tab allowed us to handle the delicate shanks and provided a mount that would sit on the skull of the mouse to hold the implants firmly in place within the brain.

The process schedule to fabricate the 3C–SiC devices began with the RCA cleaning a 50-mm, 250-μm thick Si(100) wafer. A 20-μm thick 3C–SiC heteroepitaxial film was then grown on the Si(100) wafer in our hot-wall chemical vapor deposition reactor using a previously published process (Fig. 7.8b) [85,86]. A hard mask outlining the tab structure, consisting of 500 Å of Ti followed by 1 μm of Ni deposited through electron beam evaporation, was added to the backside of the Si substrate wafer (Fig. 7.8c). The 3C–SiC surface was spin coated with AZ4620 positive photoresist followed by lithographic patterning to generate a deep reactive ion etching (DRIE) soft mask (Fig. 7.8d). The masks were defined through the removal of the excess metal and photoresist using wet chemical etching. DRIE was used to etch the 3C–SiC layer down to the underlying Si substrate (Fig. 7.8e). The wafer was then flipped over and the 3C–SiC side attached to a 50-mm handle wafer using a 1:4 blend of Crystalbond®:acetone (Fig. 7.8f). DRIE was then used to remove the underlying Si substrate, leaving the released 3C–SiC structures imbedded in the Crystalbond (Fig. 7.8g). The 3C–SiC shanks were then released with acetone. The Ti/Ni protective mask was removed using a ferric chloride solution followed with immersion in a dilute hydrofluoric acid solution. Finally, the shanks were cleaned using an O_2 plasma (Fig. 7.8h). The manufacture of the Si shanks was similar in process schedule except we utilized a high aspect ratio Bosch DRIE process, composed of several DRIE and masking cycles, to maintain sidewall integrity [87].

7.3.1.2 *In Vivo* Implantation Protocol

The University of South Florida has an animal welfare assurance #A-4100-01 on file, and animals for this experiment were maintained in accordance with the USF Institutional Animal Care and Use Committee protocol. The completed passive devices were implanted into three C57BL/6J mice within an area nearest the rostral mesencephalon, or mouse midbrain. The devices remained within the animals for 35 days. The 3C–SiC implant was located at coordinates 2.52 mm coronal, −3.05 mm sagittal, and the Si implant at −2.52 mm coronal, −3.05 mm sagittal, as measured from bregma. These locations allowed the implant to interact directly within a large portion of

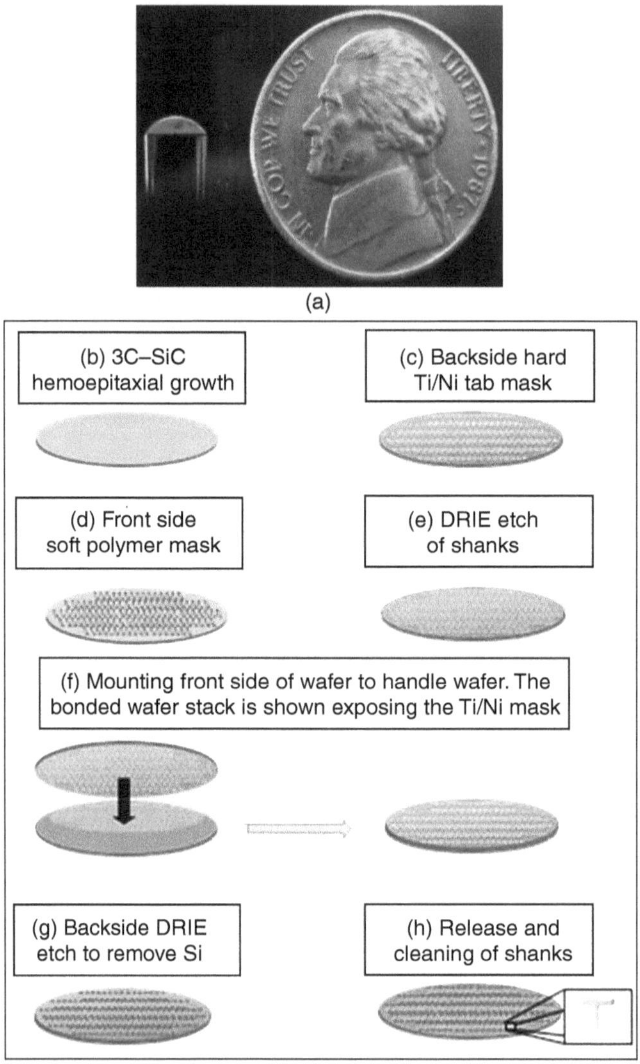

Figure 7.8 (a) A digital photograph of a completed passive implant device. The nickel provides scale. The thick Si tab allowed for us to handle the shanks and provided us the ability to affix the implants to the skull of the mouse. The process schedule for the construction of the passive 3C–SiC and Si implants used in this study. (b) A film of 3C–SiC was heteroepitaxially grown on a Si substrate via chemical vapor deposition. (c) Photolithography with metal lift-off created a hard Ni mask. (d) Photolithography using AZ4620-positive photoresist detailed the shanks and tab on the front of the wafer. (e) Deep reactive ion etching (DRIE) was used to etch the 3C–SiC (or Si) on the front of the wafer to create the shanks and tab body. (f) The wafer front side was mounted to a Si handle wafer through coating of Crystalbond®/acetone. This exposed the backside of the wafer for subsequent (g) DRIE etching to remove the excess Si substrate, leaving freestanding devices. (h) The shanks are released from the handle wafer with acetone and cleaned of the remaining metals and polymers. Images originally reported in Ref. [84].

the hippocampus. Although this area is not targeted traditionally in BMI studies, it is often a target in neuroscience and is associated with many diseases for which neural stimulation may be of assistance.

The surgical implantation was performed as follows: the mice were anesthetized with isoflurane using a calibrated vaporizer. A jeweler's drill was mounted to a digital stereotaxic manipulator arm. The origin was set to the bregma and the drill was moved into the implant position. The drill created a 1.00-mm wide hole in the mouse skull at each of the two implant locations. The drill was removed and replaced with normally closed tweezers holding the passive implant. The implant was lowered into the brain at approximately 2 mm/s to a z-depth of –6 mm, at which point the base of the tab rested on the skull of the mouse. A two-part dental epoxy was used to secure the tab to the skull and allowed to dry. The mouse scalp was sutured closed over the device and the skull, and the mouse was allowed to recover on a 37°C hot pad. After the operation, the mice were monitored daily for overall health and reactivity to ensure proper healing of the incision and the detection of any distress.

7.3.1.3 Tissue Fixation and Evaluation

At 35 days, the mice were transcardially perfused with saline (0.9% NaCl) followed by 4% paraformaldehyde. We carefully removed the skull along with the attached implants from the perfused brain. At this time we noticed that the Si implants in all animals easily withdrew from the brain, but the 3C–SiC implants were more firmly set within the brain tissue. One of the two 3C–SiC shanks broke at the base of the shank and remained within the brain of the second animal. The brains were frozen to –20°C, sliced to 20-μm thick tissue blocks within a cryostat, and the slices were placed directly onto Fisher Brand Color Frost Plus glass slides (Fisher, 12-550-17). The slices on the slides were blocked with donkey serum and sodium azide (NaN_3) in PBS. The slices were incubated overnight at 4°C after adding a solution containing NaN_3/PBS and primary antibodies anti-CD45-conjugated FTIC at 500:1 (ab25670, AbCam), 3000:1 unconjugated anti-GFAP (ab53554, AbCam), and 3000:1 unconjugated anti-MAP2 (ab11268, AbCam). CD45 is a protein marker of microglial and microphage activation, GFAP is a protein marker for astrocytes, and MAP2 is a protein marker common to the microtubules of neurons [88]. The following day, the slides were incubated in a mixture of NaN_3/PBS and secondary antibodies at room temperature at a concentration of 250:1. The secondary antibodies used were Alexa Fluor® 568 (A11057, Invitrogen) and Alexa Fluor 350

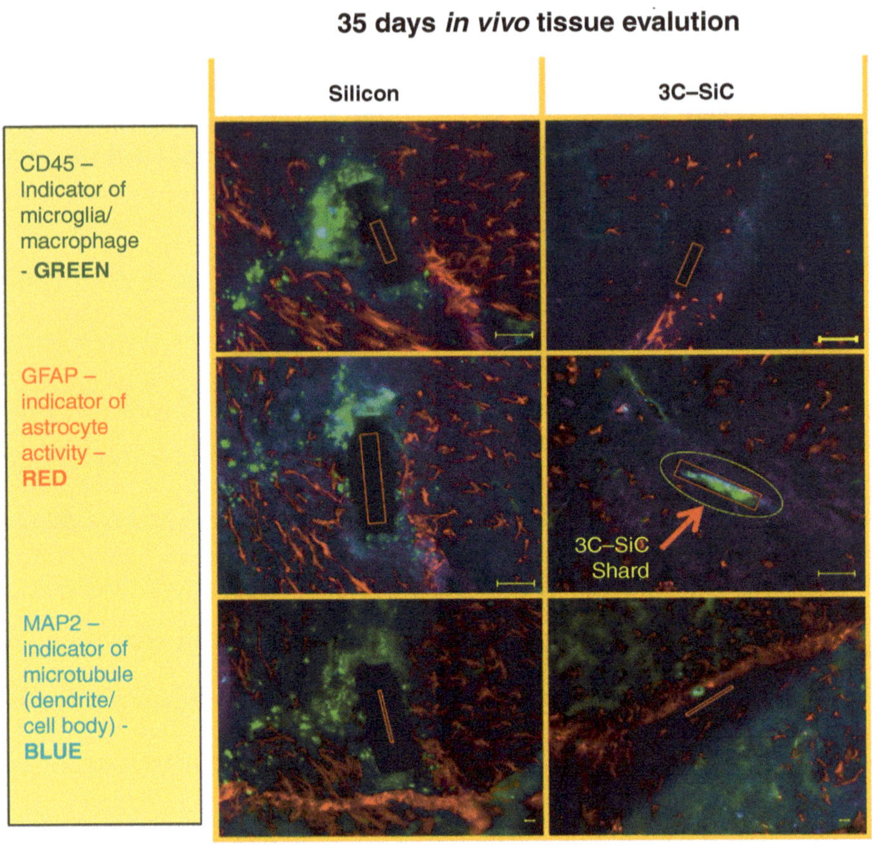

Figure 7.9 ***Selected digital photographs obtained from a Zeiss Axio Imager 2 microscope, which was used for the immunofluorescence analysis of C57BL/6J mouse hippocampal tissue.*** Each row represents a selected tissue slice from a single animal ($N = 3$). The first column shows the tissue response within the left hemisphere of the hippocampus surrounding the Si implant and the second column presents the tissue associated with 3C–SiC within the right hemisphere. The legend on the left details of the antibody stains used in the study. The orange boxes denote the approximate size of the implant at the tissue location. Scale = 50 μm. *(Image from Ref. [84].)*

(A10035, Invitrogen) for GFAP and MAP2, respectively. Fluorescent digital microscopy images were captured using a Zeiss Axio Imager 2 microscope outfitted with an ApoTome accessory.

Fig. 7.9 shows selected digital images obtained from fluorescent microscopy of the tissue surrounding the implanted materials. The first and most noticeable difference between the two materials is the concentration and frequency of both $CD45^+$ and $GFAP^+$ glial cells. The tissue that was next to the Si implants has a very large $CD45^+$ and $GFAP^+$ response and this activity radiates hundreds of microns away from the implant site. In comparison, the tissue surrounding the 3C–SiC implants displays little indication of $CD45^+$ activity, and only tissue from the third animal shows a very slight increase in $GFAP^+$ activity on only one side of the implant.

Another observation that can be made from the images is directly related to the implant location. A large area devoid of neural tissue is apparent around all of the Si implants, and the voids are often larger than 50 μm wide, a distance which is much larger than the implant thickness of ~20 μm. The left and right images are the same tissue slice for the Si and 3C–SiC implants, thus allowing for a meaningful direct comparison between the two materials within the same animal. On the other hand, the 3C–SiC material in animals 1 and 2 produced voids which are ~20 μm wide, which is the same dimension as the device thickness. The third animal tissue provided an exception with a void width larger than 50 μm, and this was the only animal, which also shows signs of increased $GFAP^+$ activity in relation to 3C–SiC.

The last observation relates to the single implant piece that remained within the brain of the second animal. We were able to slice through the 3C–SiC implant, which remained within the tissue slice. The second row and second column show this implant material imbedded within the slice. With the exception of a very small gap on the lower left side of the implant, MAP2 neural processes, as indicated by the blue haze, remains in direct contact with the 3C–SiC material and there are few $CD45^+$ and $GFAP^+$ cells within the area of the implant.

7.3.1.4 Neural In Vivo Discussion and Conclusion

Without exception, the positive reaction control of Si displayed an increase in inflammatory activity, which follows observations that have been made by many other INI device researchers [10,12,75]. This reaction, present in all of the tissue samples we examined, was marked with a large activation of microglia/macrophages in the vicinity of the device, which are, in turn, surrounded by astrocytes with elongated, interconnecting processes. It also appears with the absence of $MAP2^+$ cells that the neural processes in the vicinity of the implanted Si material have either been withdrawn or lost due to cellular death. This effect creates large voids around the device, which has the effect of increasing the distance between the neural processes and the implant surface. It was also interesting to note that there were microglia/macrophages active hundreds of microns from the implant site, although there does not appear to be any correlation with activation distance or the direction of the activation between the samples. Microglia have been credited with a sophisticated communication network through the use of chemotaxis to maintain and protect the neural environment, and the reaction to the large implant may be part of this communication, but our limited immunofluorescence analysis cannot provide adequate quantitative evidence of this assertion [89,90].

3C–SiC has shown a markedly different reaction to the materials in all three animals. The first major difference is the very limited $CD45^+$ microglia/macrophage activity in nearly all of the samples. The tissue voids are much smaller surrounding the 3C–SiC implants, as marked by the $MAP2^+$ cell processes. This is especially apparent when we examine the second animal, which has a piece of the implant material present within the slice. The neural processes are still in contact with the material and there are limited $CD45^+$ and $GFAP^+$ cells present. Although there is a large amount of green luminescence in the image, this is due to the fact that 3C–SiC produces a green photoluminescence in the presence of blue light due to the 2.3 eV bandgap [91].

For the majority of the tissue samples analyzed, the $GFAP^+$ astrocytes in the 3C–SiC tissue samples were much different in shape, size, and quantity when compared with the same cells in the Si tissue. These cells appeared in almost regular patterns through the tissue, and their processes were not extended, interconnected, or web-like. The one exception to this was in the third animal. One side of the implant shows an increased presence of $GFAP^+$ cells. As this effect did not occur in the other animals, we are not certain as to the cause of this effect. It should be noted that the active layer is very thin and could be the result of the endpoint of the acute inflammatory response to the implantation damage. A more likely cause is one side of the 3C–SiC material consists of highly defective material due to growth interface differences between the Si and 3C–SiC crystal lattices, and these defects may be causing issues (ie, roughness) that activate the neural inflammatory response [92]. In future devices, this issue will be eliminated by polishing this side of the probes to remove the rough surface.

Passive 3C–SiC neural implants have shown promise after implantation in C57BL/6J mice. 3C–SiC did not show any qualitative indication of an overwhelming positive microglial response. At the time of this chapter, we are working on producing active 3C–SiC neural probes capable of signal recording *in vivo* within active animal models. These will be used not only to produce statistical and quantitative tissue histology data, but also to evaluate long-term signal reliability.

7.3.2 Subcutaneous Evaluation of 3C–SiC and *a*-SiC

For implantable biosensors, a critical issue that limits the device longevity and functionality is biocompatibility and the implanted sensor must be able to function properly for an extended period of time. The significant risk for patients is the foreign body reaction of implantable devices. Therefore, contributing factors for this loss of functionality include the degradation and fouling of the sensor, and the changes in the tissue surrounding the sensor such as fibrosis and inflammation. One of the strategies is to use synthetic polymer polyethylene glycol (PEG) on the biosensor [93–96]. Another biocompatible material, *a*-SiC, has been used as thin film coating for many implantable biomedical devices. *a*-SiC and polycrystalline SiC (poly-SiC) have also been used for clinical studies in bone prosthetics and heart stents, which confirmed the biocompatibility of these forms of SiC [97–99]. Chapter: Amorphous Silicon Carbide for Neural Interface Applications provides an in-depth study of the use of *a*-SiC for INI applications with the results presented being wholly consistent with the data presented in this chapter.

This section further investigates SiC as an implantable sensor, realized using an RF (radio frequency) antenna [100]. Two different antenna electrode materials have been used on a semiinsulating 4H–SiC substrate: (1) an "all-SiC" sensor made of poly-3C–SiC material and (2) a titanium/gold (Ti/Au) electrode-based sensor. In this study, we have investigated comparative *in vivo* tissue histological analysis of SiC-based glucose sensors with different surface coatings for assessing biocompatibility.

7.3.2.1 Sample Preparation and Surface Modification of SiC-Based Ti/Au Antenna

Different surface coatings/surface modifications of SiC-based antenna sensors were investigated for histological tissue inflammation or fibrosis capsule development. Fig. 7.10 shows the different implanted antennas studied along with the control surfaces used during the experiments. Details of the test coupons are as follows: (a) 4H–SiC-based Ti/Au antenna sensor (glucose sensor) was prepared by depositing Ti (500 Å) and Au (4500 Å) via E-beam evaporation. (b) A SiO_2-coated Ti/Au antenna on semiinsulating 4H–SiC was prepared by depositing SiO_2 on Ti/Au antenna using the PECVD technique. A SiO_2 coating was selected to reduce direct adhesion problems with other biocompatible material coatings such as *a*-SiC

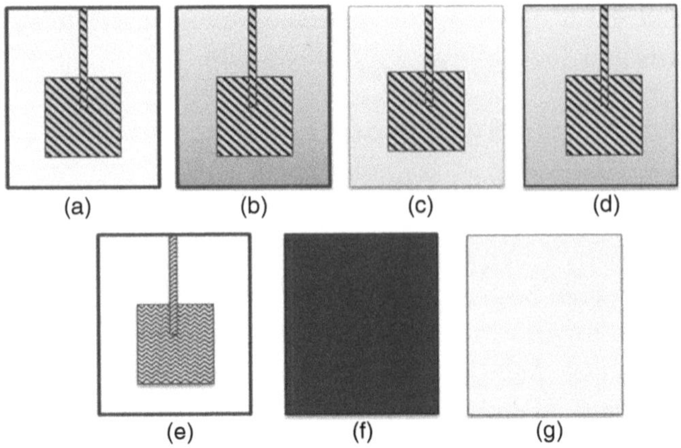

Figure 7.10 ***Sketch of the five different antenna surfaces and two controls that were subcutaneously implanted.*** (a) Ti/Au antenna on semiinsulating 4H–SiC as the base. (b) SiO_2-coated Ti/Au antenna on semiinsulating 4H–SiC. (c) PEG-coated Ti/Au antenna on semiinsulating 4H–SiC. (d) *a*-SiC-coated Ti/Au antenna on semiinsulating 4H–SiC. (e) "All-SiC" antenna, poly-3C–SiC on semiinsulating 4H–SiC. (f) Positive control Si. (g) Negative control 3C–SiC with an *a*-SiC coating on the backside [101].

and poly PEG on the Ti/Au antenna. (c) A covalently bonded PEG on oxide-coated Ti/Au antenna was one important surface modification used to increase the biocompatibility via protein inhibition. (d) *a*-SiC thin film coating of the Ti/Au antenna is another approach to make the antenna surface more biocompatible. This *a*-SiC was deposited on a SiO_2-coated Ti/Au antenna surface by using PECVD. The oxide was deposited all over the antenna surface (top layer–Au and 4H–SiC) to make sure the *a*-SiC coating was uniform. (e) "All-SiC" antenna is a poly-3C–SiC film grown on a semiinsulating 4H–SiC substrate. The heavily N_2-doped poly-3C–SiC thin film was grown using a CVD reactor and then fabricated using standard photolithography. It has been shown that SiC and its polytypes (eg, 3C–, 4H–, 6H–SiC) are more biocompatible than Si and therefore a promising candidate for bioapplications (see chapters: Cytotoxicity of 3C–SiC Investigated Through Strict Adherence to ISO 10993; Study of the Hemocompatibility of 3C–SiC and *a*-SiC Films Using ISO 10993-4). Therefore (f) Si was used as a positive control. For implantation, the Si material was diced (8.5×11.5 cm^2) and RCA cleaned as described previously. (g) 3C–SiC is a biocompatible material and it is used as a negative control. The other side was *a*-SiC coated to thus realize another "all-SiC" implant. The five antenna (a–e) reactivity to cells/tissue was compared to the two controls (f–g).

Biocompatibility and nonbiocompatibility were verified using the positive reaction control, silicon, and negative reaction control, 3C–SiC.

7.3.2.2 Device Implantation in an Animal Model

Four adult healthy mature male pigs, 20–35 kg in weight, were selected for tissue histology analysis [102]. Surgery was performed on surgical sites at the midposterior aspect shoulder (two devices), mid-thoracic pave vertebrae (two devices), midlumber pave vertebrae (two devices), and midsacral (one device). After preparation, implants were carefully placed into $1.5 \times 1\text{-cm}^2$ pockets beneath the skin (dermis) surface, as shown in Fig. 7.11. Five antennas and two controls, for a total of seven devices, were implanted into the pockets of a single animal. Four adult pigs were selected for the histological tissue/cell evaluation and surface permissiveness to cells, providing a total of 28 samples (7 different materials per 4 animals).

7.3.2.3 Implant Characterization: Atomic Force Microscopy

Using AFM, the surface quality of the devices (pre- and postimplantation) was compared. Fig. 7.12 shows the AFM surface morphology and roughness of pre- and postimplantation samples. The surface roughness of the pure SiC samples (3C–SiC and poly-3C–SiC) had similar rms values (~1 nm)

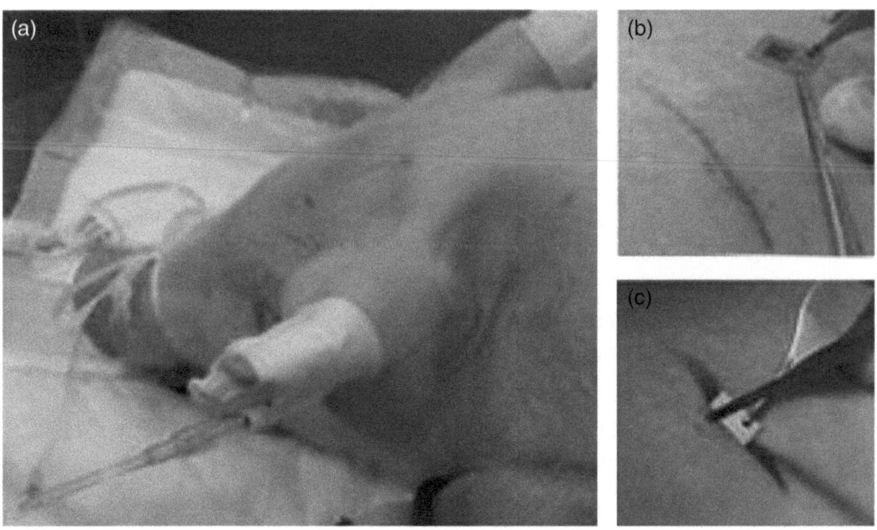

Figure 7.11 ***Subcutaneous in vivo implantation of the sensors in pig.*** (a) Shows IV and ventilation entries, while (b) shows the surgical pocket and (c) a coupon being implanted [101].

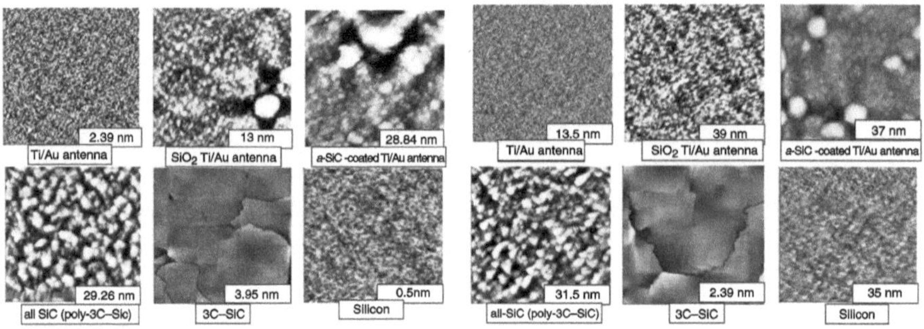

Figure 7.12 (a) Preimplant AFM micrographs and (b) postimplant AFM micrographs of Ti/Au antenna; SiO_2 Ti/Au antenna; *a*-SiC-coated Ti/Au antenna; all-SiC (poly-3C–SiC; 3C–SiC, negative control; silicon, positive control). Surface roughness rms values are located in the lower right corner of each micrograph [101].

before and after implantation. The roughness of the *a*-SiC-coated Ti/Au antenna increased from 28.8 nm to 37 nm. The roughness of the Ti/Au antenna, SiO_2-coated Ti/Au antenna, and Si were increased from preimplantation to postimplantation values by 11, 26, and 34.5 nm rms, respectively.

7.3.2.4 Implant Characterization: Light Microscopy

After 30 days postimplantation, histological sections were studied under light microscopy. After polymerization, thin (10-μm thick) translucent sections of tissue were prepared in a transversal direction to the axis of the implant using the microtome technique. The sections were stained with hematoxylin/eosin and examined using a light microscope to permit a concise description of the observed specimens and fibrous capsule formation surrounding the implants, if present. The examination was performed on 29 separate samples including 28 samples from 4 pigs and a fresh (without implant) tissue block to serve as a tissue control. A light micrograph of a fresh tissue block, without any implant, is shown in Fig. 7.13. This fresh tissue block is used as a reference, and tissue samples were compared to this healthy tissue sample.

Histological examination of a subcutaneous implant of Si (positive control) revealed that an adverse tissue response occurred to all four implants, and this response was fairly uniform at 4 weeks *in vivo*. The Si implants were surrounded by fibrous capsules; after removal of the implant the shape of the capsule was the same as the implant, which is shown in Fig. 7.14. All four Si implants had a thin circumferential fibrous capsule around them, which is shown in Fig. 7.15f. The histological tissue analysis of the SiO_2-coated

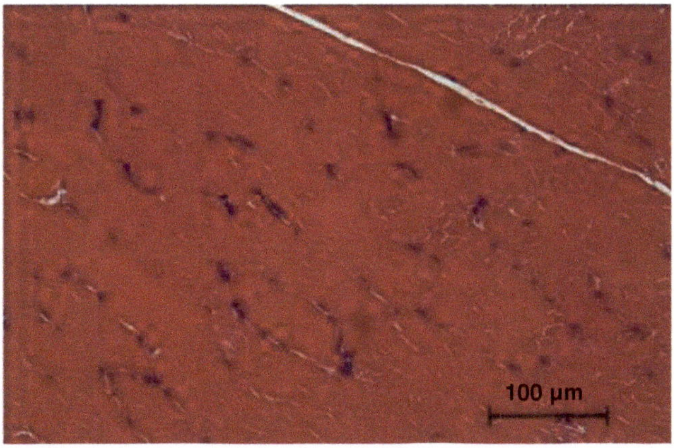

Figure 7.13 ***Histological tissue analysis of healthy tissue, without any implant, showing no signs of inflammation (original magnification ×40) [101].***

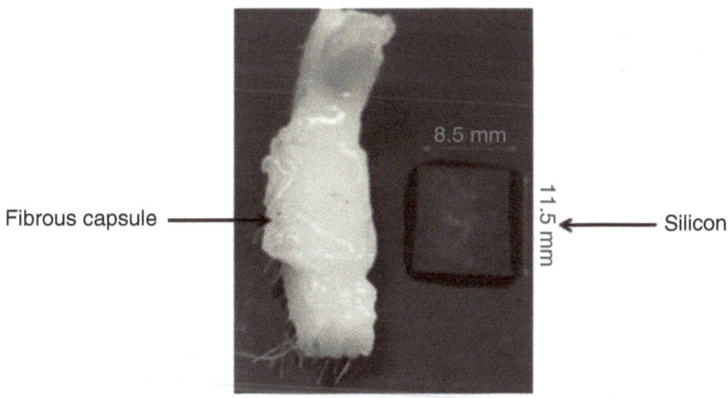

Figure 7.14 ***Photograph of the fibrous capsule after removal of an Si implant (30 days*** **in vivo** ***implantation specimen) [101].***

Ti/Au antenna specimen had similar trends as observed for the Si implants. Multiple nucleated giant cells, macrophages, fibroblasts, lymphocytes, and evenly distributed eosinophil cells were evident, which had a strong inflammatory cell infiltration. The tissue in contact with the implant began to change to fibrous tissue. In addition the surface of the implanted material (SiO_2-coated Ti/Au antenna) had a thin fibrous tissue layer shown in Fig. 7.15b, which covered the scattered connective tissues. After 30 days of subcutaneous implantation, the Ti/Au antenna demonstrated accumulation of inflammatory cells and macrophages at the implant–tissue interface (shown in Fig. 7.15a), which is a sign of an acute inflammatory response.

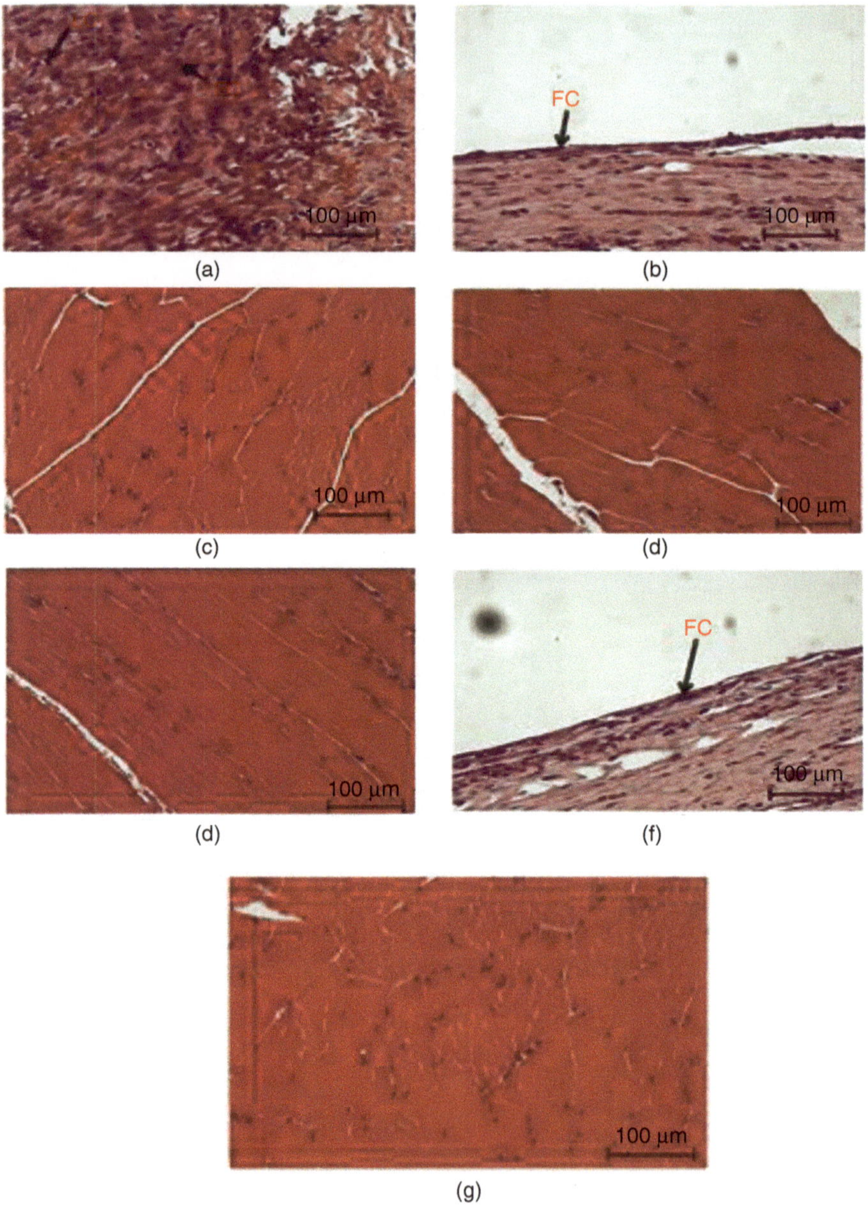

Figure 7.15 ***Seven different implant histological tissue responses.*** (a) Ti/Au antenna on SI 4H–SiC as the base; (b) SiO_2-coated Ti/Au antenna; (c) PEG-coated Ti/Au antenna; (d) *a*-SiC-coated Ti/Au antenna; (e) "all-SiC" antenna, poly-3C–SiC; (f) Si; (g) 3C–SiC with *a*-SiC coating on the backside [101].

However, there was no sign of fibrous capsule development around the embedded material after 4 weeks of implantation.

The tissue histology of the "all-SiC" antennas, which consisted of doped poly-3C–SiC as the antenna electrode and 4H–SiC (semi-insulating) as the substrate, had one of the most promising results as shown in Fig. 7.15e. There was no sign of inflammatory response for all four implants. Another promising implant was an *a*-SiC-coated Ti/Au antenna. The histopathological tissue analysis shown in Fig. 7.15d showed no sign of inflammatory response. Implants comprised of single crystal 3C–SiC film on the front side of a Si substrate, with an *a*-SiC coating on the back side, had no inflammatory response during the tissue histological examination on three different implanted samples as shown in Fig. 7.15g. There was some scar evident on the fourth tissue sample, which might be from tissue recovery after the surgery. Finally, the PEG-Ti/Au antenna showed the expected result; no sign of inflammatory response for all the samples (Fig. 7.15c) as comparable to the fresh tissue block shown in Fig. 7.13.

For this 30-day *in vivo* study, we found Si and SiO_2 to have an inflammatory response with fibrous encapsulation of mono- and multinucleated phagocyte cells surrounding the implants, as expected. The bare Ti/Au antenna also had some inflammatory signs, even though there was no encapsulation. On the other hand, the PEG-coated antenna and the "all-SiC" antenna did not have any inflammatory immune response whatsoever. At the same time, histological tissue analysis on single crystal 3C–SiC and *a*-SiC-coated samples did not show any inflammatory response. Based on the tissue histological results, the SiC-based sensors, covalently bonded PEG on the sensor, and the *a*-SiC-coated sensor can be considered as possible viable implantable sensors within the body [102].

7.4 3C–SiC FOR BMI APPLICATIONS—AN UPDATE

In the first edition of *Silicon Carbide Biotechnology*, the use of SiC in advanced BMI was discussed in detail [103]. Since the publication of the first edition, significant progress has been made in this important area. In this section we will highlight two of the major achievements. First, we present the observation that 3C–SiC does not distort images taken in rat brains via MRI, which has led to the design of 3C–SiC deep-brain stimulation implants. Second, we have been developing an ultralow-power neural recording chip for use in a wireless cortical microelectrode implant [104]. Additional work in the area of developing a SiC optrode for optogenetics applications, while

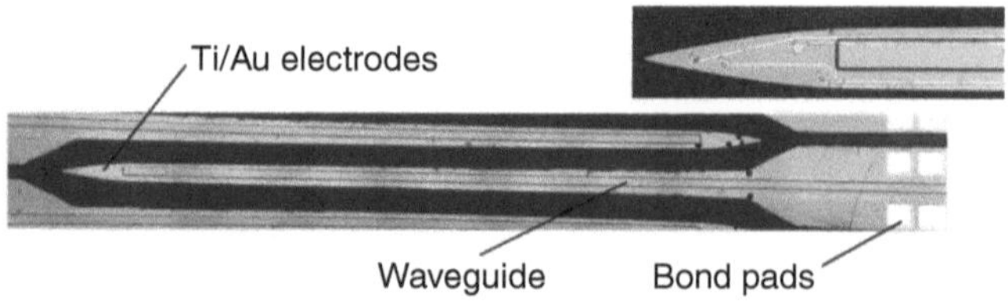

Figure 7.16 ***Optical micrograph of an SiC neural implant.*** Electrodes on the implant and the central SU-8 waveguide are visible in the photo. Inset shows detail of four Ti/Au recording electrodes at the probe tip.

not a significant activity at this time, demonstrated that we could manufacture a complex neural probe in 3C–SiC as shown in Fig. 7.16. Details of the SiC optrode development can be found in Ref. [105].

7.4.1 MRI Compatibility of 3C–SiC *In Vivo*

A major limitation facing current implantable electrode technology is the lack of compatibility with MRI. Although the metals used in neural implants are not ferromagnetic, the interaction of these metals with the extremely high-power electromagnetic fields required for high-resolution MRI, typically greater than 3T, introduces two major issues. One is the generation of image artifacts that extend well beyond the physical size of the metal contacts. In an *in vitro* study, these artifacts were found to disrupt the image 150–200 μm from the neural probe surface [106]. The second effect is that the electrons in the conductor will couple with these powerful electromagnetic fields, inducing eddy currents. The induced current in turn leads to joule heating ($Q \propto I^2R$), which flows into the surrounding tissue, leading to an elevation in the local temperature, which can increase to dangerous, and even terminal, levels [107,108]. Additionally, these MRI-induced currents have the possibility of stimulating the surrounding neural tissue [109]. Currently, the safe magnetic dose limitation for individuals with neural implants examined with MRI is 1.5 T, but this low power level may not produce the complex data necessary for accurate clinical diagnosis [107].

SiC in all of its forms has the thermal conductivity of copper at room temperature, allowing the material to transport heat very effectively [80]. Additionally, SiC has a relative magnetic permeability of 0.999, and will not readily couple with magnetic fields, unless it contains donor impurities called dopants [80]. Standard neural implantation device leads use metal conductors and dissipate heat through the conductive metal pathway; however, the leads are bounded by electrical insulation, which is normally not a good heat transporter, so the majority of the heat transfer is concentrated at the exposed end of the electrode where recording/stimulation of neural tissue occurs [108]. The greatest image artifact intensity is also produced at the exposed electrode tips [106]. Although all conductors will inductively couple with a magnetic field, SiC as a semiconductor has modifiable conductivity, which may provide a solution to this issue. Lower conductivity will still allow signal reception and transmission, albeit with some issues due to the increased electrode impedance. However, the lower conductivity decreases the production of image artifacts. The thermal properties of SiC, regardless of its electrical conductivity, effectively increases the thermal pathway across the entire device surface area, allowing effective heat dissipation over a larger area of the physiological environment, reducing small area/volume heating and thus assisting in maintaining physiologically safe temperature levels.

To illustrate the different magnetic coupling differences between metals and semiconductors, Jackeline Malheiros of the University of Saõ Paulo, Saõ Carlos, Brasil, implanted our passive 3C–SiC shanks fabricated in Section 7.3.1 alongside metal wire implants (Fig. 7.17). The passive shanks

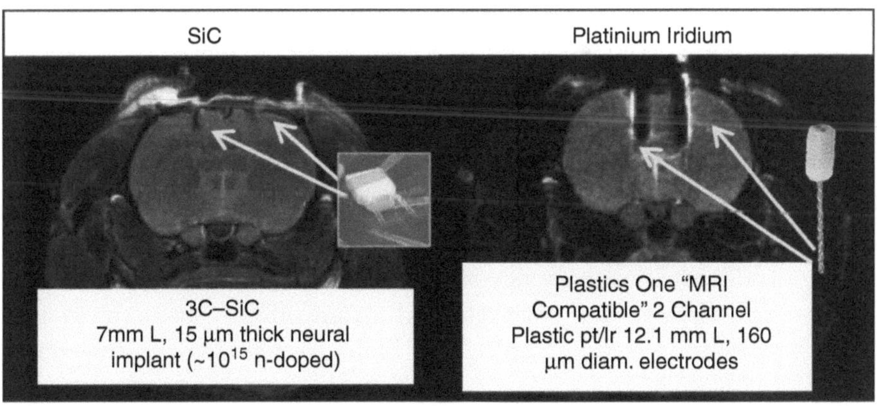

Figure 7.17 MRI images of two Sprague Dawley rats implanted with (a) dual passive probes constructed from 3C–SiC from Section 7.3.1, and (b) commercially obtained Pt/Ir alloy, dual wire MRI compatible electrodes. The images show that in contrast to 3C–SiC probes, the Pt/Ir probes are coupling with the magnetic field due to the bright image artifacts seen at the implant locations. *(Data courtesy of Dr J. Malheiros of the University of Saõ Paulo, Saõ Carlos, Brasil.)*

were generated from "as-grown" 3C–SiC, which were unintentionally doped to $\sim 10^{15}$ N_2 dopants/cm^{-3}. Commercially available, "MRI compatible" two-channel electrodes were obtained from Plastics One®. The devices were implanted in Sprague Dawley rats and the rats were placed in a 2T MRI-generated field.

The results for the metal followed prediction and published observations, as the electrons within the material coupled with the magnetic field producing a bright image artifact (Fig. 7.16 right side). The 3C–SiC, on the other hand, did not produce any noticeable coupling image artifacts. Although we had no means during the experiment to gather thermal data, the results indicate that 3C–SiC did not strongly inductively couple to the magnetic field and, therefore, should not also produce any appreciable joule heating. This may open a whole new avenue for 3C–SiC as a permanently implantable device biomaterial as implanted individuals would not have the MRI constraints required of metal-based implants.

7.4.2 Ultralow Power Recording Electronics for Wireless BMI Applications

BMI has evolved with great speed in recent years, and promised a plethora of therapeutic treatments for millions with neural damage. Unfortunately, complex BMI systems have yet to show long-term reliability, which has limited clinical testing and follow-on commercial application. Ultimately, the reliability issue has prevented BMI from assisting individuals with neurological and physical disabilities.

In order for BMI to effectively make use of the advances in robotic and computer technology, it is vital to find biological interfaces that meet the requirements of long-term electrical reliability, low-power operation, and biocompatibility. In this section we focus on two of these long-term electrical issues. First, many system components are based on a battery for power delivery, and these batteries eventually discharge and must be replaced, resulting in additional maintenance requirements. Second, many systems rely on materials that may not have reliable biocompatibility, and can possibly erode with time, demanding costly and high-risk replacement procedures [10,74].

We present a possible wireless solution for implantable microelectrode neural interfaces, which could be incorporated into a complete BMI system (Fig. 7.18). The first version of the Cortex chip, a four-channel amplifier that magnifies neural signals for processing by a microprocessor system, works within the constraints of very low-power consumption and

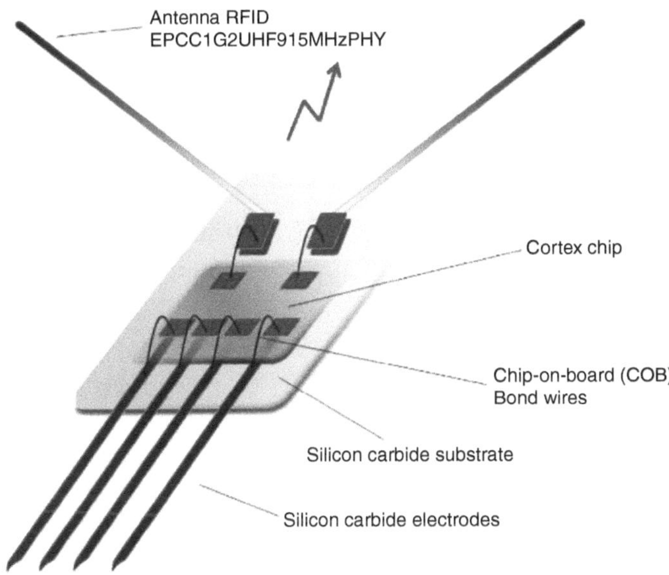

Figure 7.18 ***Conceptual diagram of the proposed BMI system, which integrates the Cortex chip, described in detail in this paper, with active 3C–SiC electrodes and an RFID wireless interface [104].***

biocompatibility. The Cortex chip relies on two features: (1) we use a chip designed to operate with a power consumption that is small enough for it to be powered by radio frequency antennas, dispensing with the need for batteries; (2) we use a probe manufactured from silicon carbide (SiC), a material that demonstrated excellent neural compatibility with murine mouse brain tissue *in vivo* (Section 7.3.1).

Many neural signal amplifiers have been developed since 2000s [110–113], but none of them meets our proposed needs. Many authors have implemented simple, operational transconductance amplifiers (OTAs) for neural signal amplification, which have mainly been based on older technologies (1500–500 nm). Improved modern designs incorporate advanced technologies, such as 0.18 μm complementary metal-oxide semiconductor (CMOS), leading to lower power consumption and smaller die area, both of which are important for implantable neural interfaces.

The recent designs unfortunately also come with a few drawbacks. Many designs [113] completely avoid the use of on-chip, and even off-chip, capacitors, which would assist to minimize power consumption. This design decision forces the use of DC coupling instead of AC coupling at the amplifier input, leading to DC signal component elimination a posteriori, which can be an issue because there might be a loss of data due to the hardware. Many others simply choose to use very large transistors [110] at the input to reduce flicker noise, which appears at very low frequencies, but this design increases the power consumption even when using more advanced technologies.

7.4.2.1 Silicon Carbide (3C–SiC) Neural Interface—The Cortex Chip

An ideal implantable BMI not only includes material neural compatibility, but it should also include the following: very low power consumption to operate without batteries, broadband data transmission, small area encapsulation with no off-chip components, and avoid high-power radio frequency operation, which is below the absorption limits of biological tissues. With these requirements in mind, we designed a special purpose OTA, which is fully discussed in the literature [104]. The amplifier has two stages, each with 14.14 of voltage gain, and an output buffer for test purposes. The gain stages, noted as gm1 and gm2, are comprised of a fully differential OTA, featuring high open loop gain, low-input referred noise, and high power supply rejection ratio (PSRR).

An off-the-shelf solution was incorporated for the wireless capability of the Cortex chip. We chose the 915 MHz PHY CM9011ff [114] from Chipus Microelectronics [104]; this choice was guided by the fact that its operation, in the range from ~1 to 3 GHz, minimizes the channel loss for edge-to-edge coupling in biological strata. In this solution, the data stream will be up to 800 kbps, with four channels using a 10-bit 20 kHz analog-to-digital conversion, in accordance to the maximum specific absorption rate (1.6 W/kg) defined by the Federal Communications Commission. In Fig. 7.19 is a photograph of the manufactured Cortex chip die (a) and packaged chip (b). Tables 7.2 and 7.3 present a summary of the results obtained for the Cortex chip in comparison with previously published results for similar circuits. Table 7.3 shows that our chip has the following advantages: better PSRR and power consumption performance, smaller area per channel, and AC instead of DC coupling.

7.4.2.2 Future Perspective—Cortex Chip

The Cortex chip, a low-power neural signal amplifier, has been fabricated using a 0.18-μm CMOS process and incorporates all the electronics integrated into an area of 0.40 mm^2. The chip was designed with four channels, and its total power consumption was only 144 μW. It was designed to match the impedance of the SiC electrodes, thus enabling better neural compatibility of the entire system. Furthermore, it is integrated with a RFID PHY

(a) (b)

Figure 7.19 (a) Cortex die photo (version with internal test pads) and (b) cortex chip encapsulated in a QFN64 surface mount package for device integration and electrical testing [104].

for wireless communication. The results showed that both local field potentials, in the range of 5 mV, and action potentials, around 100 μV, can be sampled by the system's probe without distortion. The Cortex chip has demonstrated promising results as an integrated circuit for brain–machine

Table 7.2 Summary of cortex chip performance

Regulated supply voltage	1.8 V typ. (1.6 V_{min}–2.0 V_{max})
Regulator PSRR	80 dB
Neural signal coupling	AC
Neural signal amplifier gain	46 dB (0.2–10 kHz)
Input-referred noise	36 μV_{rms} (0.2–10 kHz)
Number of channels	4
Bandwidth	0.2–10 kHz
Process	SILTERRA, CMOS 0.18 μM
Total chip area	0.447 mm^3 (0.111 mm^3/channel)
Total power	144 μW (36 μW/channel)

Table 7.3 Improved features of cortex chip

Feature	Other OTA	Cortex chip
Power (per channel)	80 μW/channel [10]	36 μW/channel
Regulator PSRR	5.5 dB [11]	64 dB
Chip area (per channel)	0.160 mm^2/channel [12]	0.111 mm^2/channel
Neural signal coupling	DC [13]	AC

interfacing, and the next stage is to incorporate 3C–SiC microelectrode probes with the chip to create an implant like the one displayed in Fig. 7.18.

7.5 CONCLUSIONS

Many materials have been used successfully to create viable, long-term implants that have provided therapeutic relief to millions of people, including orthopedic implants and dental prosthetics. However, when electrical activity is included in the device, there are many more issues that occur. Often, many of these electrical devices require adjustment, or they even fail, due to the response of the body. In this chapter, we discussed using silicon carbide as an electrical material for use in implantable devices. We examined the corrosion of *a*-SiC and 3C–SiC using physiological media and found no appreciable effect, either in the short term or in the long term through accelerated aging techniques. Consequently, 3C–SiC was tested *in vivo* in the brain of mice, and both 3C–SiC and *a*-SiC were evaluated subcutaneously in pigs. There was no appreciable immune response with either material in either animal. We demonstrated that 3C–SiC did not illicit an inductive coupling response within a rat brain when exposed to a 2T MRI system. In the end, the device is more than just a material, and we have begun combining 3C–SiC and *a*-SiC with electronics, notably the Cortex chip by Chipus. This chip will not only allow SiC electrodes to record neural signals, but it also wirelessly transmits the signals. Another advantage to this chip is that its very low power consumption allows it to receive power wirelessly, thereby eliminating the need for external or internal wiring to batteries. The advantage of 3C–SiC is evident, and in the next chapter, further advantages of *a*-SiC will become more apparent as we demonstrate its utility as a coating for neural implantation devices.

ACKNOWLEDGMENTS

The authors would like to thank Dr Christopher Locke and Dr Meralys Reyes for the development and growth of the 3C–SiC used to create the neural implantation devices in this section. Additionally, we thank Dr Joseph Register and Richard Everly for the development and growth of the *a*-SiC coatings. We would like to thank Dr Weeber for his contributions as a mentor of neuroscience to Dr Frewin, Dr Justin Trotter for assistance with the immunohistochemistry and microscope image processing, and to Dr Jennifer Daily for assistance with the neurological *in vivo* implantation surgery. Finally, we thank Dr Mario Gazziro of the UFABC, Santo André, Brasil, for his pioneering work on BMI electronics and the Brazilian funding agency CAPES, process number 400694/2012-3 (Science without Borders), for financial support for the development of the Cortex Chip. Dr Saddow is particularly grateful to Dr Gazzio and CAPEs for support during his 3-year Science without Borders project at USP in Saõ Carlos, Brasil. Funding for the neural implantation was provided in part by the National Institute of Health's Ruth L. Kirschstein National Research Service Awards (NRSA) for Individual Postdoctoral Fellows for Dr Frewin.

REFERENCES

[1] Misch CE. Contemporary implant dentistry. St. Louis, MO: Mosby Elsevier; 2007.
[2] Goodacre CJ, Bernal G, Rungcharassaeng K, Kan JY. Clinical complications with implants and implant prostheses. J Prosthet Dent 2003;90:121–32.
[3] A.U. Association. Milestones in Urology. Available from: http://www.urologichistory.museum/content/milestones/catheterization/p1.cfm; 2005.
[4] Gomez PF, Morcuende JA. Early attempts at hip arthroplasty: 1700s to 1950s. Iowa Orthopaed J 2005;25:25–9.
[5] Nuss KMR, von Rechenberg B. Biocompatibility issues with modern implants in bone – a review for clinical orthopedics. Open Orthop J 2008;2:66–78.
[6] Nacey J, Delahijnt B. The evolution and development of the urinary catheter. Aust NZ J Surg 1993;63:815–9.
[7] Franklin B. In: Pringle J, editor. Electric shocks in paralytic cases. London, Great Britain: Royal Society of Arts; 1757.
[8] Hodgkin AL, Huxley AF. A quantitative description of membrane current and its application to conduction and excitation in nerve. J Physiol Lond 1952;117:500–44.
[9] Stensaas SS, Stensaas LJ. Reaction of cerebral-cortex to chronically implanted plastic needles. Acta Neuropathol 1976;35:187–203.
[10] Rousche PJ, Normann RA. Chronic recording capability of the Utah intracortical electrode array in cat sensory cortex. J Neurosci Methods 1998;82:1–15.
[11] Biran R, Martin DC, Tresco PA. Neuronal cell loss accompanies the brain tissue response to chronically implanted silicon microelectrode arrays. Exp Neurol 2005;195:115–26.
[12] Polikov VS, Tresco PA, Reichert WM. Response of brain tissue to chronically implanted neural electrodes. J Neurosci Methods 2005;148:18.
[13] Grill WM, Norman SE, Bellamkonda RV. Implanted neural interfaces: biochallenges and engineered solutions. Annu Rev Biomed Eng 2009;11:1–24.
[14] Wise KD, Sodagar AM, Yao Y, Gulari MN, Perlin GE, Najafi K. Microelectrodes, microelectronics, and implantable neural microsystems. Proc IEEE 2008;96:1184–202.
[15] Cogan SF. Neural stimulation and recording electrodes. Annu Rev Biomed Eng 2008;10:275–309.
[16] Fournier E, Passirani C, Montero-Menei CN, Benoit JP. Biocompatibility of implantable synthetic polymeric drug carriers: focus on brain biocompatibility. Biomaterials 2003;24:3311–31.
[17] Prasad A, Xue QS, Sankar V, Nishida T, Shaw G, Streit WJ, et al. Comprehensive characterization and failure modes of tungsten microwire arrays in chronic neural implants. J Neural Eng 2012;9:056015.
[18] Lee K, Singh A, He JP, Massia S, Kim B, Raupp G. Polyimide based neural implants with stiffness improvement. Sens Actuators B 2004;102:67–72.
[19] Lee KK, He JP, Singh A, Massia S, Ehteshami G, Kim B, et al. Polyimide-based intracortical neural implant with improved structural stiffness. J Micromech Microeng 2004;14:32–7.
[20] Watkins S, Gandhi D, Rousche PJ. Biocompatibility of polyimide-based neural interfaces for chronic implant applications. J Undergrad Res 2007;1:36–40.

[21] Massey LK. Permeability properties of plastics and elastomers: a guide to packaging and barrier materials. 2nd ed Norwich, NY: William Andrew Publishing/Plastics Design Library; 2003.
[22] Minnikanti S, Diao G, Pancrazio JJ, Xie X, Rieth L, Solzbacher F, et al. Lifetime assessment of atomic-layer-deposited Al2O3–parylene C bilayer coating for neural interfaces using accelerated age testing and electrochemical characterization. Acta Biomater 2014;10:960–7.
[23] Kuhn M, Silversm DJ. Ionic contamination and transport of mobile ions in MOS structures. J Electrochem Soc 1971;118:966.
[24] Beaudoin F, Simard-Normandin M, Meunier M. Metallic contamination from wafer handling. In: Gupta DC, Bacher F, Hugh WH, editors. Silicon recombination lifetime characterization methods. West Conshohocken, PA: American Society for Testing Materials; 1988. p. 219–25.
[25] Snow EH, Grove AS, Deal BE, Sah CT. Ion transport phenomena in insulating films. J Appl Phys 1965;36:1664.
[26] Kuhn M, Silversmith DJ. Ionic contamination and transport of mobile ions in MOS structures. Solid State Sci 1971;118:5.
[27] Osburn CM, Raider SI. Effect of mobile sodium ions on field enhancement dielectric breakdown in SiO_2 films on silicon. J Electrochem Soc 1973;120:1369–76.
[28] Raider SI, Gregor LV, Flitsch R. Transfer of mobile ions from aqueous-solutions to silicon dioxide surface. J Electrochem Soc 1973;120:425–31.
[29] Carlisle EM. Silicon as an essential trace-element in animal nutrition. Ciba Found Symp 1986;121:123–39.
[30] Hench LL, Wilson J. Biocompatibility of silicates for medical use. In: Evered D, O'Connor M, editors. Ciba Foundation Symposium – silicon biochemistry, vol. 121. Chichester, UK: John Wiley & Sons; 1986. p. 231–53.
[31] O'Neil C, Jordan P, Bhatt T, Newman R. Silica and oesophageal cancer. In: Evered D, Connor M, editors. Ciba Foundation Symposium – silicon biochemistry, vol. 121. Chichester, UK: John Wiley & Sons; 1986. p. 214–30.
[32] Birchall JD, Chappell JS. The chemistry of aluminum and silicon in relation to Alzheimers disease. Clin Chem 1988;34:265–7.
[33] Birchall JD, Exley C, Chappell JS, Phillips MJ. Acute toxicity of aluminum to fish eliminated in silicon-rich acid waters. Nature 1989;338:146–8.
[34] Birchall JD. The role of silicon in biology. Chem Brit 1990;26:141–4.
[35] Birchall JD. The interrelationship between silicon and aluminum in the biological effects of aluminum. Ciba Found Symp 1992;169:50–68.
[36] Edell DJ, Toi VV, McNeil VM, Clark LD. Factors influencing the biocompatibility of insertable silicon microshafts in cerebral cortex. IEEE Trans Biomed Eng 1992;39:635–43.
[37] Fahal IH, Ahmad R, Bell GM, Birchall JD, Roberts NB. Profile of serum silicon in aluminum-overloaded patients on regular hemodialysis-treatment. J Anal Atom Spectrom 1993;8:911–3.
[38] Birchall JD. Silicon–aluminum interactions and biology. Colloid Chem Silica 1994;234:601–15.
[39] Bayliss SC, Buckberry LD, Harris PJ, Tobin M. Nature of the silicon-animal cell interface. J Porous Mater 2000;7:191–5.
[40] Hammerle H, Kobuch K, Kohler K, Nisch W, Sachs H, Stelzle M. Biostability of micro-photodiode arrays for subretinal implantation. Biomaterials 2002;23:797–804.
[41] Angelescu A, Kleps I, Mihaela M, Simion M, Neghina T, Petrescu S, et al. Porous silicon matrix for applications in biology. Rev Adv Mater Sci 2003;5:440–9.
[42] Nablo BJ, Prichard HL, Butler RD, Klitzman B, Schoenfisch MH. Inhibition of implant-associated infections via nitric oxide. Biomaterials 2005;26:6984–90.

[43] Coletti C, Jaroszeski MJ, Pallaoro A, Hoff AM, Iannotta S, Saddow SE. Biocompatibility and wettability of crystalline SiC and Si surfaces. Conf Proc IEEE Eng Med Biol Soc 2007;2007:5850–3.

[44] Frewin CL, Jaroszeski M, Weeber E, Muffly KE, Kumar A, Peters M, et al. Atomic force microscopy analysis of central nervous system cell morphology on silicon carbide and diamond substrates. J Mol Recognit 2009;22:380–8.

[45] Saddow SE, editor. Silicon carbide biotechnology: a biocompatible semiconductor for advanced biomedical devices and applications. Amsterdam: Elsevier; 2011.

[46] ISO, 10993-1. Biological evaluation of medical devices. Part 1: evaluation and testing, ed. Geneva, Switzerland: International Organization for Standardization; 2009.

[47] ISO, 10993-14. Biological evaluation of medical devices. Part 14: identification and quantification of degradation products from ceramics, ed. Switzerland: ISO Copyright Office; 2001.

[48] Celina M, Gillen KT, Assink RA. Accelerated aging and lifetime prediction: review of non-Arrhenius behaviour due to two competing processes. Polym Degrad Stabil 2005;90:10.

[49] Hukins DWL, Mahomed A, Kukureka SN. Accelerated aging for testing polymeric biomaterials and medical devices. Med Eng Phys 2008;30:1270–4.

[50] Locke CW, Frewin CL, Saddow SE. 3C-SiC epitaxial growth for advanced biomedical applications. In: La Via F, editor. Silicon Carbide Epitaxy. Kerala, India: Research Signpost; 2012. p. 225–58.

[51] Reyes M, Frewin C, Ward PJ, Saddow SE. 3C-SiC on Si hetero-epitaxial growth for electronic and biomedical applications. ECS Trans 2013;58:119–26.

[52] Zhang DQ, Alkhateeb A, Han HM, Mahmood H, McIlroy DN, Norton MG. Silicon carbide nanosprings. Nano Lett 2003;3:983–7.

[53] Jaeger RC. Introduction to microelectronic fabrication, vol. V. New Jersey: Prentice Hall; 2002.

[54] Frewin CL, Coletti C, Riedl C, Starke U, Saddow SE. A comprehensive study of hydrogen etching on the major SiC polytypes and crystal orientations. Mater Sci Forum 2009;615–617:589–92.

[55] Frewin CL, Locke C, Saddow SE, Weeber EJ. Single-crystal cubic silicon carbide: an *in vivo* biocompatible semiconductor for brain machine interface devices. In: Engineering in Medicine and Biology Society, EMBC, 2011 Annual international conference of the IEEE; 2011. p. 2957–2960.

[56] Coletti C, Jaroszeski MJ, Pallaoro A, Hoff AM, Iannotta S, Saddow SE. Biocompatibility and wettability of crystalline SiC and Si surfaces. In: Engineering in Medicine and Biology Society, 2007. EMBS 2007. 29th annual international conference of the IEEE; 2007. p. 5849–5852.

[57] Goodwin JL, Kehrli ME Jr, Uemura E. Integrin Mac-1 and β-amyloid in microglial release of nitric oxide. Brain Res 1997;768:279–86.

[58] Pocock JM, Liddle AC. Microglial signalling cascades in neurodegenerative disease. In: Lopez MNS, Castellano B, editors. Progress in Brain Research, vol. 132. Elsevier; 2001. p. 555–65.

[59] Wink DA. The chemical mechanisms in regulatory, cytotoxic, and cytoprotective roles of nitric oxide, Abstracts of papers of the American Chemical Society, vol. 215; 1998. p. U360–U360.

[60] Wink DA, Mitchell JB. Chemical biology of nitric oxide: insights into regulatory, cytotoxic, and cytoprotective mechanisms of nitric oxide. Free Radic Biol Med 1998;25:434–56.

[61] Block ML, Hong JS. Microglia and inflammation-mediated neurodegeneration: multiple triggers with a common mechanism. Prog Neurobiol 2005;76:77–98.

[62] Pestovsky O, Bakac A. Nitrous acid as a source of NO and NO_2 in the reaction with a macrocyclic superoxorhodium(III) complex. Inorg Chem 2002;41:901–5.

[63] Park JY, Lee YN. Solubility and decomposition kinetics of nitrous-acid in aqueous-solution. J Phys Chem 1988;92:6294–302.
[64] Schwartz B, Robbins H. Chemical etching of silicon. J Electrochem Soc 1961;123:1903–9.
[65] ISO, 10993-15. Biological evaluation of medical devices. Part 15: identification and quantification of degradation products from metals and alloys, ed. Switzerland: ISO Copyright Office; 2000.
[66] Coletti C, Jaroszeski MJ, Hoff AM, Saddow SE. Culture of mammalian cells on single crystal SiC substrates. MRS Proceedings, vol. 950. Cambridge University Press; 2006. p. 0950-D04.
[67] Nezafati N, Saddow SE, Frewin CL. Investigating the surface changes of silicon *in vitro* within physiological environments for neurological application. Presented at the MRS, Boston, MA; 2014.
[68] Nezafati N, Saddow SE, Frewin CL. Investigating the surface changes of silicon *in vitro* within physiological environments for neurological application. In: MRS Proceedings, Boston, MA, USA; 2014. p. 237–242.
[69] Williams KR, Gupta K, Wasilik M. Etch rates for micromachining processing – Part II. J Microelectromech Syst 2003;12:761–78.
[70] Iler RK. The chemistry of silica. Solubility, polymerization, colloid and surface properties, and biochemistry. New York/Chichester/Brisbane/Toronto: John Wiley & Sons; 1979.
[71] Vidal JJ. Toward direct brain-computer communication. Annu Rev Biophys Bioeng 1973;2:157–80. Mullins LJ, editor. Palo Alto, CA: Annual Reviews, Inc.
[72] Donoghue JP. Bridging the brain to the world: a perspective on neural interface systems. Neuron 2008;60:511–21.
[73] Lebedev MA, Nicolelis MAL. Brain-machine interfaces: past, present and future. Trends Neurosci 2006;29:536–46.
[74] Schmidt EM, Bak MJ, McIntosh JS. Long-term chronic recording from cortical neurons. Exp Neurol 1976;52:496–506.
[75] Winslow BD, Christensen MB, Yang WK, Solzbacher F, Tresco PA. A comparison of the tissue response to chronically implanted parylene-C-coated and uncoated planar silicon microelectrode arrays in rat cortex. Biomaterials 2010;31(35):9163–72.
[76] Bogdan C. Nitric oxide and the immune response. Nat Immunol 2001;2:907–16.
[77] Nablo BJ, Schoenfisch MH. *In vitro* cytotoxicity of nitric oxide-releasing sol-gel derived materials. Biomaterials 2005;26:4405–15.
[78] Long TC, Tajuba J, Sama P, Saleh N, Swartz C, Parker J, et al. Nanosize titanium dioxide stimulates reactive oxygen species in brain microglia and damages neurons *in vitro*. Environ Health Perspect 2007;115:1631–7.
[79] Frewin CL, Locke C, Saddow SE, Weeber, EJ. Single-crystal cubic silicon carbide: an *in vivo* biocompatible semiconductor for brain machine interface devices. In: Engineering in Medicine and Biology Society, EMBC, 2011 annual international conference of the IEEE, Boston, MA; 2011. p. 2957–2960.
[80] Saddow SE, Agarwal A. Advances in silicon carbide processing and applications. Boston, MA: Artech House; 2004.
[81] Asaad W, Eskandar E. The movers and shakers of deep brain stimulation. Nat Med 2008;14:17–9.
[82] ISO, 10993-1. Biological evaluation of medical devices. Part 1: evaluation and testing, ed. Geneva, Switzerland: International Standards Organization; 2009.
[83] ISO, 10993-5. Biological evaluation of medical devices. Part 5: tests for *in vitro* cytotoxicity, ed. Geneva, Switzerland: International Standards Organization; 2009.
[84] Frewin CL, Locke C, Saddow SE, Weeber EJ. Single-crystal cubic silicon carbide: an *in vivo* biocompatible semiconductor for brain machine interface devices. In: Engineering in Medicine and Biology Society, EMBC, 2011 annual international conference of the IEEE, Boston, MA; 2011. p. 2957–2960.

[85] Locke C, Anzalone R, Severino A, Bongiorno C, Litrico G, La Via F, et al. High quality single crystal 3C-SiC(111) films grown on Si(111). Mater Sci Forum 2009;615–617:145–8.

[86] Locke C, Kravchenko G, Waters P, Reddy JD, Du K, Volinsky AA, et al. 3C-SiC films on Si for MEMS applications: mechanical properties. Mater Sci Forum 2009;615–617:633–6.

[87] Laermer F, Schilp A. Method of anisotropic etching of silicon, US patent; 1999.

[88] Cosenza-Nashat MA, Kim MO, Zhao ML, Suh HS, Lee SC. CD45 isoform expression in microglia and inflammatory cells in HIV-1 encephalitis. Brain Pathol 2006;16:256–65.

[89] Kettenmann H, Hanisch UK, Noda M, Verkhratsky A. Physiology of microglia. Physiol Rev 2011;91:461–553.

[90] Hanisch UK, Kettenmann H. Microglia: active sensor and versatile effector cells in the normal and pathologic brain. Nat Neurosci 2007;10:1387–94.

[91] Botsoa J, Lysenko V, Géloën A, Marty O, Bluet JM, Guillot G. Application of 3C-SiC quantum dots for living cell imaging. Applied Physics Letters 2008;92:173902.

[92] Severino A, Frewin C, Bongiorno C, Anzalone R, Saddow SE, La Via F. Structural defects in (100) 3C-SiC heteroepitaxy: influence of the buffer layer morphology on generation and propagation of stacking faults and microtwins. Diam Relat Mater 2009;18:1440–9.

[93] Dalsin JL, Hu BH, Lee BP, Messersmith PB. Mussel adhesive protein mimetic polymers for the preparation of nonfouling surfaces. J Am Chem Soc 2003;125: 4253–8.

[94] Tsai WB, Grunkemeier JM, McFarland CD, Horbett TA. Platelet adhesion to polystyrene-based surfaces preadsorbed with plasmas selectively depleted in fibrinogen, fibronectin, vitronectin, or von Willebrand's factor. J Biomed Mater Res 2002;60: 348–59.

[95] Shen M, Horbett TA. The effects of surface chemistry and adsorbed proteins on monocyte/macrophage adhesion to chemically modified polystyrene surfaces. J Biomed Mater Res 2001;57:336–45.

[96] Hong-Sub S, Kwideok P, Ji HK, Jae-Jin K, Dong KH, Myoung-Woon M, et al. Biocompatible PEG grafting on DLC-coated nitinol alloy for vascular stents. J Bioact Compat Polym 2009;24(4):316–28.

[97] Li X, Wang X, Bondokov R, Morris J, An YH, Sudarshan TS. Micro/nanoscale mechanical and tribological characterization of SiC for orthopedic applications. J Biomed Mater Res B Appl Biomater 2005;72:353–61.

[98] Kalnins U, Erglis A, Dinne I, Kumsars I, Jegere S. Clinical outcomes of silicon carbide coated stents in patients with coronary artery disease. Med Sci Monit 2002;8: PI16–20.

[99] Santavirta S, Takagi M, Nordsletten L, Anttila A, Lappalainen R, Konttinen YT. Biocompatibility of silicon carbide in colony formation test *in vitro*. A promising new ceramic THR implant coating material. Arch Orthop Trauma Surg 1998;118: 89–91.

[100] Afroz S, Thomas SW, Mumcu G, Saddow SE. Implantable SiC based RF antenna biosensor for continuous glucose monitoring. Baltimore, MD: IEEE Sensors; 2013.

[101] Afroz S. A biocompatible SiC RF antenna for *in-vivo* sensing applications, PhD, Electrical Engineering, University of South Florida; 2013.

[102] Thomas SW, Afroz S, Saddow SE, Muffly K, Hoyte L. *In-vivo* evaluation of subcutaneous histological tissue reaction to SiC based continuous glucose sensor. J Histotechnol; In Press, 2016.

[103] Frewin CL, Abbati L, Weeber EJ, Saddow SE. SiC for brain–machine interface (BMI). In: Saddow SE, editor. Silicon carbide biotechnology: a biocompatible semiconductor for advanced biomedical devices and applications. Amsterdam: Elsevier; 2011.

[104] Gazziro M, Braga CFR, Moreira DA, Carvalho ACPLF, Rodrigues JF, JNavarro, J.S. et al. Transmission of wireless neural signals through a 0.18 μm CMOS low-power amplifier. Presented at the thirty-seventh IEEE Engineering in Medicine and Biology Society, Milan, Italy; 2015.
[105] Register J. SiC for advanced biological applications. PhD, Electrical Engineering, University of South Florida; 2014.
[106] Sadleir RJ, Grant SC, DeMarse TB, Woo EJ, Lee SY, Kim TS, et al. Study of MRI/MEA compatibility at 17.6 tesla. Int J Bioelectromagnet 2005;7:1–4.
[107] Rezai AR, Finelli D, Nyenhuis JA, Hrdlicka G, Tkach J, Sharan A, et al. Neurostimulation systems for deep brain stimulation: *in vitro* evaluation of magnetic resonance imaging-related heating at 1.5 tesla. J Magn Reson Imaging 2002;15:241–50.
[108] Elwassif MM, Kong Q, Vazquez M, Bikson M. Bio-heat transfer model of deep brain stimulation-induced temperature changes. J Neural Eng 2006;3:306–15.
[109] Tronnier VM, Staubert A, Hahnel S, Sarem-Aslani A. Magnetic resonance imaging with implanted neurostimulators: an *in vitro* and *in vivo* study. Neurosurgery 1999;44: 118–25. discussion 125–126.
[110] Harrison RR, Charles C. A low-power low-noise CMOS amplifier for neural recording applications. IEEE J Solid-State Circuits 2003;38:958–65.
[111] Holleman J, Otis B. A sub-microwatt low-noise amplifier for neural recording. 2007 annual international conference of the IEEE engineering in medicine and biology society. vols. 1–16; 2007. p. 3930–3933.
[112] Wattanapanitch W, Fee M, Sarpeshkar R. An energy-efficient micropower neural recording amplifier. IEEE Trans Biomed Circ Syst 2007;1:136–47.
[113] Muller R, Gambini S, Rabaey JM. A 0.013 mm(2), 5 mu W, DC-coupled neural signal acquisition IC with 0.5 V supply. IEEE J Solid-State Circuits 2012;47:232–43.
[114] C.M. Company. Product Outline: CM911ff – UHF Passive RFID Front-End, EPCGlobal Class 1 Gen 2/ISO18000-6C Compliant. Available from: http://www.chipus-ip.com/pdf/CM9011ff_outline_rev1v0.pdf; 2012.

CHAPTER 8

Amorphous Silicon Carbide for Neural Interface Applications

Gretchen L. Knaack*, Hamid Charkhkar, Stuart F. Cogan† and Joseph J. Pancrazio†**

*U.S. Food and Drug Administration, Center for Devices and Radiological Health, Office of Science and Engineering Laboratories, Division of Biophysics, Silver Spring, MD, United States of America

**Electrical and Computer Engineering Department, Volgenau School of Engineering, George Mason University, Fairfax, VA, United States of America

†Department of Bioengineering, Erik Jonsson School of Engineering & Computer Science, University of Texas at Dallas, Richardson, TX, United States of America

Contents

8.1 INTRODUCTION

Implantable microelectrode arrays (MEAs) offer the promise of restoring function to individuals with paralysis or amputation [1,2], of restoring vision to the blind [3,4], and of treating neurological diseases [5]. MEAs typically consist of microwire structures coated with polymeric encapsulation or silicon (Si) and thin-film polymer structures fabricated through methods common to the semiconductor device industry. While promising, neural signals recorded with these devices degrade over time and occasionally device failure is observed such that the reliability of this approach has been the subject of recent attention [6-8]. Most MEA devices are fabricated from some combination of noble metals/alloys, silicon, polymers (eg, parylene-C, silicones, polyimide), and conventional dielectric passivation (silicon oxide or silicon nitride). Silicon carbide (SiC) has emerged as a promising material for biomedical devices, which may convey advantages for MEAs. The purpose of this chapter is to discuss mechanisms involved in the degradation of MEA performance, the role of material choice in the tissue response, the biological response of neural tissue to amorphous silicon carbide (*a*-SiC) both *in vitro* and *in vivo*, and *a*-SiC as an encapsulant for MEA devices.

8.2 BIOTIC AND ABIOTIC MECHANISMS OF DEVICE FAILURE

It is believed that implanted MEA failure results, at least in part, from the brain's immune response, which activates after penetration of these devices through the blood–brain barrier [9,10]. Specifically, implantation of the device causes shearing of vasculature and disturbances to the extracellular matrix, glia, and neurons [11,12]. The wound-healing process starts immediately after the initial injury by recruiting macrophages as well as activating microglia [10,12-16]. Microglia usually exist in a ramified state until activated by breaking the blood–brain barrier [9,10,12,14,17]. Once activated, they extend processes toward the probe, exhibit amoeboid morphology, and express leukocyte-associated molecules such as CD68, which make them virtually indistinguishable from macrophages [12,17]. Activated microglia also release numerous inflammatory, cytotoxic, and neurotoxic chemicals [12]. They further exacerbate the response by secreting factors that recruit additional microglia/macrophages and astrocytes [12,15,18].

Similar to microglia, a reactive state is also induced for astrocytes, which is primarily characterized by upregulation of glial fibrillary acidic protein (GFAP) [9,10,12,15,19]. Although many investigators have reported migration and proliferation of activated astrocytes to the probe site [9,12,20], more recent live imaging did not reveal any movement of these cells and only proliferation near vasculature, but did confirm elongated processes and hypertrophy [19]. After about 4 weeks of implantation, the acute inflammatory reaction transitions into a chronic response characterized by the formation of a glial scar that encapsulates the device. The glial scar is composed primarily of reactive astrocytes, although activated microglia can also be observed [12,14-16,18,20,21]. This tissue sheath isolates the probe and is believed to increase electrical impedance [12,22,23]. This is paired with a significant loss of neurons proximal to the device [10,14]. However, the extent of this reaction varies across animals and is dependent on the size [24], shape, and insertion rate of the implant [11]. Although some elements of this response appear well characterized, others remain unclear, and a clear link between the tissue response and device performance remains to be drawn [12].

8.3 ROLE OF THE MATERIAL CHOICE IN THE TISSUE RESPONSE

There is a growing recognition that the physical and chemical properties of the device material can influence the biotic response and ultimately the device performance and longevity [25]. For example, polyimide-insulated tungsten microwire arrays chronically implanted in rats demonstrated insulation damage due to excessive oxidative stress leading to deterioration in the recording sites [22]. Furthermore, delamination of multilayer structures that constitute these implantable technologies likely contributes to device failure. For example, mechanically applied stress in parylene C generates microcracks and additional material failures arise from blistering/pore formation [26], which exposes the underlying materials to the external environment.

The material can also contribute to the biotic response, possibly potentiated with degradation processes occurring at small scales. Many of the commercially available micromachined probes are fabricated from silicon, which has the advantage of well-developed fabrication techniques, developed in the semiconductor industry to allow large-scale production of highly reproducible structures. Interestingly, silicon has been linked to cytotoxic reactions in multiple studies [7,10,27,28]. Several studies report structural degradation in commonly used materials like silicon both *in vitro* [29] and *in vivo* [22,30]. It is conceivable that the tissue reaction to a degrading material, even at small scales, could potentiate inflammatory cascades to further degrade device performance.

A hermetic coating can limit water and ion penetration of implanted materials and prevent outward dissolution of materials [31,32]. SiC is a robust material that is able to withstand high temperatures and adverse chemical conditions [32]. SiC may be produced in several different crystallographic forms, with different material properties as well as an amorphous thin film. *a*-SiC is particularly attractive as a passivation layer to protect implantable devices because of its excellent water and ion barrier properties, suitability for coating a variety of substrate materials at modestly low temperatures (<350°C) by plasma-enhanced chemical vapor deposition (PECVD), good adhesion to Si [32], low dissolution rate, and biocompatibility [31,33]. It has been used as a protective coating for orthopedic implants [34] and for coronary stents [35], where *a*-SiC is considered to reduce thrombogenicity and improve biocompatibility [36].

8.4 *IN VITRO* "NEUROCOMPATIBILITY" OF *a*-SiC

In vitro cytotoxicity testing is a vital, initial step in assessing the biocompatibility of potential implantable materials. Regulatory guidance is provided by the international standard ISO 10993, which explicitly defines testing methodologies for material screening, which begins for *in vitro* cytotoxicity testing. Consistent with the standard, prior studies concerned with implantable neurological devices have made use of mammalian fibroblasts (eg, [37,38]). In fact, it has been well established that fibroblasts tolerate silicon carbide consistent with a material that demonstrates *in vitro* biocompatibility [35,39,40]. However, as discussed in ASTM standard F2901 (2013), which pertains to the assessment of neurotoxicity of medical devices, there are limitations in the use of nonneural-derived cells for *in vitro* testing of neurological implants. Consistent with [41], F2901 recommends the use of cells derived from nervous tissue to provide a more comprehensive assessment.

The use of cortical neuronal networks cultured directly on planar MEAs has been used as a tool for assessing neuropharmacology [42–44] as well as *in vitro* material biocompatibility [45]. Neural tissue derived from primary dissection of rodents can be cultured to form functional, spontaneously active networks where action potentials, or spikes, can be monitored noninvasively (Fig. 8.1).

Spontaneous spike activity in cultured neuronal networks exhibits varying degrees of coordination across multiple neurons, or units [46–48], and is largely influenced by functional synapses [46,48,49]. With modulation of spike activity associated with exposure to a putative neuroactive compound, this approach may be a valuable tool to complement preclinical testing of therapeutics [49–51] as well as to screen for environmental toxins [43,44,52]. The technique is exquisitely sensitive to environmental perturbations, which may not be cytotoxic but certainly influence neural activity. For an application involving neural interfaces, which attempt to record activity from neuronal processes in close proximity ($\leq 100\ \mu m$) to recording microelectrode sites, it is critical that the material not only lacks cytotoxicity, but also does not diminish neural activity.

By adapting the ISO 10993-5 guidance for amount and duration of material extract exposure, the neuronal network methodology was demonstrated for a range of materials including conductors, gold and copper, as well as insulators, polyethylene (PE), and organotin containing polyvinyl chloride (PVC). As an example, the multiunit neural activity for representative

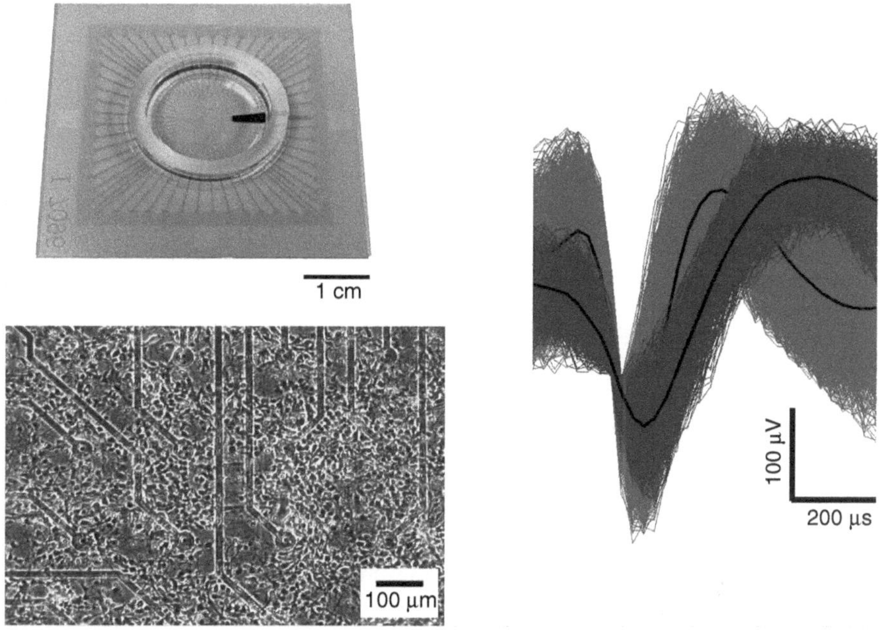

Figure 8.1 ***Commercially available planar MEAs from ALA Scientific Instruments have 60 microelectrodes with a diameter of 10 μm and interelectrode spacing of 200 μm.*** The primary neuronal culture method [42] was used to generate frontal cortex networks exhibiting stable multiunit activity after 21 days *in vitro*.

networks exposed for 24 h to extracts from appropriately sized PE and PVC samples is shown in Fig. 8.2. PVC, which is considered a positive control material, induced a significant reduction in spike activity. Data can be readily quantified by measuring the mean spike rate across each network [53], and normalizing to the corresponding baseline levels measured before exposure to the material sample.

Using this testing methodology, we examined three different packaging variations of SiC: (1) *a*-SiC deposited on Si; (2) cubic 3C–SiC on Si; and (3) *a*-SiC on 3C–SiC on Si. As shown in Fig. 8.3, none of the three variations of SiC impacted neuronal activity across the networks. A one-way analysis of variance determined that there was no significant difference in normalized spike rate with exposure to the materials.

This finding is largely consistent with the prior *in vitro* cytotoxicity assessments for SiC variants [35,54]. Furthermore, the processes and reagents used for *a*-SiC deposition and fabrication are unlikely to generate residues that would be toxic to neurons or produce alterations in neural activity.

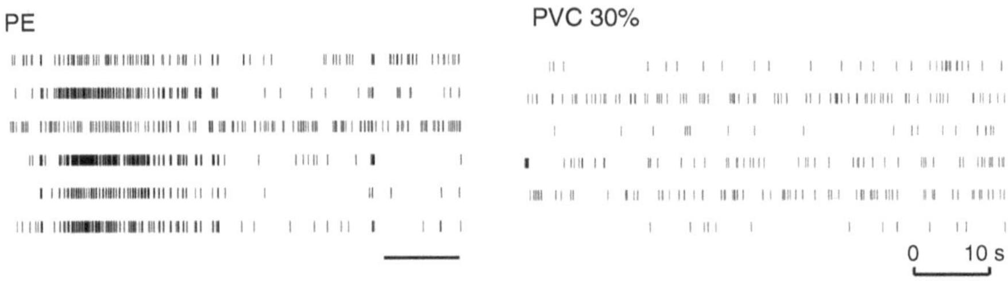

Figure 8.2 ***Raster plots showing spike activity for neuronal networks exposed to negative and positive control materials, PE and PVC.*** Extracts of materials were prepared by incubating 1 mL serum-free cell culture media for every 3 cm^2 of material sample in 35-mm diameter polystyrene Petri dishes at 37°C for 24 h under constant agitation using a laboratory shaker. The PVC extract was diluted to 30% by addition of control media; 100% extracts from the PVC completely eliminated neuronal network activity. Control extracts, which also underwent the same procedures except they were incubated with material-free cell culture media, failed to elicit any change in neuronal activity.

8.5 *IN VIVO* TISSUE RESPONSE TO *a*-SiC-COATED PROBES

Prior work by [31] showed *a*-SiC-coated microwires elicited a modest effect on the tissue viability and neuronal density after long-term implantation in the rabbit parietal cortex. Since the neuroinflammatory response is comprised of changes in multiple cell types, it is important to consider potential effects on astrocytes and microglia, as well as on neurons. Single shank, non-functional probes were either left uncoated or coated with 0.5 μm *a*-SiC using PECVD (Fig. 8.4). These test structures have dimensions typical of implantable probes, that is, 5 mm in length, 15 μm thick, and 200 μm in width at the base with the widest part of the shank in contact with cortical tissue.

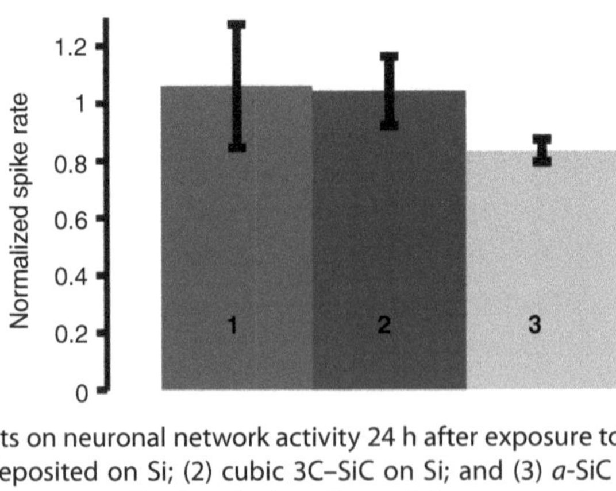

Figure 8.3 Effects on neuronal network activity 24 h after exposure to extracts derived from (1) *a*-SiC deposited on Si; (2) cubic 3C–SiC on Si; and (3) *a*-SiC on 3C–SiC on Si. The spike rates were normalized to the baseline activity measured prior to adding the extracts. Data are shown as mean normalized spike rate and SEM for $n = 3$ networks.

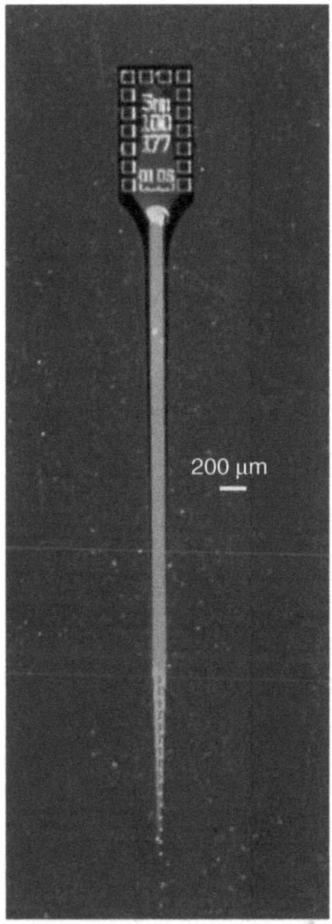

Figure 8.4 ***Photomicrograph showing nonfunctional NeuroNexus probes used for testing the tissue response resulting from 4-week implantation in the rat motor cortex.*** *(Courtesy Ms Amy Davis of NeuroNexus Technologies [Ann Arbor, MI].)*

To investigate the chronic response to *a*-SiC, implants remained in the brain for 4 weeks. Fig. 8.5 shows comparative immunohistochemically stained horizontal slices derived from the same rat implanted with a control Si probe and an *a*-SiC-coated probe. Both materials induce a loss of neurons, based on NeuN intensity proximal to the probe insertion site. Moreover, there is a visually apparent decrease in GFAP intensity, indicative of decreased levels of activated astrocytes, near the insertion site for the *a*-SiC-coated implant compared to the Si control probe. Labeling for CD68 appears to be similar for both Si control and *a*-SiC-coated probes. To process these slices, the probes were removed from the brain after sacrifice, which invariably disrupts the tissue proximal to the implant. A more comprehensive analysis would

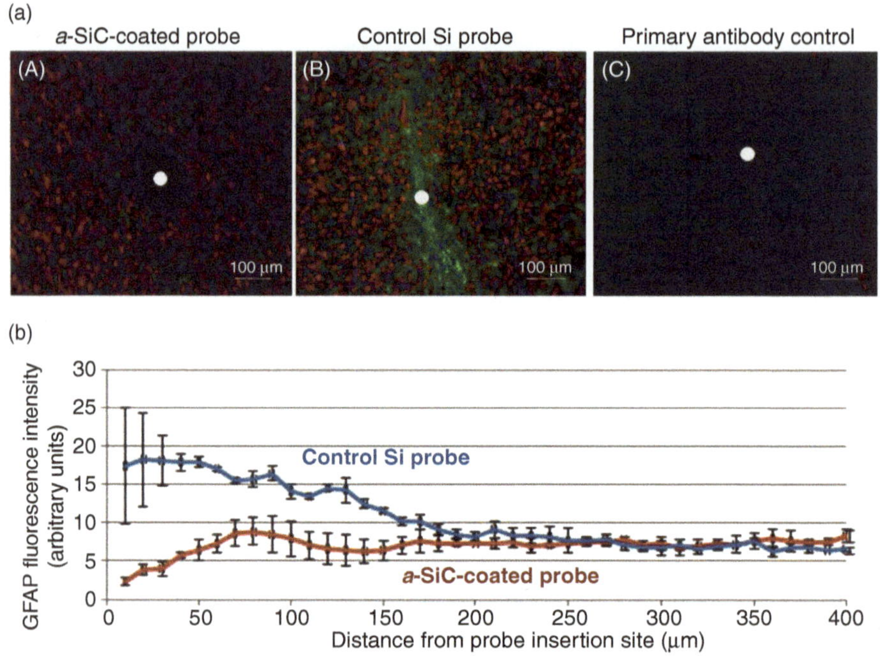

Figure 8.5 (a) Images from deep cortical tissue labeling NeuN in red (neurons), GFAP in green (astrocytes), and DAPI in blue (cell nuclei). White circles indicate center of probe locations. (A) is from tissue implanted with an *a*-SiC-coated probe, (B) from a control Si probe, and the (C) is the no primary control indicating little or no nonspecific antibody binding. (b) Immunohistochemistry summary data comparing results from an *a*-SiC-coated (red) and a control Si (blue) probes after implantation for 4 weeks. Data are mean ± SEM for $n = 2$ slices.

rely on longer implantation durations and the use of probe capturing techniques [21] whereby sagittal slices of the implanted cortical tissue are prepared for immunohistochemical analysis with the probe remaining in place.

8.6 SUMMARY

The findings from the *in vitro* and *in vivo* studies suggest that *a*-SiC is sufficiently tolerated by central nervous system tissue to serve as a passivation layer for implanted neural devices. Test structures coated with *a*-SiC exhibit a reduced inflammatory effect based on the decreased level of activated astrocyte labeling along the shank after 4 weeks of implantation. As an encapsulation layer to create a hermetic and insulating barrier between a device and the surrounding biological milieu, *a*-SiC has features that offer promise for implantable neural interfaces [31,55]. These features include, but are not limited to, the ability to withstand high temperatures and adverse chemical conditions. The material acts as a diffusion barrier, requires a relatively low temperature for deposition, and exhibits a slow dissolution rate.

There are some limitations to the use of *a*-SiC deposited by the PECVD process. Future MEAs are likely to have complex three-dimensional structures that will require conformal coatings with good step coverage. It is yet to be demonstrated that this can be achieved with PECVD *a*-SiC while preserving the required barrier properties and biocompatibility. In addition, the fabrication process of multilayered devices necessitates management of the different coefficients of thermal expansion as well as intrinsic stresses. Alterations in hydrogen content can influence intrinsic stress and the coefficient of thermal expansion of *a*-SiC deposited by PECVD [56], allowing physical properties of the material to be tailored to intended applications.

It is important to recognize that the value of *a*-SiC coating is not limited to stiff Si structures which exhibit a modulus of approximately 10 GPa. There is a body of research indicating that probe flexibility may reduce the mechanical mismatch that exists between brain tissue and the implant to ultimately diminish the chronic inflammatory response. Capadona and coworkers have demonstrated a bioinspired, mechanically dynamic nanocomposite material for intracortical probes [57,58]. Implantation of microelectrodes, which are comprised of the nanocomposite, exhibit a modulus that rapidly transitions from 5 GPa to 12 MPa upon implantation in rat brain [57]. These devices showed a diminished inflammatory response compared to a stiff wire [57], and result in unappreciable neuron loss at the device–tissue interface as long as 4 months after implantation [59]. Shape memory polymer technology may offer a manufacturable route to creating intracortical implants that soften with insertion into the brain to minimize the mechanical mismatch [60, 61]. Since *a*-SiC can be deposited at low temperatures through PEVCD, thin coatings of this versatile material may be applied to a variety of flexible substrates. The combination may produce structures which are less susceptible to corrosion and degradation without significantly affecting device stiffness. The result would be a novel class of implant technologies with improved biocompatibility.

ACKNOWLEDGMENTS

This work was sponsored by the Defense Advanced Research Projects Agency (DARPA) MTO under the auspices of Dr Jack Judy through the Space and Naval Warfare Systems Center, Pacific Grant/Contract No. N66001-12-1-4026.

REFERENCES

[1] Hochberg LR, Serruya MD, Friehs GM, Mukand JA, Saleh M, Caplan AH, Branner A, Chen D, Penn RD, Donoghue JP. Neuronal ensemble control of prosthetic devices by a human tetraplegia. Nature 2006;442:164–71.

[2] Velliste M, Perel S, Spalding MC, Whitford AS, Schwartz AB. Cortical control of a prosthetic arm for self-feeding. Nature 2008;453:1098–101.

[3] Lorach H, Goetz G, Mandel Y, Lei X, Kamins TI, Mathieson K, Huie P, Dalal R, Harris JS, Palanker D. Performance of photovoltaic arrays *in-vivo* and characteristics of prosthetic vision in animals with retinal degeneration. Vision Res 2015;111(Pt B):142 148.

[4] Normann RA, Greger B, House P, Romero SF, Pelayo F, Fernandez E. Toward the development of a cortically based visual neuroprosthesis. J Neural Eng 2009;6:035001.

[5] Park Yun S, et al. Early detection of human epileptic seizures based on intracortical local field potentials. IEEE Conference: 6th International IEEE EMBS Conference on Neural Engineering (NER). San Diego, CA; 2013.

[6] Kane SR, Cogan SF, Ehrlich J, Plante TD, McCreery DB, Troyk PR. Electrical performance of penetrating microelectrodes chronically implanted in cat cortex. IEEE Trans Biomed Eng 2013;60(8):2153–60.

[7] Rousche PJ, Normann RA. Chronic recording capability of the Utah intracortical electrode array in cat sensory cortex. J Neurosci Methods 1998;82:1–15.

[8] Williams JC, Rennaker RL, Kipke DR. Long-term neural recording characteristics of wire microelectrode arrays implanted in cerebral cortex. Brain Res Protoc 1999;4:303–13.

[9] Fawcett JW, Asher RA. The glial scar and central nervous system repair. Brain Res Bull 1999;49:377–91.

[10] Winslow BD, Tresco PA. Quantitative analysis of the tissue response to chronically implanted microwire electrodes in rat cortex. Biomaterials 2010;31:1558–67.

[11] Bjornsson CS, et al. Effects of insertion conditions on tissue strain and vascular damage during neuroprosthetic device insertion. J Neural Eng 2006;3:196–207.

[12] Polikov VS, Tresco PA, Reichert WM. Response of brain tissue to chronically implanted neural electrodes. J Neurosci Methods 2005;148:1–18.

[13] Kozai TD, Vazquez AL, Weaver CL, Kim SG, Cui XT. *In vivo* two-photon microscopy reveals immediate microglial reaction to implantation of microelectrode through extension of processes. J Neural Eng 2012;9:1–17.

[14] Potter KA, Buck AC, Self WK, Capadona JR. Stab injury and device implantation within the brain results in inversely multiphasic neuroinflammatory and neurodegenerative responses. J Neural Eng 2012;9:1–14.

[15] Stice P, Muthuswamy J. Assessment of gliosis around moveable implants in the brain. J Neural Eng 2009;6:1–10.

[16] Woolley AJ, Desai HA, Otto KJ. Chronic intracortical microelectrode arrays induce non-uniform, depth-related tissue responses. J Neural Eng 2013;10:1–11.

[17] Kozai TD, et al. *In vivo* two-photon microscopy reveals immediate microglial reaction to implantation of microelectrode through extension of processes. J Neural Eng 2012;9:1–17.

[18] Szarowski DH, et al. Brain response to micro-machined silicon devices. Brain Res 2003;983:23–35.

[19] Bardehle S, et al. Live imaging of astocyte response to acute injury reveals selective juxtavascular proliferation. Nat Neurosci 2013;16:580–6.

[20] Fitch MT, Doller C, Combs CK, Landreth GE, Silver J. Cellular and molecular mechanisms of glial scarring and progressive cavitation: *in vivo* and *in vitro* analysis of inflammation-induced secondary injury after CNS trauma. J Neurosci 1999;19:8182–98.

[21] Woolley AJ, Desai HA, Steckbeck MA, Patel NK, Otto KJ. In situ characterization of the brain-microdevice interface using device capture histology. J Neurosci Methods 2011;201:67–77.

[22] Prasad A, Xue Q-S, Sankar V, Nishida T, Shaw G, Streit WJ, Sanchez JC. Comprehensive characterization and failure modes of tungsten microwire arrays in chronic neural implants. J Neural Eng 2012;9(5):056015.
[23] Wang C, et al. Characteristics of electrode impedance and stimulation efficacy of a chronic cortical implant using novel annulus electrodes in rat motor cortex. J Neural Eng 2013;10:1–19.
[24] Stice P, Gilletti A, Muthuswamy J. Thin microelectrodes reduce GFAP expression in the implant site in rodent somatosensory cortex. J Neural Eng 2007;4:42–53.
[25] Fournier E, Passirani C, Montero-Menei CN, Benoit JP. Biocompatibility of implantable synthetic polymeric drug carriers: focus on brain biocompatibility. Biomaterials 2003;24(19):3311–31.
[26] Massey LK. Permeability properties of plastics and elastomers: a guide to packaging and barrier materials. Norwich, NY, USA: Plastics Design Library/William Andrew Pub; 2002.
[27] Hench LL, Wilson J. Biocompatibility of silicates for medical use. In: Evered D, O'Connor M, editors. Ciba Foundation Symposium – silicon biochemistry. Chichester, UK: John Wiley & Sons; 1986.
[28] O'Neil C, Jordan P, Bhatt T, Newman R. Silica and oesophageal cancer. In: Evered D, O'Connor M, editors. Ciba Foundation Symposium – silicon biochemistry. Chichester, UK: John Wiley & Sons; 1986.
[29] Alexander GB, Heston WM, Iler RK. The solubility of amorphous silica in water. J Phys Chem 1954;58:453–9.
[30] Barrese JC, Rao N, Paroo K, Triebwasser C, Vargas-Irwin C, Franquemont L, Donoghue JP. Failure mode analysis of silicon-based intracortical microelectrode arrays in non-human primates. J Neural Eng 2013;1(0):1–23.
[31] Cogan SF, Edell DJ, Guzelian AA, Liu YP, Edell R. Plasma-enhanced chemical vapor deposited silicon carbide as an implantable dielectric coating. J Biomed Mater Res A 2003;67(3):856–67.
[32] Zorman CA. Silicon carbide as a material for biomedical microsystems. DTIP of MEMS & MOEMS; 2009.
[33] Iliescu C, Chen B, Poenar DP, Lee YY. PECVD amorphous silicon carbide membranes for cell culturing. Sensor Actuat B Chem 2008;129(1):404–11.
[34] Sella C, Lecoeur J, Sampeur Y, Catania P. Corrosion resistance of amorphous hydrogenated SiC and diamond-like coatings deposited by rf plasma-enahanced chemical vapor deposition. Surf Coat Tech 1993;60:577–83.
[35] Amon M, Bolz A, Schaldach M. Improvement of stenting therapy with a silicon carbide coated tantalum stent. J Mat Sci Mater Med 1996;7:273–8.
[36] Hansi C, Arab A, Rzany A, Ahrens I, Bode C, Hehrlein C. Differences of platelet adhesion and thrombus activation on amorphous silicon carbide, magnesium alloy, stainless steel, and cobalt chromium stent surfaces. Catheter Cardiovasc Interv 2009;73(4):488–96.
[37] Lee K, Singh A, He J, Massia S, Kim B, Raupp G. Polyimide based neural implants with stiffness improvement. Sens Actuators B 2004;102:67–72.
[38] Massia SP, Holecko MM, Ehteshami GR. *In vitro* assessment of bioactive coatings for neural implant applications. J Biomed Mater Res 2003;68A:177–86.
[39] Kotzar G, Freas M, Abel P, Fleischman A, Roy S, Zorman C, Moran JM, Melzak J. Evaluation of MEMS materials of construction for implantable medical devices. Biomaterials 2002;23(13):2737–50.
[40] Naji A, Harmand MF. Cytocompatibility of two coating materials, amorphous alumina and silicon carbide, using human differentiated cell cultures. Biomaterials 1991;12(7):690–4.
[41] Harry GJ, Billingsley M, Bruinink A, Campbell IL, Classen W, Dorman DC, Galli C, Ray D, Tilson HA. *In vitro* techniques for the assessment of neurotoxicity. Environ Health Perspect 1998;106(Suppl. 1):131–58.
[42] Knaack GL, Charkhkar H, Hamilton FW, Peixoto N, O'Shaughnessy TJ, Pancrazio JJ. Differential responses to ω-agatoxin IVA in murine frontal cortex and spinal cord derived neuronal networks. Neurotoxicology 2013;37:19–25.

[43] Pancrazio JJ, Gray SA, Shubin YS, Kulagina DS, Shaffer KM, Eisemann K, et al. A portable microelectrode array recording system incorporating cultured neuronal networks for neurotoxin detection. Biosens Bioelectron 2003;18:1339–47.

[44] Pancrazio JJ, Keefer EW, Ma W, Stenger DA, Gross GW. Neurophysiological effects of chemical agent hydrolysis products on cortical neurons *in vitro*. Neurotoxicology 2001;22:393–400.

[45] Charkhkar H, Frewin C, Nezafati M, Knaack GL, Peixoto N, Saddow SE, Pancrazio JJ. Use of cortical neuronal networks for *in vitro* material biocompatibility testing. Biosens Bioelectron 2014;53:316–23.

[46] Gross GW, Azzazy HME, Wu MC, Rhoades BK. The use of neuronal networks on multielectrode arrays as biosensors. Biosens Bioelectron 1995;10:553–67.

[47] Selinger JV, Pancrazio JJ, Gross GW. Measuring synchronization in neuronal networks for biosensor applications. Biosens Bioelectron 2004;19:675–83.

[48] Wagenaar DA, Pine J, Potter SM. An extremely rich repertoire of bursting patterns during the development of cortical cultures. BMC Neurosci 2005;7:11–29.

[49] Johnstone AFM, Gross GW, Weiss DG, Schroeder OHU, Gramowski A, Shafer TJ. Microelectrode arrays: a physiological based neurotoxicity testing platform for the 21st century. Neurotoxicology 2010;31:331–50.

[50] Shimono K, Baudry M, Panchenko V, Taketani M. Chronic multichannel recordings from organotypic hippocampal slice cultures: protection from excitotoxic effects of NMDA by non-competitive NMDA antagonists. J Neurosci Methods 2002;120:193–202.

[51] Xia Y, Gopal KV, Gross GW. Differential acute effects of fluoxetine on frontal and auditory cortex networks *in vitro*. Brain Res 2003;973:151–60.

[52] Kulagina NV, O'Shaughnessy TJ, Ma W, Ramsdell JS, Pancrazio JJ. Pharmacological effects of the marine toxins, brevetoxin and saxitoxin, on murine frontal cortex neuronal networks. Toxicon 2004;44:669–76.

[53] McConnell ER, McClain MA, Ross J, LeFew WR, Shafer TJ. Evaluation of multi-well microelectrode arrays for neurotoxicity screening using a chemical training set. Neurotoxicology 2012;33(5):1048–57.

[54] Mahmoodi M, Ghazanfari L. Fundamentals of biomedical applications of biomorphic SiC. In: Gerhardt R, editor. Properties and applications of silicon carbide. In-Tech; 2011. p. 297–345.

[55] Hsu JM, Tathireddy P, Rieth L, Normann AR, Solzbacher F. Characterization of *a*-SiC(x):H thin films as an encapsulation material for integrated silicon based neural interface devices. Thin Solid Films 2007;516(1):34–41.

[56] King SW, Ross L, Li H, Xu GH, Bielefeld J, Atkins RE, Henneghan PD, Davis K, Johnson DC, Lanford WA. Influence of hydrogen content and network connectivity on the coefficient of thermal expansion and thermal stability for *a*-SiC:H thin films. J Non-Crystal Solids 2014;389:78–85.

[57] Harris JP, Hess AE, Rowan SJ, Weder C, Zorman CA, Tyler DJ, Capadona JR. *In vivo* deployment of mechanically adaptive nanocomposites for intracortical microelectrodes. J Neural Eng 2011;8(4):046010.

[58] Hess AE, Capadona JR, Shanmuganathan K, et al. Development of a stimuli-responsive polymer nanocomposite toward biologically optimized, MEMS-based neural probes. J Micromechan Microeng 2011;21(5). Article Number: 054009.

[59] Nguyen JK, Park DJ, Skousen JL, Hess-Dunning A, Tyler DJ, Rowan SJ, Weder C, Capadona JR. Mechanically-compliant intracortical implants reduce the neuroinflammatory response. J Neural Eng 2014;11:056014.

[60] Ware T, Simon D, Arreaga-Salas DE, Reeder J, Rennaker R, Keefer EW, Voit W. Fabrication of responsive, softening neural interfaces. Adv Funct Mater 2012;22(16):3470–9.

[61] Ware T, Simon D, Liu C, Musa T, Vasudevan S, Sloan A, Keefer EW, Rennaker RL, Voit W. Thiol-ene/acrylate substrates for softening intracortical electrodes. J Biomed Mater Res Part B 2014;102(1):1–11.

CHAPTER 9

SiC Nanowire-Based Transistors for Electrical DNA Detection

Edwige Bano*, Louis Fradetal, Maelig Ollivier*,†, Ji-Hoon Choi*,† and Valérie Stambouli‡**
*IMEP-LAHC Minatec, Grenoble, France
**IMEP-LAHC and LMGP Minatec, Grenoble, France
†LTM-CNRS Minatec, Grenoble, France
‡LMGP Minatec, Grenoble, France

Contents

9.1 INTRODUCTION

It is well known that DNA is the molecule which carries genetic codes. The sensitive and selective detection of DNA sequences is central in modern bioanalysis. The fast and direct detection of small quantities of DNA is important in many fields. In the healthcare field, DNA biomarkers enable early medical diagnosis of certain serious diseases such as cancer [1]. Elsewhere, DNA detection is used in forensic science as well as in the food industry, environmental protection, and biodefense to evidence the presence of pathogenic viruses or genetically modified organisms [2]. For all these application fields, in situ, rapid, and easy methods to detect and quantify DNA are necessary. Presently, the usual method for the detection of nucleic acid sequences relies on sophisticated labeling fluorescent detection technology [2]. This method requires one to tag the target DNA sequence, which is then detected by a fluorophore. Upon the specific recognition with DNA probes, which are immobilized on a solid surface, a fluorescent signal is measured. However, this fluorescence method suffers from several drawbacks: it is centralized in laboratories, it is time consuming, and its cost is high due to complex and expensive optical detection instrumentation.

Alternatively, label-free optical and nonoptical methods such as electrical methods are under extensive development, notably to perform direct, low-cost, and in situ detection [3]. In this case, and in the absence of fluorescent visualization, the detection signal relies entirely on a variation of a physical parameter, such as electrical conductance or capacitance, of the supporting material, which, in turn, should be thoroughly elaborated and optimized in order to produce the maximum signal possible.

Nanowire-based field effect transistors (NWFETs) have been under intense development during the past decade in view of electrical label-free biosensing. Indeed, they can achieve very sensitive (up to fm), real time, label-free electrical detection of several kinds of charged biochemical species, such as DNA, proteins, and ions [4–6]. Most of these NWFETs are based on silicon (Si) [6], mainly due to their compatibility with the microelectronics industry.

Other nanowire (NW) materials offering excellent electronic properties, as well as inherent chemical robustness, nontoxicity, and biocompatibility, are under investigation. For instance, some recent studies focus on NWFET devices based on GaN NWs [7], doped graphene [8,9], or carbon nanotubes (NTs) [10,11] in view of DNA detection. Silicon carbide (SiC) is another promising semiconductor for these applications. Its wide bandgap (2.36 eV for the 3C–SiC polytype) and relatively high electron mobility (900 $cm^2/V\,s$ for 3C–SiC) allow for the realization of efficient field effect transistors (FETs) [12,13]. Mostly due to its chemical inertness, SiC is a highly stable and biocompatible material [14]. SiC is already used for many biomedical applications ranging from the coating for implants [15,16] to myocardial heart probes, bone prosthetics, coronary heart stents, and biomimetic structures [17]. SiC is also being developed as an electrode for brain–machine interfaces [18] and to enhance cellular reproduction in tissue regeneration [19]. Coupling both NW and SiC properties, SiC NWFETs could potentially be an excellent candidate for nanoelectronics in harsh or biological environments [20–22].

In the field of sensing and detection, SiC is already used under several morphologies according to the species to be detected and the way of detection. Planar SiC FETs are used to achieve electrical detection of gases [23–25]. SiC nanoparticles are used for *in vivo* pH measurement using fluorescence [26]. Yang et al. used nanocrystalline SiC electrodes for electrochemical biosensing applications [27]. Williams et al. selectively functionalized SiC NWs for protein immobilization but without reaching the step of electrical detection of bioanalytes [28].

On this basis, our aim is to fabricate SiC NWFETs and to evaluate their potential in order to envisage their use as label-free electrical DNA biosensors. This chapter reports the fabrication of SiC NWFETs for DNA biosensing involving DNA grafting and hybridization. Early results are reported on the electrical detection of DNA grafting and hybridization.

To realize SiC NW-based devices, such as NWFETs, it is essential to develop single crystalline SiC at the nanoscale. One-dimensional (1D) nanomaterials have a great interest for nanoelectronic devices because of their remarkable physical and electrical properties as well as their high surface-to-volume ratio. Semiconductor NWs have been the topic of extensive research as critical building blocks for many applications, such as electronics [29], photonic devices [30], and sensors [31], due to their more interesting electrical and optical properties in contrast to their bulk material forms. However, it is clear that SiC 1D nanostructures are not easy to produce without any structural defects whatever the elaboration process used. Currently, a common technique for the growth of SiC NWs is the bottom-up approach, such as vapor–liquid–solid (VLS) [32] or vapor–solid (VS) methods [33]. Critical problems of these methods are that the as-grown SiC NWs present a high density of structural defects, such as stacking faults, and unintentional high n-type doping levels. These characteristics lead to low electrical performance (such as weak gate effects and low mobility) of the related devices and limit their practical applications.

In this chapter, first we report two alternative methods to achieve SiC NW growth, such as VLS or VS, that will be detailed in chapter: Silicon Carbide Based Nanowires for Biomedical Applications. Here, using the top-down approach, these alternative methods are the inductively coupled plasma (ICP) etching method and the carburization method. This section will be followed by a description of the technological process that we have developed to build SiC NWFETs.

Then, we report the DNA functionalization process that we have developed on the various as-grown SiC nanostructures. The validation of

DNA functionalization and hybridization on these nanostructures has been achieved using characterization techniques such as fluorescent microscopy, X-ray photoelectron spectroscopy (XPS), and atomic force microscopy (AFM). Finally we report the preliminary results on electrical DNA detection using SiC NWFETs.

9.2 ELABORATION OF SiC NANOSTRUCTURES

In this section, we focus on two innovative methods that we developed to obtain high-quality SiC 1D nanostructures for their integration into nanoFETs. Both are based on a top-down approach using an ICP etching process. As this process offers a highly anisotropic profile, high etch rate, and less etch damage compared with other dry etching methods, it is considered as a possible solution to achieve high crystalline nanostructures with low structural defects and controlled doping level.

The first method gives rise to SiC nanopillars (NPs) obtained by ICP etching of a SiC epitaxial layer. The second method provides Si–SiC core–shell NWs by carburation of etched Si NWs in a chemical vapor deposition (CVD) reactor.

Physical characterization techniques such as AFM, scanning electron microscope (SEM), dual focused ion beam–scanning electron microscope (FIB-SEM), and transmission electron microscopy (TEM) are used to characterize and optimize the grown nanostructures.

9.2.1 SiC Nanopillars by Inductively Coupled SF_6/O_2 Plasma Etching

9.2.1.1 Introduction

The technology of single crystalline SiC growth has been developed over the past few years and has shown commercial availability of SiC substrates of ever-increasing diameter and quality [17]. Therefore, if the top-down approach is used on a single crystalline SiC substrate, it is possible to achieve highly crystalline SiC nanostructures with low structural defects with controlled doping levels. Moreover, it also permits one to control precisely the geometry of the nanostructures, such as diameter and density.

Due to the strong bond energy between the silicon and carbon atoms (4.25 eV) in the SiC lattice, plasma-based dry etching, such as ICP etching, is a promising method for top-down processing of SiC since it can be carried out at much higher plasma density ($>10^{11}$ cm^{-3}) than in reactive ion etching (RIE) systems ($>10^{9}$ cm^{-3}) [34], and with a lower chamber pressure (<10 mTorr compared to 10–300 mTorr). High plasma density allows higher sputtering rates, which are able to break Si–C bonds and the operation at low pressure improves the directionality of the ion flux, which enables the achievement of highly anisotropic etch profiles. Moreover, the plasma density and sputtering energy can be independently controlled, thereby increasing flexibility in optimizing the etch process [35,36].

Many efforts have been made in order to etch SiC with plasma methods, mainly for the patterning of microelectromechanical systems (MEMS) applications [36,37], or the realization of via holes or trenches for power microwave devices [38,39]. Two papers have reported the fabrication of nanometer-scale SiC structures by the top-down approach. The first one [40] has shown nanoelectromechanical systems devices based on top-down SiC NWs with a diameter of 55 nm by etching heteroepitaxially grown 3C–SiC layers on Si. However, the growth of SiC on Si substrate leads to a low crystalline quality due to the large lattice mismatch (almost 20%) and the difference of thermal expansion coefficient (almost 8%) between the 3C–SiC layer and the Si substrate. The second study [41] has reported NP structures of SiC obtained by dry ICP-RIE with a self-assembled silicon dioxide (SiO_2) mask. The maximum achieved pillar height cannot exceed 600 nm due to the low selectivity of SiO_2 with respect to SiC. In addition, self-assembled methods could not fully control pitch distance, mask diameter, and mask materials. Therefore, it is still worthwhile investigating the etching of features for nanoscale single crystalline SiC pillars using a dry etching method from the practical application viewpoint.

In the following section, SiC NP ICP etching is reported using an SF_6/O_2-based gas mixture. The etching profile evolution has been studied according to the etch process parameters.

9.2.1.2 Experiment

Experiments were carried out in a commercial high density plasma-etching chamber from Applied Materials Inc., which has been presented elsewhere [42]. The wafers used in this study were 8 degrees off-axis 4H–SiC(0001), which is fabricated by physical vapor transport (PVT) methods [43]. The doping level of the 4H–SiC wafer was 10^{17} cm^{-3} with n-type. We fabricated an additional marker to identify the crystallographic orientation of [11–20], while the (0001) direction is tilted to 8 degrees from the surface normal. Many different fluorinated-based gases, such as SF_6 [36,38,44], NF_3 [34], and CHF_3 [45], with the addition of O_2 have been considered to achieve high-rate ICP etching of SiC. Among these, we have employed SF_6 (sulfur hexafluoride) for SiC etching due to the relatively obtained high etch rate [44].

Regarding the etch masks, two different sizes of circular patterns (115 and 370 nm diameters) were prepared with the same pitch distance (7 μm) using electron beam lithography (JEOL JSM-7401F). Various metal (Ni, Al, and Cu) masks of 110 nm thickness were deposited on the Si face of the SiC substrate by e-beam evaporation. The patterned samples were loaded on a He backside-cooled chuck biased with 13.56 MHz of power. Total gas flow rate, ICP coil power, substrate bias voltage, chamber pressure, and etching times remained constant at 50 sccm, 1500 W, 150 V, 6 mTorr, and 180 s, respectively, unless otherwise mentioned. SEM was used to examine the morphology of etched samples, while initial mask pattern size and thickness after lift-off processing measured by tapping-mode AFM, Veeco Dimension 3100.

9.2.1.3 Results and Discussion

Effect of Metal Mask Species

A masking material with high selectivity with respect to SiC in the SF_6/O_2 etching process is needed to achieve maximum height of the SiC NPs. To examine suitable mask materials, ICP etching was carried out at 1500 W of ICP coil power with an etching gas mixture of SF_6:O_2 of 40:10 sccm with different metal masks (Cu, Ni, and Al) for an etching time of 3 min.

SiC NPs with the Cu mask show a roughness on the sidewalls (see Fig. 9.1a). This might be attributed to the irregularly formed etch by-products on the mask. These products are formed during the etching process via Cu reacting with F or O radicals. These nonvolatile reaction products (Cu_2F or Cu_2O) formed on the mask exhibit high etch selectivity and sputter-resistant properties, which is why Cu has been evaluated as the best candidate materials for the pattern mask materials of MEMS devices [37]. However, Cu metal masks are not suitable for etching at the nanoscale. Indeed, irregularly formed products on the mask modify the mask geometry, which induces roughness on the sidewalls of etched pillars (Fig. 9.1a). SiC pillars with an Al mask presented a higher roughness on the sidewalls than that with the Cu mask along the vertical direction (Fig. 9.1c). It might be due to mask damage caused by energetic ion bombardment. Ni masks are much more sputter resistant and protect the SiC beneath the mask without irregularly forming a volatile metal–fluoride layer. Consequently, etching profiles of SiC NPs processed with Ni masks show clear sidewall surfaces on the NPs with uniform mask shape (Fig. 9.1b).

Fig. 9.1f shows a high density of residue (grass-like structure) on the substrate, which is believed to be due to a micromasking effect caused by the redeposition of sputtered material from the Al mask [46]. The mask material

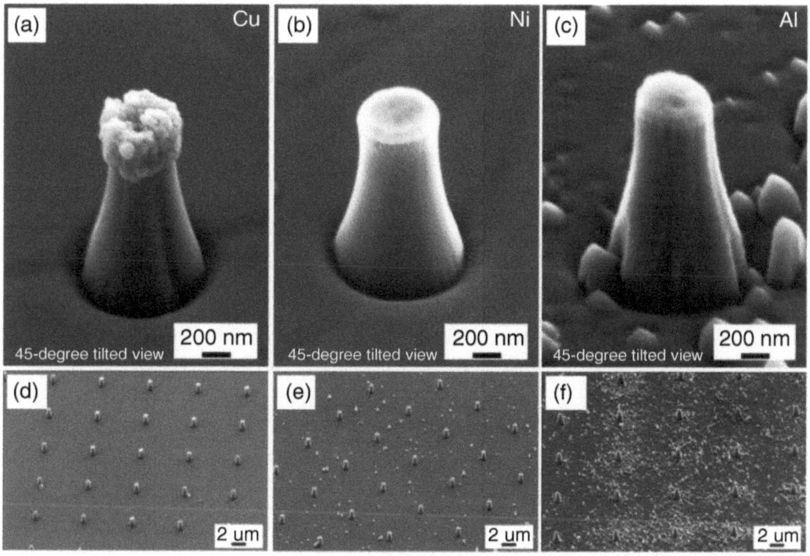

Figure 9.1 ***SEM images of SiC NPs with (a) and (d) Cu, (b) and (e) Ni, and (c) and (f) Al metal mask (45-degree tilted view) (initial mask size = 370 nm).***

with a relatively low selectivity (such as Al) is more likely to be sputtered from the mask onto the SiC surface by the energetic ion bombardment than other masks with a relatively high selectivity, such as Cu and Ni [47]. Consequently, sputtered mask material on the surface leads to a micromasking effect that eventually results in rough and textured surfaces with grass-like features. The observed micromasking effect decreased in the following order, Al, Ni, and Cu (see Fig. 9.1d–f), which shows the reverse tendency of sputter resistance (Cu > Ni > Al). As a result, the higher the sputter resistance of the mask the cleaner the SiC substrate surface compared with lower sputter resistance of the mask.

Even though the Cu mask shows the clearest surface, Cu is not a suitable mask material due to the induced roughness on the pillars. Based on these results, Ni was chosen as the mask material for the etching of NPs because of the observed clear sidewall surface and the relatively low density of micromasking.

Effect of SF_6/O_2 Flow Rate Ratio

The etching behavior depends strongly on the composition of the etch gases. Specially, in the case of an SF_6 plasma, oxygen concentration is known to influence the profile of SiC NPs, such as anisotropy, surface cleanness, metal mask selectivity, and etch rate by controlling the neutral fluorine

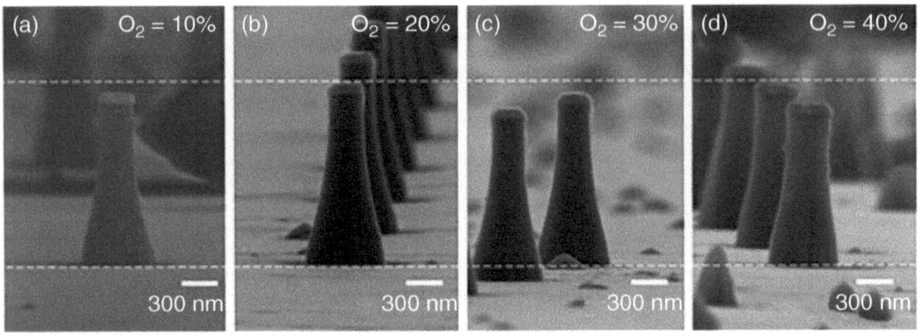

Figure 9.2 ***Side-view SEM images of ICP-etched SiC NPs at different O_2 concentrations of (a) 10%, (b) 20%, (c) 30%, and (d) 40%, respectively (initial mask size = 370 nm and etching time = 180 s).***

density [48]. To examine the oxygen effect, ICP etching was carried out with different oxygen concentrations from 10 to 40 vol%, while all other experimental conditions were kept constant.

Fig. 9.2 shows SEM images of NP arrays fabricated with different O_2 concentrations. The pillar height initially increases with O_2 concentration, and reaches a maximum value of 1.58 ± 0.02 μm at around 20% of O_2 concentration. Then, it decreases with increasing O_2 concentration, as shown in Fig. 9.3. The optimum O_2 concentration has been determined at the aspect ratio maximum. The aspect ratio (H/D) is simply calculated by dividing the height (H) by the diameter (D) of the pillar at the base. The aspect ratio shows a maximum value of 2.4 at 20% of O_2 concentration. The optimum O_2 concentration (20%) and the observed trend are comparable to results

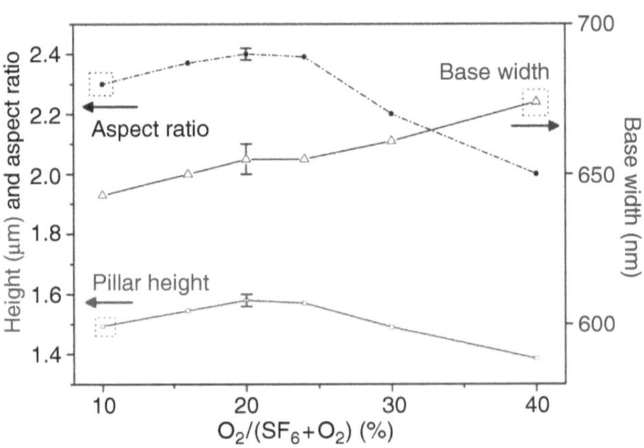

Figure 9.3 ***Height and base width of the pillars, and corresponding aspect ratio with different O_2 concentrations (initial mask size = 370 nm and etching time = 180 s).***

reported elsewhere [49]. The addition of O_2 can keep the F radical concentration high, which is responsible for the etching of SiC, by reacting with unsaturated CF_n and SF_n bonds so that F radicals do not combine with them. In addition, F radicals from the plasma preferentially etch Si rather than C [36]. As a result, a carbon-rich layer is formed on the SiC substrate, which is one of the limiting factors for SiC etching. The presence of appropriate O_2 level provides the driving force for removing the carbon by forming volatile etches products, such as CO, CO_2, and CF_2 [44]. Consequently, those two factors are believed to contribute to the increase of the observed etch rate. However, further addition of oxygen decreases the etch rate due to the dilution of the F concentration.

It was found that the O_2 concentration also affects the morphology (or vertical profile) of the etched NPs. We believe that a higher density of F radicals induces a more vertical profile of the NPs. Besides, the base diameter slightly increases with increasing O_2 concentration due to the deposition of a passivation layer during the etching process, such as SiF_xO_y.

Broader base diameter of the NPs than the size of mask could be explained by several factors, such as a mask shadowing effect [36,50], deposition of a passivation layer, or mask erosion. In the ideal case, incident ions arrive at the substrate vertically during the etching process, but in a practical case, incident ions show a Gauss angle dispersion due to scattering with other particles [50]. Therefore, incident ions are reduced in the vicinity of the base due to the shadowing effect of the pattern and pillar. Simultaneously, the deposition of a passivation layer (SiF_xO_y) on the SiC NPs during the etching process also contributes to tapered NPs since the deposition rate decreases from the bottom to the top of the pillars.

Generally, these two effects are enough to describe the slope of structures with microscale patterns. However, in the case of nanoscale etch patterns and long etching time, mask erosion also needs to be considered. Moreover, if the pillar height is high enough to induce a nonuniform charge distribution within the etching features, the deflected ions can also significantly influence the tapered profile of the pillars. This will be discussed in more detail in the next paragraphs.

Fig. 9.2d shows a deposited layer on the sidewalls of the SiC NPs. It is believed to be due to the redeposition of nonvolatile species including etch products [51]. The excess inclusion of O_2 (above 20% of O_2 concentration) tends to react with the Si and form SiO_x or SiF_xO_y on the surface, which has lower etch rates than SiC (the SiO_2 etch rate is about 90% of the SiC etch rate for SF_6/O_2 [52]). Meanwhile, insufficient O_2 concentration (below

Figure 9.4 ***Side-view SEM images of SiC NPs at different bias voltages of (a) 100 V, (b) 150 V, (c) 200 V, and (d) 300 V, respectively (initial mask size = 115 nm, pressure = 6 mTorr, and etching time = 180 s).***

20%) induces a carbon-rich layer and nonvolatile carbon-related products (CF_x) on the substrate. At the optimum O_2 concentration (20%), clean wall surfaces are observed due to the balance between the etching by F radicals and the deposition by O radicals.

Effects of Bias Voltage and Chamber Pressure

The effects of bias voltage and process pressure have been investigated using a small mask circular pattern (115 nm diameter) to fabricate SiC NPs. This size is the smallest circular pattern size that we could realize with a metal (Ni) deposition of 110 nm. The circular pattern below 100 nm disappears during the lift-off process due to low adhesion between Ni and SiC substrate.

Fig. 9.4 shows SEM images of etched SiC NPs as a function of bias voltage for 180 s of etching time. Higher bias voltage increases the directionality of ion and ion bombardment energy, so it provides more efficient bond breaking of SiC. As a result, the height and aspect ratio of the SiC NPs is gradually increased with an increase of bias voltage, as shown in Fig. 9.5a. Pillar height and aspect ratio reach the value of 2.2 μm and 7.4 at 300V bias voltage, respectively. Moreover, the size of the observed pillar base microtrench increases with increasing bias voltage due to enhanced directionality of energetic ions. On the other hand, mask thickness and diameter

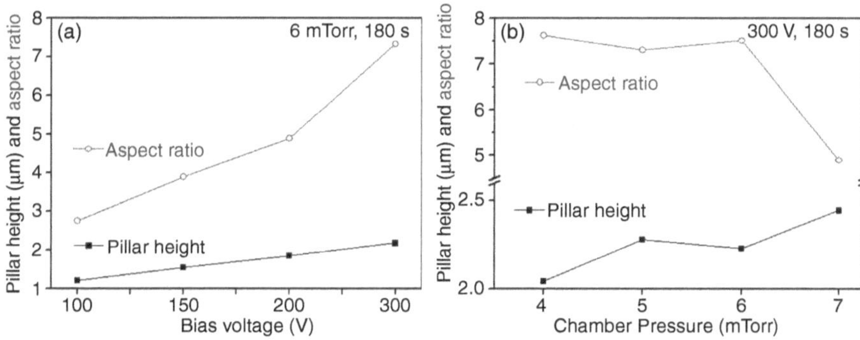

Figure 9.5 ***Height and aspect ratio of the pillars as a function of (a) bias voltage and (b) chamber pressure (initial mask size = 115 nm and etching time = 180 s).***

considerably decrease with increasing bias voltage, which is induced by enhanced physical sputtering. As a result, the initial mask size (115 nm) is considerably reduced to 60 nm for 300 V of bias voltage, as shown in Fig. 9.4d.

Fig. 9.6 shows SEM images of the etched SiC NPs with a small mask circular pattern (115 nm) as a function of chamber pressure at 300 V bias voltage and 180 s etching time. There are no big differences between 5 mTorr and 6 mTorr in terms of pillar height and aspect ratio, as depicted in Fig. 9.5b. The pillar height at 4 mTorr is slightly decreased due to the

Figure 9.6 ***Side-view SEM images of NPs at different chamber pressures of (a) 4 mTorr, (b) 5 mTorr, (c) 6 mTorr, and (d) 7 mTorr, respectively (initial mask size = 115 nm, bias voltage = 300 V and etching time = 180 s).***

decrease of the ion density, but there is no very obvious change of the aspect ratio.

At a pressure of 7 mTorr, the pillar height slightly increases due to the increase of the ion density, while the aspect ratio was considerably reduced from 7.4 at 6 mTorr to 4.9 at 7 mTorr (Fig. 9.5b). Increasing the pressure also increases the number of colliding atoms, so the mean free path of the ions decreases [44]. It reduces the directionality of the ions and finally results in a tapered profile with low aspect ratio.

Effects of bias voltage and chamber pressure have been investigated to elaborate the SiC NP formation using a 115-nm diameter circular mask pattern. The obtained SiC pillars at an optimum bias voltage (300 V) and chamber pressure (6 mTorr) show a height of 2.2 μm and aspect ratio of 7.4. The etched SiC NPs at these conditions have a minimum diameter of 60 nm. The length of the pillars (below 100 nm in diameter) was around 800 nm, which is long enough to reveal the electrical properties of the SiC nanoFET that will be presented later. Mask erosion and a tapered profile are inevitable for the etching of hard-to-etch materials like SiC using the ICP etching process but these optimized parameters have resulted in reasonable feature definition.

An alternative etching technique, such as capacitively coupled plasmas, might be a better choice than the ICP methods just described in terms of the anisotropic profiles, due to the characteristic of higher ion energies with low density [53]. But this technique also has some drawbacks. For example, metal mask thickness can be eroded much faster than for the ICP process due to the relatively high-energy ion bombardment. Finally, it leads to lower pillar height than in the case of ICP etching. Therefore, there is a need for further comparative studies according to the different etching techniques to elaborate the high aspect ratio of SiC NPs.

Etching Profile Evolution Over-Etching Time

Due to mask erosion during the ICP process, pillar height is strongly limited by the lateral mask size as well as vertical mask thickness. Therefore, we have to take into account the mask erosion for the design of the mask pattern size in order to obtain long SiC NPs. In this section, we have investigated the etching behavior using a large circular pattern (370 nm diameter) as a function of etching time, not only to get longer SiC NPs, but also to monitor the etch characteristic parameters, such as mask size, mask thickness, pillar height, and sidewall bowing.

Fig. 9.7 shows the morphology of the etched SiC NPs according to etching time at the optimum O_2 concentration (20 %) and 150 V bias

Figure 9.7 ***SEM images of SiC NPs for different etching times of (a) 360 s, (b) 480 s, (c) 660 s, and (d) 840 s, respectively (initial mask size = 370 nm).***

voltage. The height of the SiC NPs was found to increase depending on the etching time. For a quantitative analysis, the mask size, mask thickness, and pillar height are plotted as a function of etching time, as shown in Fig. 9.8.

The aspect ratio of the pillars increases with etching time (see Fig. 9.7). The maximum ratio was 6.45 for an 840-s etch time. Long etching times lead to high etching depths with high anisotropy, but it also gives an undercut etching profile at 840 s (see Fig. 9.7d), due to high chemical etching by direct exposure of F radicals from the plasma. Minimum diameter of SiC NPs at 840 s is around 60 nm at the undercut point. Therefore, it is obviously necessary to balance the physical (related to ion bombardment) and

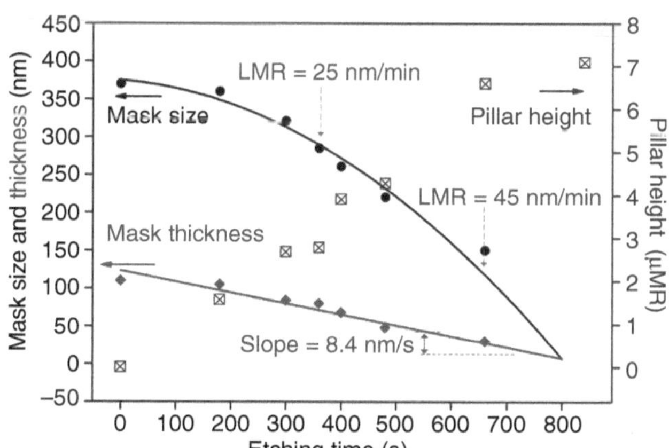

Figure 9.8 ***Mask size and thickness, and corresponding pillar height as a function of etching time.*** LMR, lateral mask erosion rate from the fitted curve.

chemical (related to F radicals) contributions in order to get straight pillar features.

The etch rate is calculated by the pillar height over-etching time. It remains almost constant (550 nm/min) during etching, and then begins to decrease (165 nm/min) in the region between 660 and 840 s when the metal mask is completely removed. The etch selectivity of mask materials (Ni) with respect to the SiC is around 65.

Mask thickness at the center is linearly reduced with time as shown in Fig. 9.8, so the vertical etch rate of the mask remained constant throughout the etching process (8.4 nm/min). Besides, the lateral etch rate of the mask increases evidently with increasing etching time. Vertical and lateral etch rate of the mask was fitted to a linear and a second-order polynomial for the experimental data, respectively. Mask erosion behavior seems to be quite well described by fitting the curve with R^2 values for mask size and thickness of 0.93 and 0.97, respectively.

To analyze mask profiles in more detail, the Ni mask part of each SiC NP is magnified, as shown in Fig. 9.9. Mask erosion is a little bit faster at the edges due to strong physical sputtering of energetic ions, which is enhanced by the local distortion of the electric field [47]. Therefore, mask faceting is observed due to the ion bombardment effect. Consequently, mask slope angle is progressively decreased from 55 degrees at 36 s to 35 degrees at 660 s, as shown in Fig. 9.9. This can explain the increasing of lateral mask erosion rate over-etching time, because mask size is more rapidly decreased at lower mask slope under the same sputtering condition. Lateral mask erosion rate from the fitted curve increased from 25 to 43 nm/min, and the corresponding mask slope decreased from 55 to 35 degrees, when the etching time increased from 360 to 660 s. This result confirms the angle-dependent sputtering yield of the mask, showing an increase in mask erosion rate at lower mask slope [54].

As the etching proceeds, the edge of the mask pattern is etched away. The SiC below the etched Ni mask is no longer protected and begins to be etched. Finally, when the metal mask is completely removed, SiC NPs begin to develop a slight convex shape at the top of the pillar, as shown in Fig. 9.9e.

As noted previously, mask erosion can be one of the reasons for the observed vertically tapered sidewalls of SiC NPs since reduced mask thickness exposes more of the energetic ion flux on the top part of the pillars than the bottom part. As the mask pattern size is shrinking at the nanoscale, mask erosion and undercutting play a crucial role for determining the maximum height and etching profile of SiC NPs.

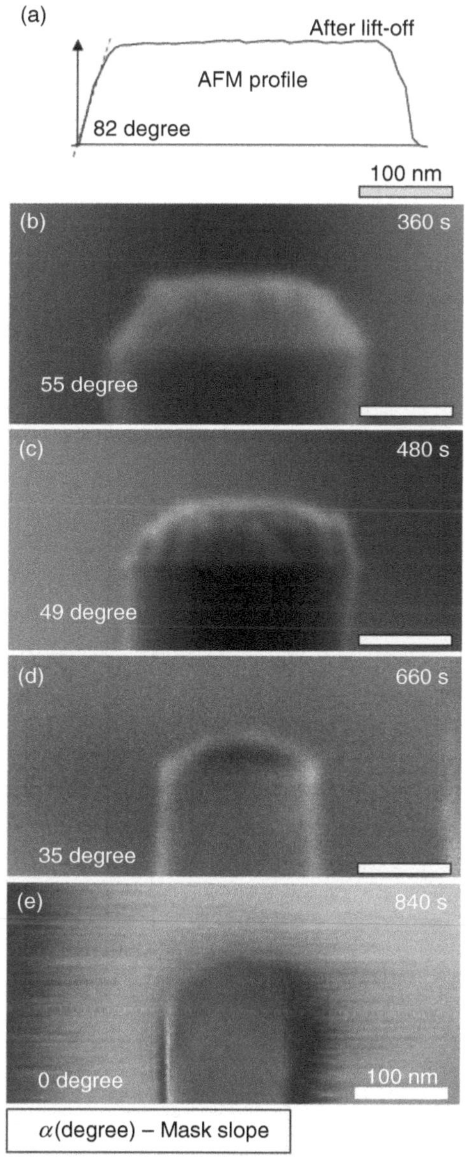

Figure 9.9 (a) AFM profile after lift-off, and (b–e) magnified SEM images of the Ni mask after etching for (b) 360 s, (c) 480 s, (d) 660 s, and (e) 840 s. 100 nm scale.

9.2.1.4 Conclusion

The etching characteristics of 4H–SiC(0001) have been investigated using ICP etching in an SF_6/O_2 plasma to elaborate SiC NPs. The etching profile of the SiC NPs and etch rate are controlled by O_2 concentration in the etching gas. The effects of bias voltage and chamber pressure have also been investigated using a circular mask pattern of 115 nm. The obtained SiC NPs at an optimum bias voltage (300 V) and pressure (6 mTorr) exhibited a high aspect ratio (7.4) with an etching depth of 2.2 μm. Etching profile and mask evolution were monitored versus etching time using a circular mask of 370 nm diameter. As the etching progressed, the pillar height and aspect ratio increased until the mask removal point. The vertical etch rate of the mask remained constant throughout the etching process, but the lateral etch rate of the mask gradually increased with increasing etching time. This "top-down" approach with dry plasma etching has shown the capability of achieving single crystalline SiC NPs with high aspect ratio (6–7) and a large etch depth (>7 μm) [55]. We also reported elsewhere [56] that hexagonal facets of the SiC NPs have been obtained for long etching times because of the crystallographic structure of the *a*-SiC phase. Thanks to the mask erosion and undercutting, the minimum diameter of etched pillars can be shrunk down to 60 nm. This will allow for the study of the electrical properties of single crystalline SiC nanostructures after integration into nanoFETs (Section 9.3). Due to their shape, this also opens new possibilities to apply these NPs for widespread applications, such as biomedical probes [57].

9.2.2 Si–SiC Core–Shell Nanowires by Carburization

9.2.2.1 Introduction

As introduced in Section 9.1, high-quality 1D SiC nanostructures are needed to realize SiC NW-based nanoelectronic devices or sensors, such as FETs, to be used in harsh and/or biological environments. We have presented in Section 9.2.1 a "top-down" approach leading to SiC NPs. The second alternative method developed in this section gives rise to Si–SiC core–shell NWs. The idea is to benefit from the excellent electronic transport into the Si core NW and from the biocompatibility of the SiC shell all around the Si NW for new integration prospects for *in vivo* biosensors. In this section, we present innovative Si–SiC core–shell NWs, which are elaborated from previously etched Si NWs by a two-step process:

1. In situ deoxidation of Si NWs under H_2 flux;
2. Growth of a SiC shell by Si carburization under CH_4 or C_3H_8.

Each step of the synthesis has been characterized either by SEM or TEM. The dependence of the thickness of the SiC shell with the carburization time has also been studied and is presented next.

Figure 9.10 ***SEM image of etched Si NWs.*** (a) Cross-section view. (b) Detail of the top of one Si NW, tilted at 30 degrees.

9.2.2.2 Experiment

For this study, Si NWs are obtained by the top-down approach using a SF_6/O_2 plasma etching of an exactly <001>-oriented Si substrate [58]. The as-formed NWs are 3.4 µm long with a diameter of 200 nm and are spaced with a pitch of 200 nm (Fig. 9.10a). These Si NWs display a nonnegligible roughness (Fig. 9.10b) due to the etching process itself. However, these etched Si NWs present major advantages such as single crystallinity, absence of metal catalysts, and high reproducibility of the fabrication process.

The aim of this set of experiments is the study of the dependence of the thickness of the SiC layer with the carburization time. Concerning the in situ deoxidation, a high H_2 flow and the high Si ad-atom mobility at high temperature lead to the smoothing of the Si NW sidewalls, accompanied by a slight diameter reduction (Fig. 9.13b). These smoothed Si NWs are excellent starting points for the carburization process.

The carburation step is performed in an alumina hot-wall CVD furnace at 1100°C and atmospheric pressure. The main flow is composed of a mixture of H_2 diluted into Ar with a ratio of 50 sccm:200 sccm. The removal of native silicon oxide on the Si NWs is processed directly inside the reactor thanks to the reductive properties of H_2.

Methane is used as a carbon precursor and injected into the reactor during the temperature ramp from 800°C to the dwell with a flow of 0.5 sccm. This flow is then increased to 4 sccm at the temperature plateau, which lasts from 1 to 60 min. This high dilution of the precursor gas into the carrier gases allows for the limiting of the carbon deposit in favor of the carburization of silicon. The system is naturally cooled down after the specified growth time.

A dual FIB-SEM (FEI Helios 450S) is used to prepare the thin lamella for TEM characterizations of the as-grown SiC layer both on a control planar Si(100) surface and on carburized Si NWs. Standard thin films are grown on Si(100) and this baseline process is used to study the SiC layer growth directly on the carburized silicon NWs. This process consists in a fib-cut perpendicular to the *z*-axis of the NWs leading to thin lamella of sliced NWs.

First, the sample is coated either with a polymethylmethacrylate (PMMA) resin or an e-beam-induced platinum deposit in order to protect and maintain the carburized Si NWs during the focused ion beam (FIB) processing. Then, a standard FIB lamella is processed with a volume of $10 \times 20 \times 5$ μm. Thanks to a micromanipulator, the thin lamella is moved to a grid and fixed with an ion beam-induced platinum deposit at 90 degrees for specimen thinning. A second ion beam induced platinum (Pt) deposition is done on the top of the lamella. This Pt layer is useful to evacuate the charges and, also, to control the speed of the thinning. The lamella is thinned until a thickness of 50 nm, and then characterized by TEM.

9.2.2.3 Results and Discussion

In this section we will deal with the carburization of Si NWs under CH_4 flow, which has been studied as a function of the time of carburization. Si NWs carburized under CH_4 (Fig. 9.13c) have the same roughness before and after carburization (Fig. 9.10), and the overall morphology is kept, such as length and diameter.

The thickness of the 3C–SiC layer, measured by TEM, is about $\varepsilon \approx 3$ nm and is completely sealing the Si NW surface (Fig. 9.11a). This thickness is in accordance with the thickness measured on planar Si(100) as shown in Fig. 9.11b. Therefore the Si–SiC core–shell NWs have been well elaborated. The 3C–SiC layer surrounding the Si NWs is single crystalline and no voids were observed at the interface of the Si–SiC contrary to previous studies conducted on Si substrates [59,60].

In Fig. 9.11c, the thickness of the 3C–SiC layer as a function of the squared root of the carburization time is reported. It appears to be a linear relationship between the thickness and the square root of process time, which is typical for a diffusion-limited process. In the case of the carburization of Si, it has been shown that Si is the diffusing species through the SiC layer [60]. In a previous study conducted at low pressure, this phenomenon has been used to produce SiC NTs from carburization of Si NWs [61].

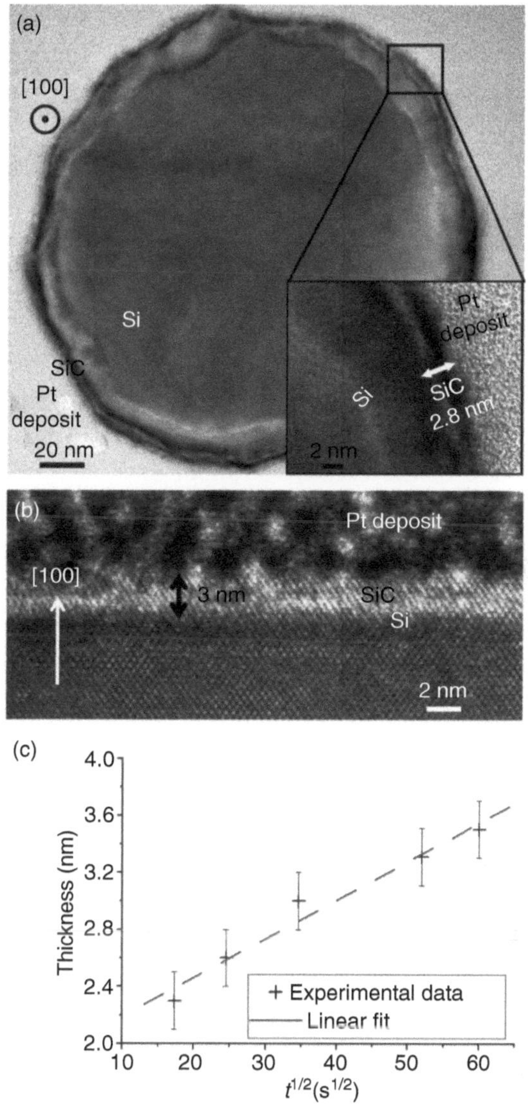

Figure 9.11 (a) Bright-field TEM image in <001> zone axis of a carburized Si NW during 10 min of SiC film growth at 1100°C. The thickness of the SiC layer is about 3 nm and crystallized in the cubic (3C–SiC) phase. (b) Bright-field TEM image of a 3C-SiC layer obtained by carburization of a Si(100) surface for 10 min at 1100°C. The 3C-SiC thickness is comparable to the shell thickness on NWs. (c) Thickness of the 3C-SiC layer on top of Si NWs as a function of carburization time, for a temperature plateau of 1100°C. A thicker layer of about 20 nm appears on the image which is only due to the initial roughness.

According to the second Fick's law, and the Einstein–Smoluchowski's relation, the depth of the Si surface affected by the carburization is given by Eq. (9.1) in first approximation [62]:

$$\varepsilon \approx \sqrt{2D_{Si}t} + b \tag{9.1}$$

where ε is the thickness of the 3C–SiC shell, t is the carburization time at 1100°C, D_{Si} is the apparent diffusion coefficient of Si through the SiC layer, and b is the SiC thickness elaborated from 800 to 1100°C. Applying Eq. (9.1) the calculation of the diffusion coefficient of Si into SiC at 1100°C under CH_4 is $D_{Si} = 3.7 \cdot 10^{-18}$ cm^2/s. This value is in accordance with previous studies conducted under similar experimental conditions [63,64].

The study and the estimation of the apparent diffusion of Si through 3C–SiC in the case of the carburization of Si NWs was of interest because carburization takes place here on many different crystallographic planes contrary to previous studies conducted on bulk Si(100) surface [59,60]. Moreover, the control of the out-diffusion of Si allows the realization of various nanostructures: Si–SiC core–shell NWs, SiC NWs, or SiC NTs [61].

In order to check the SiC shell conformity and to be sure that our future application, that is, nanobiosensors based on Si–SiC core–shell NWs, is possible, a simple chemical etching of the SiC layer was done. Carburized and noncarburized Si NWs were etched with a potassium hydroxide (KOH)-diluted-in-water solution at 50°C for 2 min [65]. As shown in Fig. 9.12, the Si–SiC core–shell NWs have the same morphology before (Fig. 9.12a) and after the chemical etching (Fig. 9.12b) whereas the noncarburized Si NWs have totally disappeared after etching (Fig. 9.12d).

9.2.2.4 Conclusion

Si–SiC core–shell NWs have been successfully elaborated from etched Si NWs by a process including in situ deoxidation of Si NWs at high temperature under H_2 flow and the subsequent carburization of Si. The carburization process is limited by the out-diffusion of Si through the first SiC layer. It has been possible to calculate the apparent diffusion coefficient of Si through SiC at 1100°C under CH_4 in the case of the Si NW carburization. After carburizing the smooth Si NWs obtained after the in situ deoxidation step, no morphological change is visible (Fig. 9.13d). Besides that, an accurate control of Si out-diffusion via the control of the total pressure inside the reactor can support the growth of either Si–SiC core–shell NWs [66] or SiC NTs [61,67].

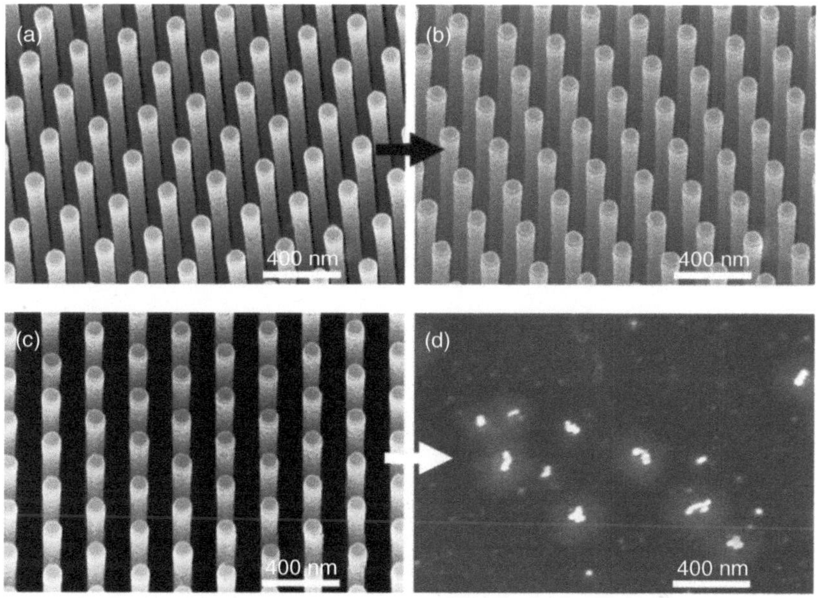

Figure 9.12 SEM images of: (a) carburized Si NWs before the KOH etching; (b) carburized Si NWs after the KOH etching; (c) noncarburized Si NWs before the KOH etching; (d) noncarburized sample after the KOH etching. Note the Si NWs have disappeared without the protective SiC coating.

Finally, a perfectly conformal 3C–SiC coating of ~2.5 nm thick was obtained on Si NWs during the carburization step. The conformality was tested by chemical etching in hot KOH. This 3C–SiC layer was observed to act as a protective shell against harsh environments such as acid or basic solutions for the Si core of the nanostructure. These innovative nanostructures have potential applications in the nanobiosensors field thanks to the high biocompatibility of the 3C–SiC [17] and the very good and well-known electronic properties of Si.

9.3 TECHNOLOGICAL PROCESS OF nanoFETs

9.3.1 Introduction

Only a few works have been reported on nanoFETs based on SiC nanostructures, and the experimental electrical performance of such nanoFETs is still very low compared to theoretical expectations [68]. For instance, low values of effective mobility with nanoFETs based on VLS SiC NWs are reported in the literature: 16 cm^2/V s for electrons [12,13] and 6 cm^2/V s for holes [69]. In this section, we report our results of the electrical performance

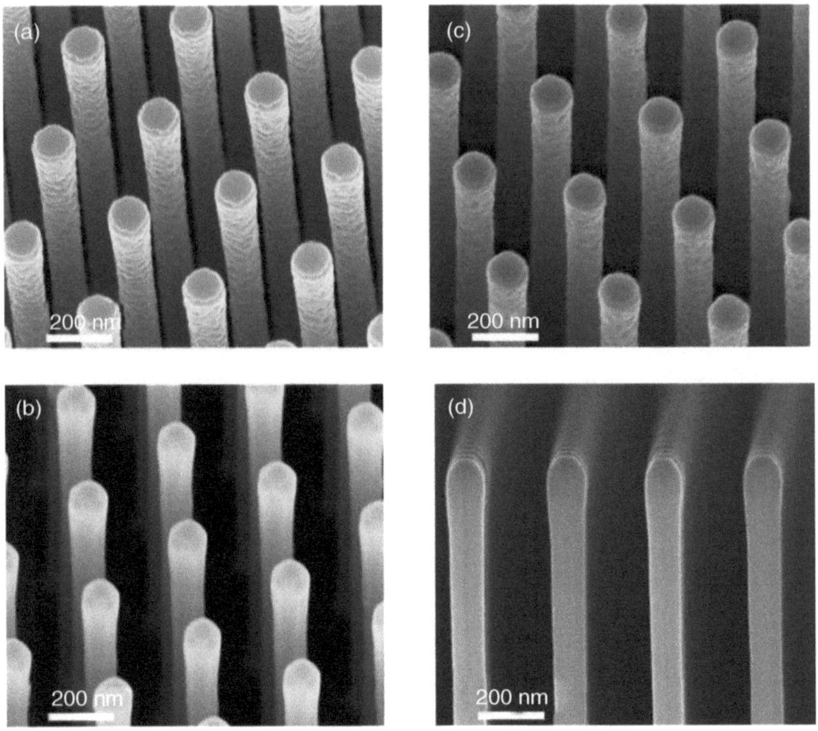

Figure 9.13 ***SEM images of etched Si NWs.*** (a) As prepared. (b) After in situ deoxidation under H_2 at 1000°C during 5 min; sidewalls have been smoothed. (c) NWs in (a) after carburization under CH_4. (d) NWs in (b) after carburization.

of nanoFETs using innovative 1D SiC nanostructures, presented in Section 9.2, as the transport channel such as: SiC NPs, Si–SiC core–shell NWs, and SiC NTs.

Backgated nanodevices, based on these nanostructures, are realized with the upstream facilities of the Advanced Technology Platform (PTA) in Minatec using deposition and etching, and also advanced e-beam lithography and lift-off techniques for achieving optimized microcontacts.

9.3.2 Experiment

Five different kinds of SiC-based 1D nanostructures were used for this study (Fig. 9.14):

1. 5-μm length Si NWs (210-nm diameter) obtained by the top-down approach, consisting of a plasma etching process of Si(100) wafers described elsewhere [58]. These Si NWs are used as reference material for nanoFETs fabrication;

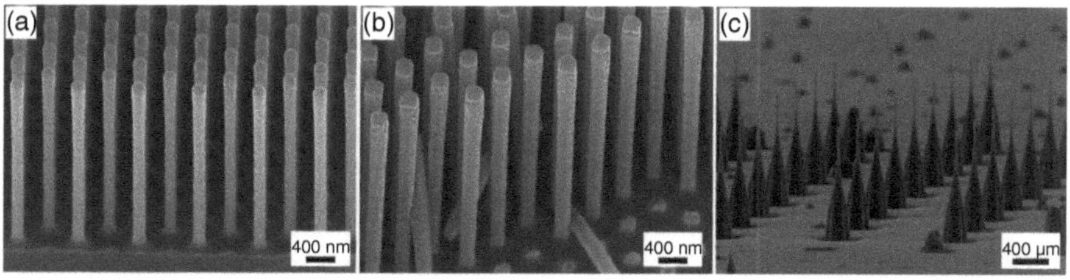

Figure 9.14 SEM images of (a) Si NWs obtained by plasma etching process. Si–SiC core–shell NWs have exactly the same morphology. (b) SiC NTs obtained by carburization of Si NWs. (c) Typical SiC NPs obtained by a plasma-etching process.

2. 5-μm length Si–SiC core–shell NWs (210-nm diameter) obtained by a combined top-down and bottom-up approach: first, Si NWs are obtained by a plasma etching process, followed by a carburization route at atmospheric pressure, leading to a single crystalline, 3-nm thick 3C–SiC shell described in detail in Section 9.2.2 and elsewhere [66]. As a consequence, the morphology of Si NWs and Si–SiC NWs is identical (see Fig. 9.14a);
3. 5-μm length 3C–SiC NTs (250-nm diameter) obtained by a combined top-down and bottom-up approach: first, Si NWs are obtained by a plasma etching process, followed by a carburization route where the pressure is monitored in order to empty out Si through a SiC shell [67].
4. 620-nm length 3C–SiC NPs (120-nm diameter) obtained by a top-down approach, consisting of a plasma etching process of a 3C–SiC epilayer with an n-doping level of about $2–8 \times 10^{16}$ cm^{-3} described in detail in Section 9.2.1 and elsewhere [55,56];
5. 570-nm length 4H–SiC NPs (110-nm diameter) obtained by a similar top-down approach, consisting of a plasma etching process of 4H–SiC wafers with an n-doping level of about 4×10^{15} cm^{-3} also described elsewhere [55,56] and in Section 9.2.1.

These nanostructures have been integrated into back-gate FETs by the following two different processes:

1. Simultaneous process (Fig. 9.15): In this first process, described elsewhere [70], a p^{++} Si(100) substrate covered with a 100-nm thick Si_3N_4 layer is used, acting as a gate oxide, where nanostructures are dispersed. Optical lithography is then used to define drain and source contacts with a spacing of 1 μm, and the nanostructures are placed in between. After development, contacts are metalized (Ni 80 nm/Al 120 nm) using an e-beam evaporator. A lift-off step is then performed to remove residual resist and an O_2 plasma step is used to clean the wires from all

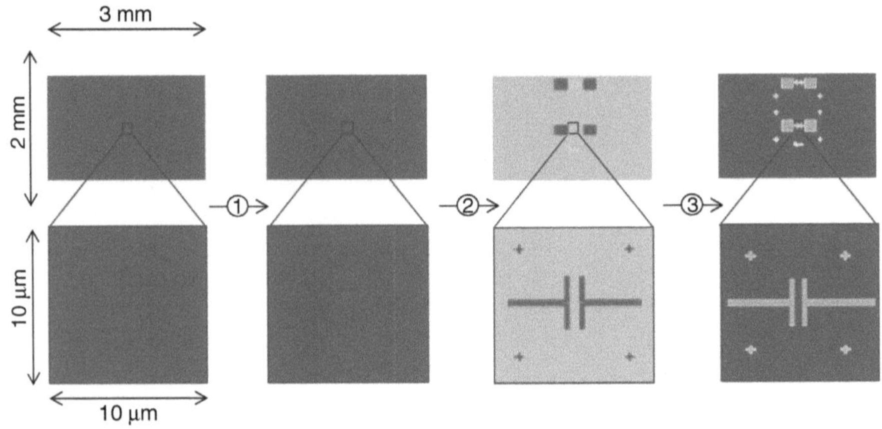

Figure 9.15 ***Fabrication steps for microcontacts in the simultaneous process.*** (1) Dispersion of the nanostructures, (2) design of the macro- and microcontacts by optical lithography, and (3) contact metallization.

resist waste, which could perturb electrical transport. This process was used for nanostructures (1), (2), and (3). SEM micrographs of the as-fabricated nanoFETs are shown in Fig. 9.18a–c, respectively.

2. Sequential process (Figs. 9.16 and 9.17): For nanostructures (4) and (5), the p^{++} Si(100) substrate is covered with a 80-nm thick SiO_2 layer. Because of the small length of these nanostructures, optical lithography is only used to define a window for the macrocontacts but is not usable as a microcontact. Therefore, an e-beam lithography step is necessary to define drain and source microcontacts to connect the nanostructure.

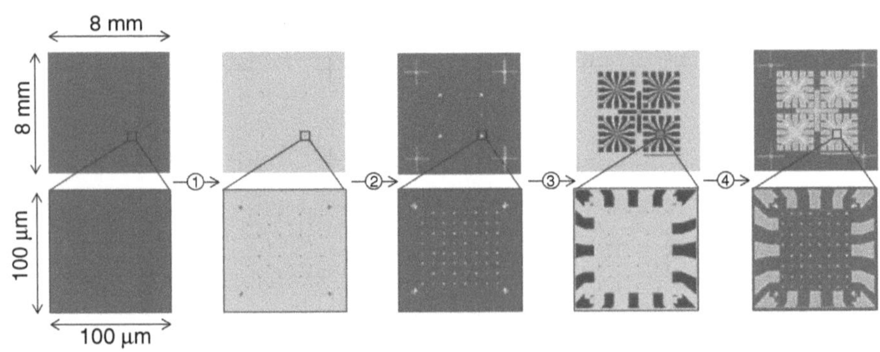

Figure 9.16 ***Fabrication steps for microfabrication, showing alignment markers and macrocontacts in the sequential process.*** (1) Design of the alignment markers by e-beam lithography, (2) metallization of the markers, (3) design of the macrocontacts by optical lithography, and (4) macrocontact metallization.

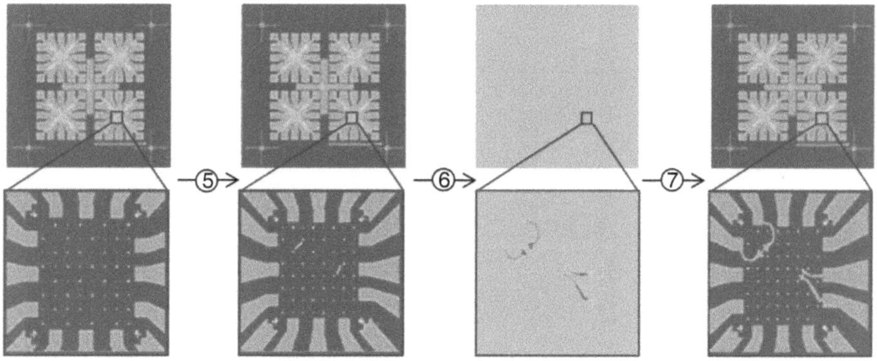

Figure 9.17 ***Fabrication steps for microcontacts in the sequential process.*** (5) Dispersion of the nanostructures, (6) design of the microcontacts by e-beam lithography, (7) macrocontact metallization.

Following the integration process, these two devices are annealed at 650°C for 30 s in an N_2 ambient within a rapid thermal annealing furnace to obtain ohmic contacts and low contact resistance. SEM images of the as-fabricated 3C–SiC NPFET and 4H–SiC NPFET are shown in Fig. 9.18d and e, respectively.

The drain current (I_{DS}) of the nanoFETs as a function of the source–drain voltage (V_{DS}) was measured at room temperature and in ambient air using a semiconductor parameter analyzer (Keithley 4200), the results of which are presented next.

9.3.3 Results and Discussion

In Fig. 9.19, I_{DS}–V_{DS} characteristic curves with gate voltage (V_G) varying in the range of +20 to −20 V are reported for the Si NWFET, Si–SiC core–shell NWFET, and SiC NTFET, respectively. It clearly shows that each nanoFET works and that a gate effect is obtained, with an n-type behavior. However, current levels measured even for high gate voltages are really low compared to the literature on SiC nanoFETs [13] and Si ones [71].

It is possible to estimate an apparent carrier mobility, $\mu_{app,}$ in these nanoFET thanks to the cylinder–plate capacitance model in Eq. (9.2) [72,73]:

$$\mu_{app} = \frac{g_m L^2}{C_{gate} V_{DS}} \quad (9.2)$$

where g_m is the transconductance, which can be obtained from the slope of the I_{DS} versus V_G plot, L is the length of the channel, C_{gate} is the gate capacitance, and V_{DS} is the drain–source voltage. The gate capacitance C_{gate} can be

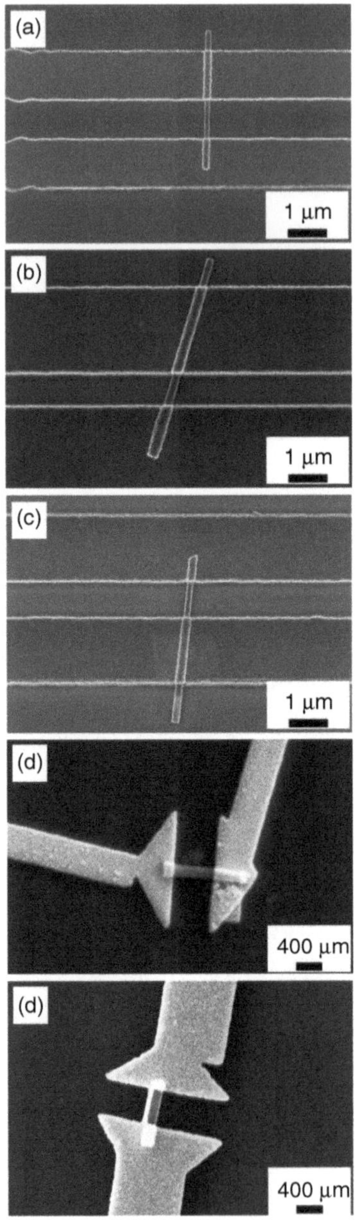

Figure 9.18 SEM images of (a) a Si NWFET, (b) a Si–SiC core–shell NWFET with drain and source contacts spacing of 1 μm, (c) a SiC NTFET with drain and source contacts spaced of 1 μm, (d) a 3C–SiC NPFET with drain and source contacts spaced at 570 nm, and (e) a 4H–SiC NPFET with drain and source contact spacing of 620 nm. (a–d) Drain and source contact spacing of 1 μm.

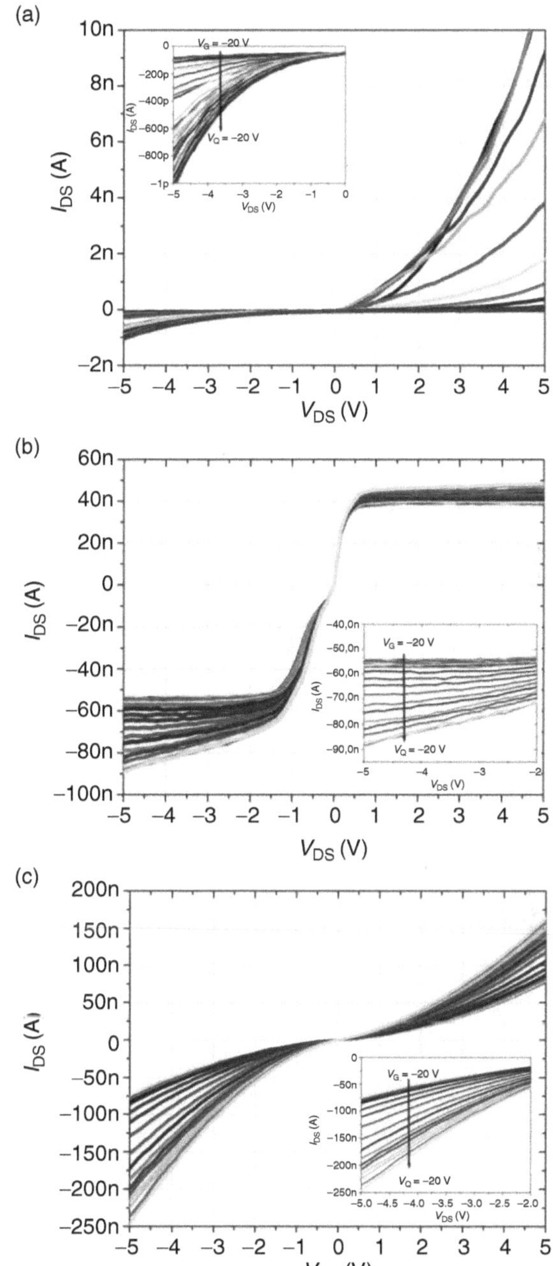

Figure 9.19 ***Experimental characteristics of the nanoFETS showing their I_{DS}–V_{DS} characteristic curves, for V_G values of +20 V to −20 V, in steps of 2 V.*** (a) Si NWFET; (b) Si–SiC core–shell NWFET; (c) SiC NTFET.

estimated by the metallic cylinder on an infinite metal plate model as described by Eq. (9.3), where h is the insulator thickness and r is the radius of the nanostructure with a quasicircular cross-sectional approximation [73]:

$$C_{gate} = \frac{2\pi\varepsilon_0\varepsilon_r L}{\cosh^{-1}\left(\frac{h+r}{r}\right)} \tag{9.3}$$

The extracted apparent mobility is really low, around 0.01 $cm^2/V\ s$ for these three different nanoFETs.This can be explained by the absence of contact rectification due to contact annealing since, with the same integration process, electrical performance enhancement of Si nanoFETs has already been demonstrated thanks to metallic contact silicidation [70].

Fig. 9.20 shows results of electrical measurements of I_{DS}–V_{DS} characteristic curves for (a) 3C–SiC NPFET and (b) 4H–SiC NPFET for V_G varying between +3 V and −3 V in steps of 1.5 V. First, a gate effect is directly observable and the current levels are higher than in Fig. 9.19 because of the silicidation of contacts during annealing. Moreover, the extracted apparent carrier mobilities of these two nanoFETs are 232 and 53 $cm^2/V\ s$, respectively, which are the best reported results for SiC nanoFETs [74] compared to those reported in the literature [12,13].This can be partly explained by the high crystalline quality of both 3C–SiC and 4H–SiC materials constituting the NWs. Indeed, in the case of SiC, the top-down approach leads to very low crystalline defects densities in the NWs compared to those obtained by a bottom-up approach. Last, but not least, the association of a low doping level of 3C–SiC and 4H–SiC (of about 10^{15} at cm^{-3}) [75] with good crystalline quality also guarantees these high measured mobilities. Moreover, these first experimental measurements on SiC nanoFETs tend to confirm that the electron mobility is higher in 3C–SiC than in 4H–SiC.

9.3.4 Conclusion

In this section, experimental results from the integration of innovative 1D SiC nanostructures are reported based on SiC NPs, Si–SiC core–shell NWs, and SiC NTs incorporated into nanoFETs. The top-down study leading to 3C–SiC NPs has also displayed evidence of the highest experimental electron mobility already reported for a SiC nanoFET. In the near future, thanks to the optimization of contact quality through silicidation, electrical performance of these transistors will be greatly enhanced.

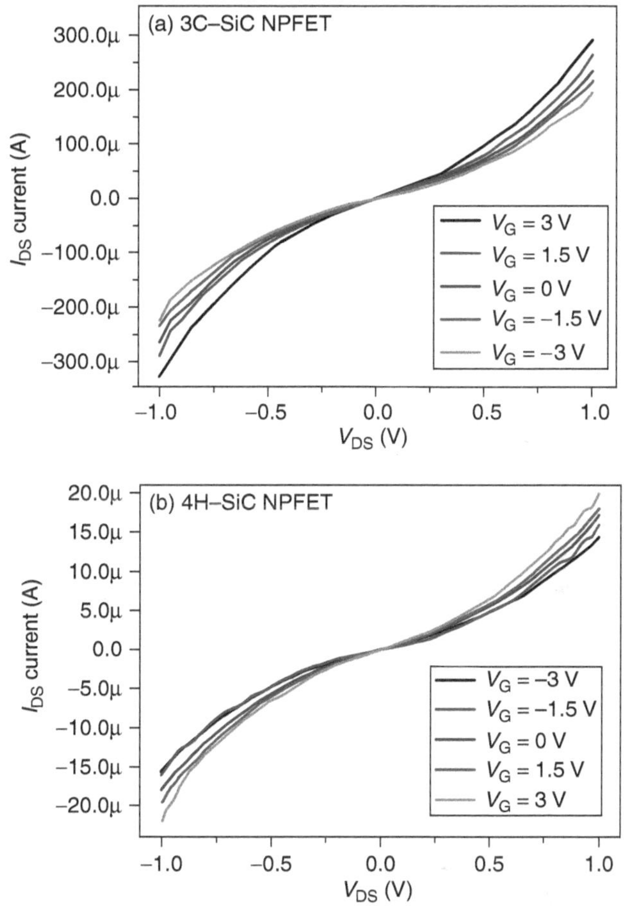

Figure 9.20 ***Experimental characteristics of the nanoFETS showing I_{DS}–V_{DS} characteristics curves for V_G values from +3 V to −3 V in steps of 1.5 V.*** (a) 3C–SiC NPFET; (b) 4H–SiC NPFET.

9.4 FUNCTIONALIZATION AND DNA HYBRIDIZATION

9.4.1 Introduction

Generally speaking, the fundamental DNA structure can be simply perceived as a double helical structure: two single DNA strands twisted around each other in the form of a double helix (Fig. 9.21). Single DNA strands consist of a long chain of repeating subunits, which are called nucleotides (Fig. 9.21). The latter are composed of three parts: a five-carbon cycle sugar, named as ribose, a phosphate group, and a nucleobase (adenine, guanine, cytosine, and thymine). The phosphate groups are negatively charged at

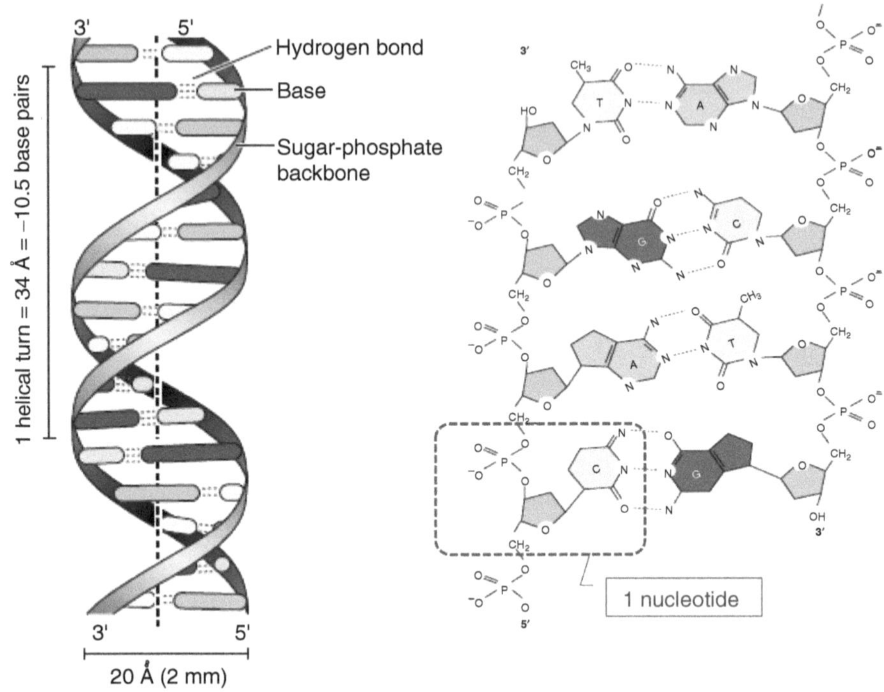

Figure 9.21 ***Schematic model of the DNA double helix showing the composition of the nucleotide subunit.***

neutral pH, their pK_a being near 0 (refer, for instance, to http://en.wikibooks.org/wiki/Structural_Biochemistry/Nucleic_Acid/DNA/DNA_structure among others). That is why DNA is a negatively charged molecule and, more precisely, an anionic polymer (at pH = 7).

When immobilized on the surface of a semiconductor NW, the DNA's negative charge induces a change in the conductance of the NW by the field effect mechanism. Depending on the doping type of the NW, the conductance change is not the same. In the case of an n-type semiconductor NW, the DNA charge depletes the whole volume of the NW and reduces its conductance. On the contrary, in the case of a p-type semiconductor NW, an accumulation effect is expected under the surface, which increases the conductance [3–5]. To the best of our knowledge, no study has been reported on label-free electrical DNA detection using SiC NWFET-based devices.

To achieve SiC NWFETs for label-free electrical DNA detection with high detection performance, it is crucial to control and optimize the DNA functionalization process on the SiC NW. To be efficient, the functionalization process must provide covalent DNA probe grafting, which is both localized and selective. On the one hand, it should take into account the global morphology and geometrical configuration of the sensing area represented by both the NW and its immediate environment. The NW surrounding involves electric metallic contacts or other NWs. On the other hand, the process of sensor fabrication is issued from microelectronic techniques, which are not always compatible with biochemical functionalization steps. Some adaptation must be carried out. Finally, the lack of adapted nanocharacterization techniques makes difficult the analyses of the results of the DNA functionalization steps.

For these reasons, prior to performing the DNA probe covalent grafting and hybridization on single SiC NWs, it is important to check the functionalization efficiency on other kinds of SiC nanostructures. This enables one to perform physicochemical characterization such as XPS and fluorescence microscopy. In this aim, we have used SiC structures such as (1) planar SiC surfaces [76], (2) top-down SiC NP arrays [66,76], and (3) core–shell Si–SiC NP arrays [56,77]. Then, the functionalization process is applied on a single NP similar to a transistor channel [77]. For this specific application, the functionalization process is locally performed thanks to e-beam lithography. Due to the size of the area of interest, fluorescence microscopy is combined with AFM measurements to characterize the functionalized transistor.

In the following, first we describe the DNA functionalization process, which leads to the covalent DNA probe grafting and hybridization on SiC wafers. Then, we use this process to functionalize different SiC nanostructures. In both cases, we report the corresponding characterization results to validate the functionalization process [76,77]. Finally, we report the early results on label-free electrical detection of DNA using SiC NWFETs [78].

9.4.2 DNA Functionalization Process

The DNA functionalization process used to covalently graft DNA probes on both SiC wafers and different SiC nanostructures up to DNA hybridization is presented in Fig. 9.22. The method is adapted from the one that we have already used on silicon dioxide and metal oxide [79,80]. The first step is the hydroxylation of the surface. This step creates hydroxyl groups, which are at the origin of the surface hydrophilicity. The hydroxylation can be achieved by several chemical treatments. Some of them involve hydrofluoric acid [81]. In our case, we use oxygen/air plasma, which is more convenient. The next step is the silanization, which enables us to deposit a layer of an organosilane on the SiC surface in order to create a covalent link between SiC and the DNA probes. The aminopropyltriethoxysilane (APTES, ROTH with purity over 97%) is terminated by amino groups

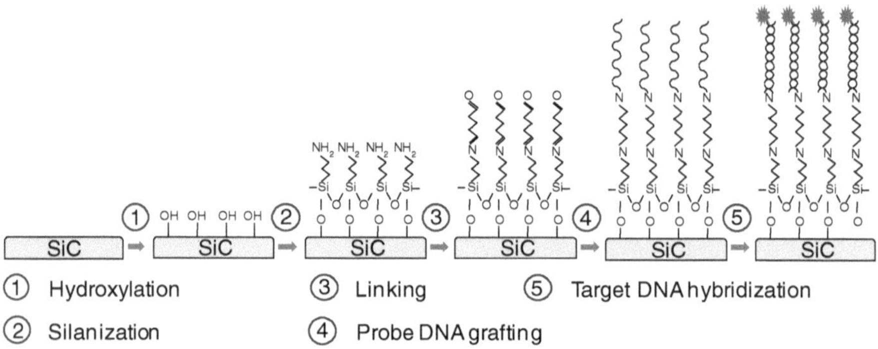

Figure 9.22 ***Flowchart of the chemical process to achieve DNA hybridization on an SiC surface.***

that enable the covalent bond with NH_2-terminated DNA probes via a cross-linker. The triethoxy function of APTES allows reticulation and is less sensitive to hydrolization than methoxy groups. The grafting of APTES is performed by the CVD technique. Due to the lower viscosity of vaporized APTES compared to the liquid version, the CVD method is expected to give better results than the liquid method with nanostructures. The samples are placed inside a sealed chamber with nitrogen and 150 μL of liquid APTES. The vaporization temperature is set to 80°C for 1 h. Then the samples are rinsed with pure ethanol and deionized water, and then annealed at 110°C for 1 h. To make a chemical bond between the amine functions of both APTES and DNA strands, the samples are immersed in a solution of 10% glutaraldehyde (grade I, 50% concentrated, provided by Sigma-Aldrich) in water. After rinsing, DNA probe grafting on the SiC surface is carried out by depositing 2 μL drops of a solution of the DNA probes ($C = 10$ μmol/L) diluted in a phosphate buffer. The probe strand sequence is 5′-NH_2-C_6-TTTTTGATAAACCCACTCTA-3′. After 2 h, the N═C bonds are stabilized through a reduction step in a sodium borohydride solution (0.1 mol/L) by modifying the imine bonding into an amine one. The DNA target molecules are either complementary single strands of DNA that fully hybridize with the DNA probes to create double strand DNA, or noncomplementary sequences. The complementary target strand sequence is 5′-CATAGAGTGGGTTTATCCA-3′ while the noncomplementary strand is 5′-CATCTCACTAACGCGGTCA-3′. These target strands are labeled with a fluorescent cyanine Cy3 molecule. Complementary, or noncomplementary, hybridization is performed with a 2 μmol/L DNA target solution for 45 min at 42°C under a wet atmosphere. The samples are then rinsed with saline solution to remove all of the unbound targets from the surface and dried with nitrogen. The complementary DNA targets can be removed thanks to a DNA denaturation process by immersing the chip in sodium hydroxide (NaOH) ($C = 0.1$ μM) solution during 45 min and afterward the hybridization step could be reiterated.

9.4.3 Characterization of the DNA Functionalization Process on Planar SiC and Different SiC Nanostructures

The planar SiC surfaces come from a 6H–SiC(0001) on-axis single crystal wafer provided by Cree and polished by Novasic™ (Fig. 9.23a). The SiC nanostructures are NPs obtained by a top-down method using ICP etching. Compared to the bottom-up approach, top-down gives shorter structures but in a better-organized array with a lower density of structural defects and controlled doping level.

The fabrication of the SiC NP array is achieved by etching of a bulk substrate of 4H–SiC provided by Cree. A Ni mask patterned by e-beam lithography defines both the position and diameter of the pillars. The etching step is performed at low pressure in a reactor with a mixture of SF_6 and O_2 gases as described previously. More details about this process are given in [56]. The term "nanopillars" is due to their dimensions, which are 1.6 μm in height with 0.8 μm diameter. The 5 μm pitch between the pillars corresponds to a density of 4×10^4 pillars/mm^2. A tilted SEM image and a schematic view of the pillars are given in Fig. 9.23b.

The second array of nanostructures is composed of core–shell Si–SiC NPs obtained by carburization of etched Si NPs. The carburization is conducted in vapor phase with methane as the carbon source [66]. The 3-nm thick SiC shell totally covers the Si core and both are single crystalline. The NP height is 0.5 μm for 200 nm diameter. Compared to the previous array (Fig. 9.23b and e), the density is 10 times superior, 4×10^5 pillars/mm^2, with a distance of 0.5 μm between pillars. The difference of morphology can be easily visualized from the 45-degree tilted SEM image and its schematic view (Fig. 9.23c). The NWFET is made with a horizontal SiC NW flattened on a silicon dioxide surface. The NW is 1 μm in length for 50 nm of diameter. It is connected at both extremities by nickel–gold contacts.

9.4.3.1 Planar SiC

First, the functionalization is performed on a planar 6H–SiC surface. During this process, drops of DNA probes are deposited on the surface. After DNA hybridization, rinsing, and drying, homogeneous typical fluorescent

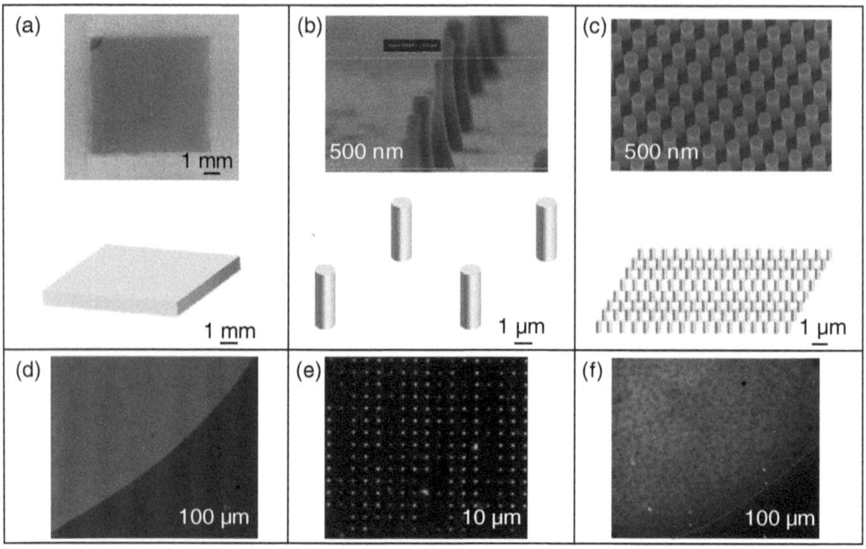

Figure 9.23 (a) Optical image of a 6H–SiC substrate and its schematic representation, (b) SEM view of SiC NP and its schematic representation, (c) SEM view of core–shell NP and its schematic representation, (d) epifluorescence microscopy image of a hybridized DNA drop at the edge of a SiC substrate, (e) epifluorescence microscopy image of functionalized SiC NPs after DNA hybridization, (f) epifluorescence microscopy image of functionalized core–shell Si–SiC NPs after DNA hybridization.

spots are observed using fluorescence microscopy. In Fig. 9.23d, the edge of a fluorescent spot is clearly visible. Such a spot indicates the specific hybridization of complementary DNA strands with the immobilized DNA probe strands. It validates the efficiency of the functionalization process involving APTES in the vapor phase on a planar SiC surface.

To further investigate the DNA presence, physicochemical analyses were carried out using XPS. This technique characterizes the chemical state of the extreme surface (first 10 nm). Due to the natural surface charging that occurs with these types of materials, the analyses have been performed using a flood gun. The overall charge compensation gives a C1s atmospheric carbon contamination peak at 284.6 eV.

The DNA single-strand molecule is composed of successive nucleotides containing a phosphate group, a deoxyribose, and a nucleobase (Fig. 9.21). While the APTES molecule just contains one kind of nitrogen atom, each DNA strand contains various kinds of nitrogen atoms provided by the different nucleobases as well as specific phosphorous atoms. Three different samples issued from a highly polished on-axis 6H–SiC wafer are analyzed: a bare SiC substrate, a silanized SiC substrate, and a SiC substrate after DNA hybridization.

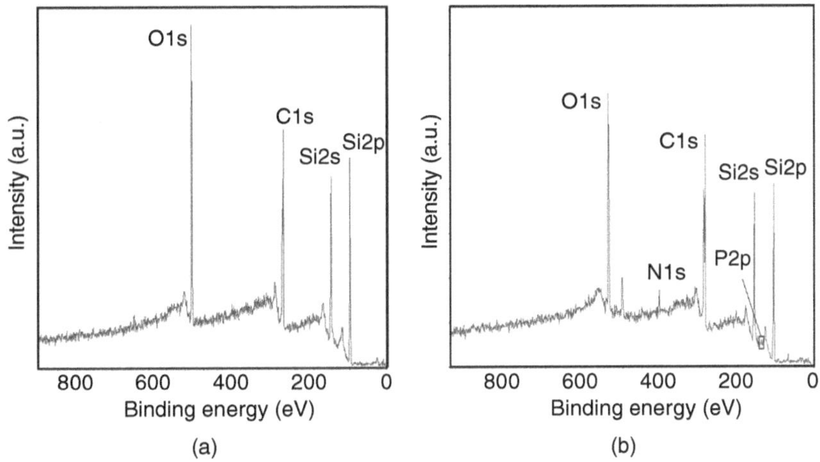

Figure 9.24 XPS surveys of (a) bare 6H–SiC and (b) biomodified 6H–SiC sample with hybridized DNA.

The general XPS survey of the bare sample (Fig. 9.24a) exhibits mainly the presence of silicon, carbon and oxygen, the latter coming from organic contamination. In the case of the silanized sample, if similar elements are found (C, Si, and O), an additional nitrogen peak is detected which confirms the presence of APTES on the SiC surface. In the case of the SiC sample modified with hybridized DNA, the XPS general survey (Fig. 9.24b) exhibits the presence of carbon and silicon coming from both the SiC substrate as well as organic molecules. Besides, other chemical species are detected such as nitrogen and especially phosphorus. Fig. 9.25a and c presents the N1s high-resolution spectra in the case of both silanized samples and after DNA

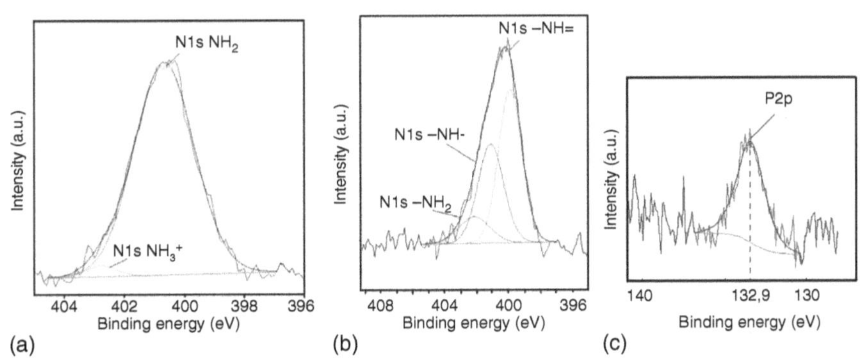

Figure 9.25 XPS high-resolution spectra of (a) N1s in the case of a silanized SiC surface, (b) N1s in the case of a hybridized DNA on SiC surface, and (c) P2p in the case of hybridized DNA on the SiC surface.

hybridization while Fig. 9.25b presents the P2p high-resolution spectrum. The P2p peak located at a binding energy of 132.7 eV is characteristic of phosphate groups in the DNA molecule. The N1s peak obtained after SiC silanization is constituted of both mainly free amine, $—NH_2$, and to a lesser extent a protonated amine, $—NH_3^+$. On the contrary, after DNA hybridization, the N1s peak (Fig. 9.25b) can be fitted using three peaks: —N═, —NH—, and $—NH_2$. The distance between these peaks is fixed as follows: 1.22 eV between —N= and —NH— and 1.04 eV between —NH— and $—NH_2$ which is a typical result for a surface covered by DNA [82,83]. Consequently, the result of the chemical analysis of the surface by XPS demonstrates the presence of DNA on the SiC surface after hybridization.

9.4.3.2 SiC Nanopillar Array

On the basis of the previous and highly successful results, the functionalization process was transferred onto different SiC NP arrays. Fig. 9.23e shows a top view epifluorescence microscopy image of SiC NPs obtained by plasma etching after the DNA functionalization process and DNA hybridization. From this image, an array of fluorescent spots is clearly observed. Each spot corresponds to one NP. The fluorescence is more intense on and around the NPs than on the planar region due to their high surface/volume ratio. By considering that the NPs have cylinder shapes the sum of the lateral and top pillar areas is equal to 10 μm^2 while the base area is 2 μm^2.

From this top-view observation, the repartition of DNA molecules along the NPs could not be determined. That is why the confocal laser-scanning microscope was used to take images on different focusing plans. An image set is taken from the ground plane up to 4 μm with a step of 0.25 μm. By assembling the 17 pictures, a 3D stereoscopic fluorescent view of the surface can be built (Fig. 9.26a). In this set, while keeping the focusing plane away from the substrate, it can be observed that the fluorescence intensity of the substrate decreases and the one of the pillar stays constant. Although pillars are 1.6 μm in height, the fluorescence is visible up to 4 μm away from the surface because of vertical light scattering. This representation shows the regularity and the important intensity of the pillar fluorescence. This result is a visual proof of the interest of nanostructures for nanosensor sensitivity improvement. However, the question is: are the DNA molecules uniformly distributed along the NP side or are they more concentrated at the top or at the bottom of the NP?

Simulations were conducted to further investigate the distribution of DNA molecules along the NP. To do this, the pillar was meshed into 17 cylindrical finite elements. The fluorescence intensity inside an element is

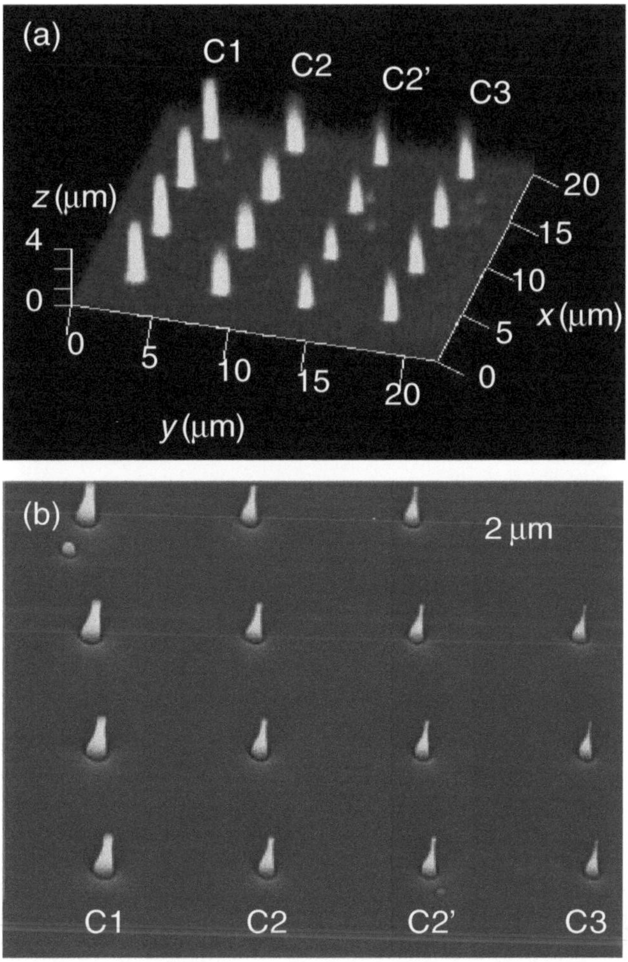

Figure 9.26 (a) Three-dimensional reconstructed view of fluorescent DNA on NPs obtained by confocal scanning laser fluorescence microscopy, and (b) SEM view of the SiC NPs.

considered as a constant that can be calculated from the intensity provided by the DNA of the element plus the sum of intensities coming from the neighboring elements. To calculate the fluorescence intensity contribution of DNA from other elements, the Bouguer law Eq. (9.4) is used.

$$I = a + \frac{b}{(c - z)^2} \tag{9.4}$$

where a is the background noise intensity in arbitrary units, b the intensity when the distance between the source and the detector is zero in arbitrary units, c the distance offset in micrometer, and z the vertical coordinate in micrometer.

This law describes the intensity detected by a sensor following the distance between the source and the sensor. The fluorescence intensity is considered to be similar for all fluorophores. This implies that the constant b is assimilated to the quantity of grafted DNA in the element. The fluorescence intensity is measured for the 17 slices of the stack. The evolution of the intensity is investigated as a function of z: for different nanostructures and for an area without NPs where all the DNA molecules are located in the $z = 0$ element. Collected data are used to determine the a, b, and c parameters. An algorithm is then used to find the b parameter of all the elements in order to fit the measured data on a NP. From this method, it is possible to approximate the most probable repartition of DNA molecules along the NP.

Values of a, b, and c were then deduced from calculation as $a = 50$ a.u., $b = 45750$ a.u., and $c = 3.82$ μm. Resulting simulations show the fluorescent intensity as a function of z in the case of three hypotheses (Fig. 9.27a). In the first hypothesis, DNA molecules are mainly located on the pillar base. In this case, the intensity continuously decreases with z and the curve is concave (Fig. 9.27a, case 1). In the case of a localization of the DNA molecules on the top of the NP (Fig. 9.27a, case 2) the curve is composed of two concave parts. Finally, if DNA molecules are equally distributed along the structure, the corresponding curve exhibits for low z a convex part, and for high z a concave one (Fig. 9.27c, case 3). To compare these calculated behaviors to the measured ones, the geometrical differences between the pillars are to be taken into account. Indeed, after the e-beam lithography step, three main types of pillars can be distinguished as shown in Fig. 9.26a and b. The first type is represented by a pillar from the C1 column on Fig. 9.26a and b. These pillars are well defined and are the tallest one. Pillars from the C2 and C2′ lines are smaller and exhibit a more slender top than the previous one. C3-type pillars are blurring and needle shaped. Depending on the pillar geometry, the fluorescence intensity is plotted as a function of z in Fig. 9.27b in the case of each pillar type. The three curves present a convex part for lower z and a concave one for high z. The curve of C3-type pillar is considered as an intermediate case between case 1 and case 3 because its convex part is very limited. Finally, it can be concluded that the DNA molecules should be located all along the NP for types C1 and C2. The particular morphology of the top of the C3 type of pillar leads to nonuniform DNA grafting.

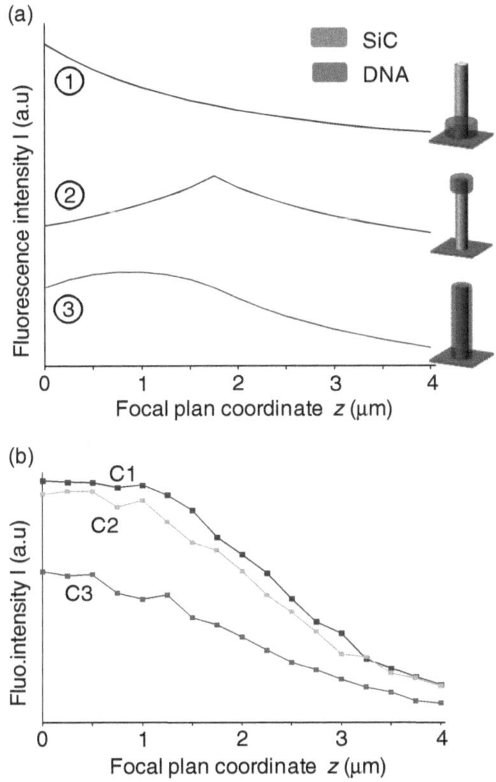

Figure 9.27 (a) Calculated distribution of the fluorescence intensity as a function of the focal plane coordinate for different locations of DNA along the SiC NP. (1) DNA molecules located around the pillar base. (2) DNA molecules located on top of the pillar. (3) DNA molecules located along the pillar side. (b) Measured fluorescence intensity data obtained in the case of C1, C2, and C3 type pillars (see text and Fig. 9.26 for C1, C2, C3 definition).

9.4.3.3 Si–SiC Core–Shell Nanowire Array

The second array of nanostructures is composed of core–shell Si–SiC NPs. An SEM view of this array is shown in Fig. 9.23c. Fig. 9.23f is a top-view image of the array after functionalization and DNA hybridization obtained by epifluorescence microscopy. In contrast with the discrete distribution of fluorescence obtained in the case of the previous array (Fig. 9.23b), a unique fluorescent spot covering the entire array is observed. The fluorescence intensity is more homogeneous at the center of the spot, whereas a gradient in intensity is observed all around the center (area between continuous and dotted white lines in Fig. 9.23f). Such a gradient in fluorescence is due to the spreading of the DNA probe liquid droplet caused by the capillary effect.

The comparison between this result and the previous one highlights the importance of the global nanostructure morphology on the functionalization process efficiency. The pitch distance seems to be an important parameter. Depending on the shape of the nanostructure, the DNA molecules could be concentrated around the pillars or ignore them. The first result shows that a more complete and systematic study concerning the influence of both size and pitch of a nanostructure array on functionalization is necessary.

9.4.3.4 SiC Nanowire-Based Sensor

Finally, the process is applied on one single NP used as a transistor channel – the SiC NWFET-based sensor. For this specific application, the functionalization process is locally performed thanks to e-beam lithography. Due to the size of the area of interest, fluorescence microscopy is combined with AFM measurements to characterize the functionalized transistor [77].

In this case, the functionalization of the nanostructure is made after the device fabrication process because of the 600°C temperature required for nickel–gold contact annealing. The FET fabrication process is fully described in [84]. The DNA probes must be localized accurately on the NW to guarantee the best sensitivity. Indeed, to detect the lowest level of concentration, all the DNA targets must hybridize in the sensitive area. To precisely localize the DNA probes, the functionalization process is coupled with lithography and lift-off steps (Fig. 9.28). Prior to silanization, an electrosensitive resist, PMMA diluted at 4%, is spin coated on the bare SiC sample. Then, a 1-μm wide area is opened by electronic lithography and development along the NW. The parameters used for spin coating are 4000 revolutions per second for 1 min and an exposure dose of 220 $\mu As/cm^2$. The solution used for development is a mixture of methylisobutylketon and isopropanol (3:1). Then, the functionalization process is started with silanization followed by glutaraldehyde grafting. The sample is immersed in acetone for 20 min to remove the resist. After lift-off, the functionalization is ended by DNA grafting. In this study, electronic lithography is chosen

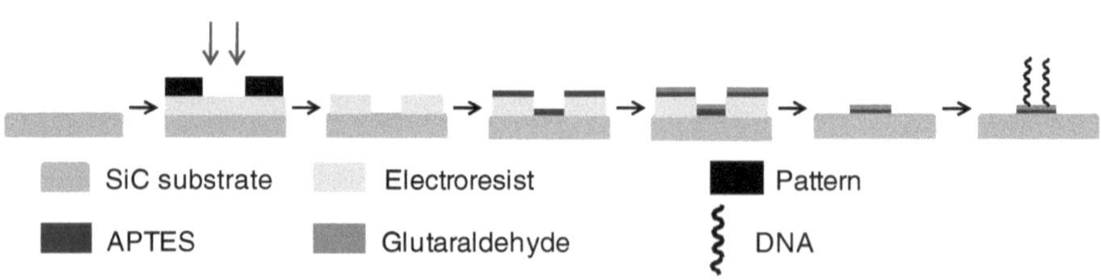

Figure 9.28 ***Process of localized functionalization of SiC surface.***

because it is possible to use the same pattern twice in the process. Optical lithography can also be used to create DNA patterns [85] but the resolution is limited. Moreover, e-beam lithography allows one to specifically functionalize areas smaller than 0.5 μm^2. But in the present case, a wider area is processed to keep the NW away from the border effects. Unlike the process described in Ref. [86], lift-off is performed after glutaraldehyde grafting because the molecule is very reactive and will be physisorbed on the substrate if the resist is removed before. Lift-off is facilitated by the large difference between the thickness of the PMMA and APTES layers, which are 270 nm versus 1 nm. The immersion in acetone does not modify the properties of the organic layer.

In Fig. 9.29a is a detailed view of the e-beam lithography pattern used during the SiC NWFET realization for contacting the NW to the electrodes.

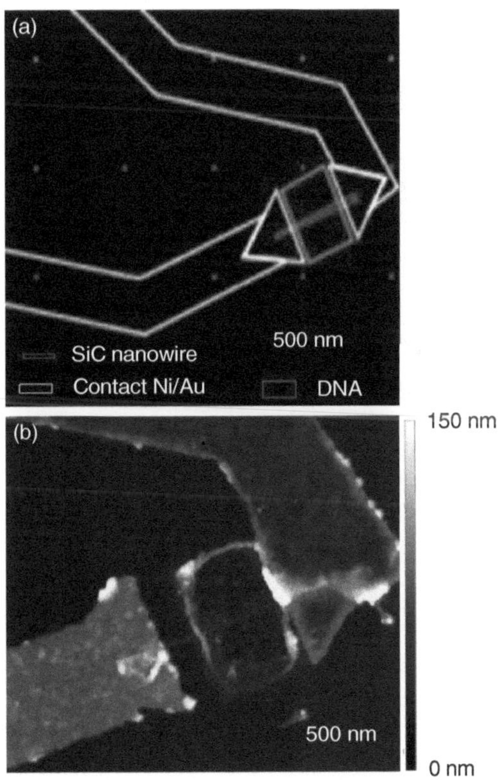

Figure 9.29 (a) Detailed view of a pattern used for SiC NWFET fabrication and functionalization, (b) AFM topography mode image of the same area after functionalization and DNA hybridization.

The same pattern was used again during the functionalization to circumscribe DNA grafting and hybridization around the NW. The area to be functionalized (in green on the picture) is 1 μm in length, the same length as the NW, and 2 μm in width. Following the functionalization process and DNA hybridization step, the resulting NWFET is observed using AFM (Fig. 9.29b). The NW is still connected to the metal electrodes as is clearly seen at the center of the image. The functionalized area around the NW can be distinguished from two features. First, its surrounding 100-nm high edges are composed of resist and polymerized APTES residues resulting from border effects. Second, from depth profiling measurement, it can be deduced that the center of this area is at a higher level than the external ground substrate. The difference of level is 10 ± 3 nm, corresponding to the thickness of APTES and DNA layers. It can be observed that the center exhibits a homogeneous and smooth surface state, which is suitable for DNA immobilization.

9.4.4 Conclusion

Thanks to different characterization techniques such as fluorescence, XPS, and AFM, we could validate the covalent DNA probe grafting and hybridization on different SiC nanostructures. This functionalization combined with a lithography process allowed the realization of DNA functionalized SiC NWFETs. Further experiments, which are described in the following section, consist in the electrical characterization of such systems to perform electrical label-free DNA hybridization detection.

9.5 ELECTRICAL DETECTION OF DNA

This section reports early results on the electrical detection of DNA grafting and hybridization using SiC NWFET devices as resistors [78]. We remind the reader that DNA is negatively charged due to nonprotonated phosphate groups (Fig. 9.21).

Here, for electrical DNA detection, SiC NWFETs are fabricated using 3C–SiC NWs prepared by vapor phase epitaxy [87]. The NW length ranges from 5 to 10 μm for a diameter between 20 and 50 nm. Then, the NWs are connected using e-beam lithography. A chip is prepared with several NWFETs, which are electrically characterized measuring the device I_d–V_d and I_d–V_g characteristics. A typical I_d–V_d curve of NWFETs with Schottky contacts is reported in Fig. 9.30. The I–V curve exhibits a symmetric shape with no hysteresis. The polarization of the back-gate electrode induces a weak but significant change of the I_d current.

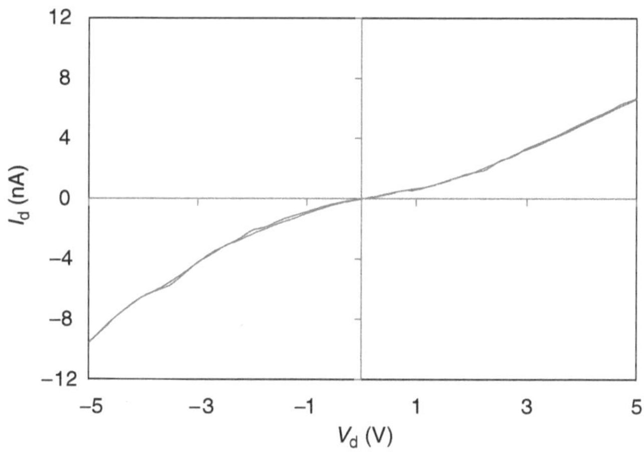

Figure 9.30 $I_d = f(V_d)$ ***characteristic after the fabrication of the SiC NWFET*** ($V_g = 0$ V).

After SiC NWFET fabrication, the functionalization process is performed resulting in covalent DNA probe molecule grafting on the SiC NWs. More particularly two SiC NWFETs are selected for DNA detection (Fig. 9.31). Thanks to e-beam lithography and the lift-off technique, only one of these SiC NWFETs is functionalized (sensor) whereas the other one is not functionalized, even though both are located on the same chip. The nonfunctionalized one acts as a reference NWFET for the electrical DNA detection measurement.

Prior to the functionalization process, e-beam lithography was used to open a cavity (1 × 2 μm^2) around the SiC NWFET sensor while the other one (reference SiC NWFET) was still covered by the resist. As detailed in Section 9.4.2, the functionalization process includes a silanization step with APTES deposited in the vapor phase and glutaraldehyde

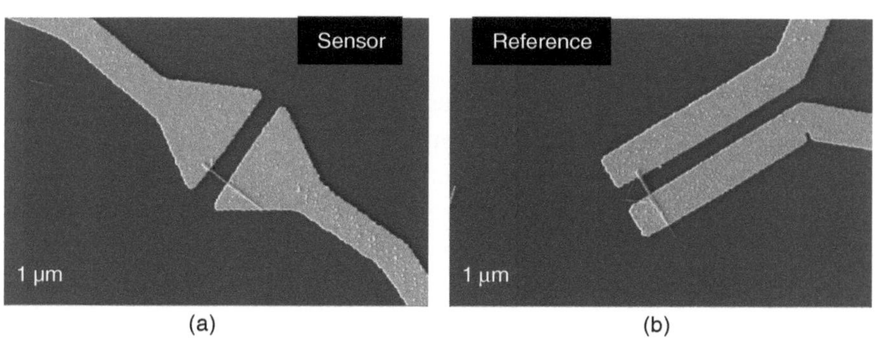

Figure 9.31 ***SEM images of two SiC NWFETs: sensor (a) and reference (b) devices.***

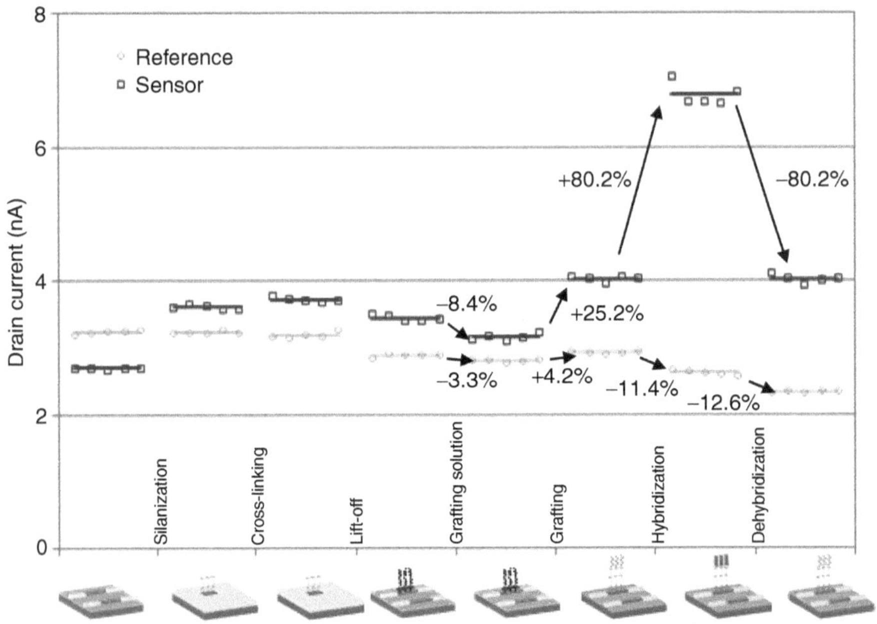

Figure 9.32 ***Plot of the drain current evolution of both, the functionalized SiC NWFET and the unfunctionalized (control) NWFET after different steps.*** The drain voltage is equal to 5 V while the back gate is grounded.

as a cross-linker. Afterward, a lift-off is performed by immersing the chip in acetone. Then, the last step is the grafting of the DNA probes. The response of both NWFETs was tested at each process step as well as by performing (1) hybridization with complementary DNA targets, and (2) dehybridization (or denaturation) achieved by immersing the chip in NaOH solution.

Between each step of the functionalization processes, the chip is dried and I_d–V_d characteristics of both FETs measured five times. Fig. 9.32 shows the respective evolution of the drain current I_d for both NWFETs after these different steps for a value of $V_d = 5$ V and $V_g = 0$ V. To quantify its variation between each step, the change of the mean current value in percentage is calculated by taking a common reference, which is the current mean value obtained after lift-off.

Apart from silanization, from cross-linking to lift-off, no significant change in the current was observed. Then, the grafting buffer solution (phosphate solution containing no DNA probe) deposited on both NWFETs does not modify the current significantly. The weak current variation proves that the phosphate solution, which is used for further DNA probe grafting, is not effective on the current. On the other hand, after grafting and hybridization of the DNA probes, the mean current strongly increases for the functionalized NWFET sensor by 25% and 80%, respectively, while keeping relatively constant values (4% and 10%) for the nonfunctionalized NWFET. After DNA dehybridization, the change of the current obtained for the functionalized sensor is opposite to the change obtained after DNA hybridization (–80%), while it is still –10% in the case of the nonfunctionalized NWFET. The symmetric results obtained in the case of the functionalized SiC NWFET prove (1) the reversibility of the dehybridization reaction and (2) the reliability of the sensor. In addition, the increase of the current upon DNA grafting and hybridization seems to indicate a p-doping characteristic of the SiC NWs.

All of these results prove the specificity, reversibility, stability, and reliability of the SiC NWFET sensor to detect DNA. These results are preliminary but they provide the basis for promising devices that fully exploit the excellent properties of SiC. Further investigations will have to include more statistical measurements and characterization at lower DNA concentrations as well as endurance in terms of long-time measurements, influence of parameters such as temperature, and chemical inertness for sensing in biological liquids.

9.6 SUMMARY

In the first part of this chapter we described the physical and morphological characteristics of different SiC nanostructures, such as SiC NPs and Si–SiC core–shell NWs. These nanostructures have been elaborated respectively using two alternative methods: ICP etching and a carburization method. Then a technological process to fabricate SiC NWFETs was described. In the second part, the chapter focused on the surface modification of the as-grown SiC nanostructures in view of label-free electrical DNA detection. A functionalization process has been developed to covalently graft DNA probes onto SiC surfaces. DNA grafting and hybridization have been validated using various characterization techniques: fluorescence microscopy, XPS, and AFM. Finally, preliminary results of label-free electrical DNA detection using SiC NWFETs were presented. The obtained current evolution showed the specificity, reversibility, stability, and reliability of the SiC NWFET sensor. These results provide the basis for promising sensing devices that fully exploit the excellent properties of SiC.

ACKNOWLEDGMENTS

We want to thank several colleagues for providing us samples: Thierry Baron, Mickael Martin, and Bassem Salem (LTM, Grenoble) for the Si NWs; Giovanni Attolini (IMEN, Parma) for VLS SiC NWs growth by VLS; Didier Chaussende (LMGP, Grenoble) for the bulk SiC sample grown by PVT; and Anne Henry (IFM, Linköping) for the SiC–3C epilayer. The authors want also to thank Laurence Latu-Romain (LTM, Grenoble) who cosupervised the PhD thesis of Maelig Ollivier and Ji-Hoon Choi; and Arnaud Mantoux (SiMaP, Grenoble) for his support in the carburization process. Finally, the authors thank the members of the technical staff of the PTA facilities at Grenoble for their technical support and also: Gérard Ghibaudo, Mireille Mouis (IMEP-LAHC), and Konstantinos Zekentes (FORTH) for fruitful discussions.

This work was supported by Grenoble INP BQR (Bonus-Quality-Research) funds and the European Network for Excellence NANOFUNCTION (EC contract 257375).

REFERENCES

[1] Beaver JA, Jelovac D, Balukrishna S, Cochran R, Croessmann S, Zabransky DJ, Park BH. Detection of cancer DNA in plasma of early stage breast cancer patients. Clin Cancer Res 2014;20(10):2643.

[2] Sassolas A, Leca-Bouvier BD, Blum L. DNA biosensors and microarrays. Chem Rev 2008;108:109.

[3] Hunt HK, Armani AM. Label free biological and chemical sensors. Nanoscale 2010; 2:1544.

[4] Cui Y, Wei Q, Park H, Lieber CM. Nanowire nanosensors for highly sensitive and selective detection of biological and chemical species. Science 2001;293:1289.

[5] Patolsky F, Zheng G, Lieber CM. Nanowire-based biosensors. Nat Protoc 2006;1:1711.

[6] Chen K-I, Li B-R, Chen Y-T. Silicon nanowire field effect transistor-based biosensors for biomedical diagnosis and cellular recording investigation. Nano Today 2011;6:131.

[7] Chen CP, Ganguly A, Lu CY, Chen TY, Kuo CC, Chen RS, Tu WH, Fischer WB, Chen KH, Chen LC. Ultra-sensitive in situ label free DNA detection using GaN nanowire base extended gate field effect transistor sensor. Anal Chem 2011;83:1938.

[8] Balasubramanian K, Kern K. Label free electrical biodetection using carbon nanostructures. Adv Mater 2014;26:1154.

[9] Dubuisson E, Yang Z, Loh KP. Optimizing label free DNA electrical detection on graphene platform. Anal Chem 2011;83:2452.

[10] Baek Y-K, Yoo SM, Kim JH, Jung DH, Choi YK, Kim YS, Lee SY, Jung HT. The effect of network density on the DNA-sensing performance of single-walled carbon nanotubes. J Phys Chem C 2009;113:21566.

[11] Gui EL, Li LJ, Zhang K, Xu Y, Dong X, Ho X, Lee PS, Kasim J, Shen ZX, Rogers A, Mhaisalkar SG. DNA sensing by field-effect transistors based on networks of carbon nanotubes. J Am Chem Soc 2007;129:14427.

[12] Seong H-K, Choi H-J, Lee S-K, Lee J-I, Choi D-J. Optical and electrical transport properties in silicon carbide nanowires. Appl Phys Lett 2004;85:1256.

[13] Zhou W, Liu X, Zhang Y. Simple approach to β-SiC nanowires: synthesis, optical, and electrical properties. Appl Phys Lett 2006;89:223124.

[14] Deeken CR, Esebua M, Bachman SL, Ramshaw BJ, Grant SA. Assessment of the biocompatibility of two novel, bionanocomposite scaffolds in a rodent model. J Biomed Mater Res B Appl Biomater 2011;96B:351.

[15] González P, Serra J, Liste S, Chiussi S, León B, Pérez-Amor M, Martínez-Fernández J, de Arellano-López AR, Varela-Feria FM. New biomorphic SiC ceramics coated with bioactive glass for biomedical applications. Biomaterials 2003;24:4827.
[16] Santavirta S, Takagi M, Nordsletten L, Anttila A, Lappalainen R, Konttinen YT. Biocompatibility of silicon carbide in colony formation test *in vitro*. Arch Orthop Trauma Surg 1998;118:89.
[17] Saddow SE, Editor Silicon carbide biotechnology: a biocompatible semiconductor for advanced biomedical devices and applications. Waltham, MA: Elsevier; 2012.
[18] Frewin CL, Locke C, Saddow SE, Weeber EJ. Single-crystal cubic silicon carbide: an *in vivo* biocompatible semiconductor for brain machine interface devices, Conference Proceedings of Engineering in Medicine and Biology Society, EMBC, 2011 Annual International Conference of the IEEE, p. 2957; 2011.
[19] Deeken CR, Fox DB, Bachman SL, Ramshaw BJ, Grant SA. Characterization of bionanocomposite scaffolds comprised of amine-functionalized gold nanoparticles and silicon carbide nanowires crosslinked to an acellular porcine tendon. J Biomed Mater Res B Appl Biomater 2011;97:334.
[20] Zekentes K, Rogdakis K. SiC nanowires: material and devices. J Appl Phys 2011;44: 133001.
[21] Godignon P, Martin I, Gabriel G, Gomez R, Placidi M, Villa R. New generation of SiC based biodevices implemented on 4″ wafers. Mater Sci Forum 2010;645–648: 1097.
[22] Senesky DG, Jamshidi B, Cheng KB, Pisano AP. Harsh environment silicon carbide sensors for health and performance monitoring of aerospace systems: a review. IEEE Sens J 2009;9:1472.
[23] Lloyd Spetz A, Huotari J, Bur C, Bjorklund R, Lappalainen J, Jantunen H, Schütze A, Andersson M. Chemical sensor systems for emission control from combustions. Sens Actuators B Chem 2013;187:184.
[24] Andersson M, Pearce R, Lloyd Spetz A. New generation SiC based field effect transistor gas sensors. Sens Actuators B Chem 2013;179:95.
[25] Kanungo J, Anderson M, Darmastuti Z, Basu S, Käll P-O, Ojamäe L, Spetz AL. Development of SiC-FET methanol sensor. Sens Actuators B Chem 2011;160:72.
[26] Guo J, Xiong S, Wu X, Shen J, Chu PK. In situ probing of intracellular pH by fluorescence from inorganic nanoparticles. Biomaterials 2013;34:9183.
[27] Yang N, Zhuang H, Hoffmann R, Smirnov W, Hees J, Jiang X, Nebel CE. Nanocrystalline 3C-SiC electrode for biosensing applications. Anal Chem 2011;83:5827.
[28] Williams EH, Schreifels JA, Rao MV, Davydov AV, Oleshko VP, Lin NJ, Steffens KL, Krylyuk S, Bertness KA, Manocchi AK, Koshka Y. Selective streptavidin bioconjugation on silicon and silicon carbide nanowires for biosensor applications. J Mater Res 2013;28:68.
[29] Zhong ZH, Wang DL, Cui Y, Bockrath MW, Lieber CM. Nanowire crossbar arrays as address decoders for integrated nanosystems. Science 2003;302:1377.
[30] Huang MH, Mao S, Feick H, Yan HQ, Wu YY, Kind H, Weber E, Russo R, Yang PD. Room-temperature ultraviolet nanowire nanolasers. Science 2001;292:1897.
[31] Patolsky F, Zheng GF, Hayden O, Lakadamyali M, Zhuang XW, Lieber CM. Electrical detection of single viruses. Proc Natl Acad Sci 2004;101:14017.
[32] Seong HK, Choi HJ, Lee SK, Lee JI, Choi DJ. Optical and electrical transport properties in silicon carbide nanowires. Appl Phys Lett 2004;85:1256.
[33] Bechelany M, Brioude A, Stadelmann P, Ferro G, Cornu D, Miele P. Very long SiC-based coaxial nanocables with tunable chemical composition. Adv Funct Mater 2007;17:3251.
[34] Wang JJ, Lambers ES, Pearton SJ, Ostling M, Zetterling CM, Grow JM, Ren F, Shul RJ. Inductively coupled plasma etching of bulk 6H-SiC and thin-film SiCN in NF_3 chemistries. J Vacuum Sci Technol A 1998;16:2204.

[35] Jiang LD, Plank NOV, Blauw MA, Cheung R, van der Drift E. Dry etching of SiC in inductively coupled Cl2/Ar plasma. J Phys D Appl Phys 2004;37:1809.
[36] Plank NOV, Blauw MA, van der Drift EWJM, Cheung R. The etching of silicon carbide in inductively coupled SF_6/O_2 plasma. J Phys D Appl Phys 2003;36:482.
[37] Kim DW, Lee HY, Park BJ, Kim HS, Sung YJ, Chae SH, Ko YW, Yeom GY. High rate etching of 6H–SiC in SF_6-based magnetically-enhanced inductively coupled plasmas. Thin Solid Films 2004;447–448:100.
[38] Chabert P. Deep etching of silicon carbide for micromachining applications: etch rates and etch mechanisms. J Vacuum Sci Technol B 2001;19:1339.
[39] Leerungnawarat P, Hays DC, Cho H, Pearton SJ, Strong RM, Zetterling CM, Östling M. Via-hole etching for SiC. J Vacuum Sci Technol B 1999;17:2050.
[40] Feng XL, Matheny MH, Zorman CA, Mehregany M, Roukes ML. Low voltage nanoelectromechanical switches based on silicon carbide nanowires. Nano Lett 2010;10:2891.
[41] Kathalingam A, Kim MR, Chae YS, Sudhakar S, Mahalingam T, Rhee JK. Self assembled micro masking effect in the fabrication of SiC nanopillars by ICP-RIE dry etching. Appl Surf Sci 2011;257:3850.
[42] Cunge G, Inglebert RL, Joubert O, Vallier L, Sadeghi N. Ion flux composition in HBr/Cl_2/O_2 and HBr/$Cl_2/O_2/CF_4$ chemistries during silicon etching in industrial high-density plasmas. J Vacuum Sci Technol B 2002;20:2137.
[43] Chaussende D, Ucar M, Auvray L, Baillet F, Pons M, Madar R. Control of the supersaturation in the CF-PVT process for the growth of silicon carbide crystals: research and applications. Crystal Growth Des 2005;5:1539.
[44] Khan FA, Adesida I. High rate etching of SiC using inductively coupled plasma reactive ion etching in SF6-based gas mixtures. Appl Phys Lett 1999;75:2268.
[45] Ning J, Gong QC, Sun GS, Liu ZL. The ICP etching technology of 3C-SiC films. J Phys Conf Ser 2006;34:511.
[46] Voss LF, Ip K, Peartona SJ, Shul RJ, Overberg ME, Baca AG, Sanchez C, Stevens J, Martinez M, Armendariz MG, Wouters GA. SiC via fabrication for wide-band-gap high electron mobility transistor/microwave monolithic integrated circuit devices. J Vacuum Sci Technol B 2008;26:487.
[47] Zeze DA, Forrest RD, Carey JD, Cox DC, Robertson ID, Weiss BL, Silva SRP. Reactive ion etching of quartz and Pyrex for microelectronic applications. J Appl Phys 2002;92:3624.
[48] Cho H, Lee KP, Leerungnawarat P, Chu SNG, Ren F, Pearton SJ, Zetterling CM. High density plasma via hole etching in SiC. J Vacuum Sci Technol A 2001;19:1878.
[49] Khan FA, Roof B, Zhou L, Adesida I. Etching of silicon carbide for device fabrication and through via-hole formation. J Electron Mater 2001;30:212.
[50] Kanechika M, Sugimoto N, Mitsushima Y. Control of shape of silicon needles fabricated by highly selective anisotropic dry etching. J Vacuum Sci Technol B 2002;B 20:1298.
[51] Voss LF, Ip K, Pearton SJ, Shul RJ, Overberg ME, Baca AG, Sanchez C, Stevens J, Martinez M, Armendariz MG, Wouters GA. SiC via fabrication for wide-band-gap high electron mobility transistor/microwave monolithic integrated circuit devices. J Vacuum Sci Technol 2008;B 26:487.
[52] Kim DW, Lee HY, Kyoung SJ, Kim HS, Sung YJ, Chae SH, Yeom G. Magnetically enhanced inductively coupled plasma etching of 6H-SiC. IEEE Trans Plasma Sci 2004;32:1362.
[53] Yamaguchi T, Komuro T, Koshimizu C, Takashima S, Takeda K, Kondo H, Ishikawa K, Sekine M, Hori M. Direct current superposed dual-frequency capacitively coupled plasmas in selective etching of SiOCH over SiC. J Phys D Appl Phys 2012;45: 025203.
[54] Rakhshandehroo MR, Pang SW. Fabrication of Si field emitters by dry etching and mask erosion. J Vacuum Sci Technol B 1996;14:612.

[55] Choi JH, Latu-Romain L, Bano E, Dhalluin F, Chevolleau T, Baron T. Fabrication of SiC nanopillars by inductively coupled SF_6/O_2 plasma etching. J Phys D Appl Phys 2012;45:235204.
[56] Choi JH, Latu-Romain L, Bano E, Henry A, Lee WJ, Chevolleau T, Baron T. Comparative study on dry etching of alpha- and beta-SiC nano-pillars. Mater Lett 2012;87:9–12.
[57] Yakimova R, Petoral RM Jr, Yazdi GR, Vahlerg C, Lloyd Spetz A, Uvdal K. . J Phys D Appl Phys 2007;40:6435.
[58] Martin M, Avertin S, Chevolleau T, Dhalluin F, Ollivier M, Baron T, Joubert O, Hartmann JM. . J Vacuum Sci Technol B 2013;31(4):041806.
[59] Scholz RF, Werner P, Gosele U, Tan TY. The contribution of vacancies to carbon outdiffusion in silicon. Appl Phys Lett 1999;74(3):392–4.
[60] Li JP, Steckl AJ. Nucleation and void formation mechanisms in SiC thin film growth on Si by carbonization. J Electrochem Soc 1995;142(2):634–41.
[61] Latu-Romain L, Ollivier M, Mantoux A, Auvert G, Chaix-Pluchery O, Sarigiannidou E, Bano E, Pelissier B, Roukoss C, Roussel H, Dhalluin F, Salem B, Jegenyes N, Ferro G, Chaussende D, Baron T. From Si nanowire to SiC nanotube. J Nanopart Res 2011;13:5425–33.
[62] Graul J, Wagner E. Growth mechanism of polycrystalline β-SiC layers on silicon substrate. Appl Phys Lett 1972;21(2):67–9.
[63] Ferro G, Monteil Y, Vincent H, Thevenot V, Tran MD, Cauwet F, Bouix J. Atomic force microscopy growth modeling of SiC buffer layers on Si (100) and quality optimization. J Appl Phys 1996;80:4691–702.
[64] Cimalla V, Wohner T, Pezoldt J. The diffusion coefficient of silicon in thin SiC layers as a criterion for the quality of the grown layers. In: Carter CH, Devaty RP, Rohrer GS, editors. Silicon Carbide and Related Materials – 1999 PTS, 1 & 2, vol. 338-3. Zurich-Uetikon, Switzerland: Trans Tech Publications Ltd; Material Science Forum 2000. p. 321–4. .
[65] Deguel GR, Kendall DL, Galeazzi R. Anisotropic etching of silicon at different KOH/H_2O concentrations. J Electrochem Soc 1984;131(8):C320.
[66] Ollivier M, Latu-Romain L, Martin M, David S, Mantoux A, Bano E, SouliÃ¨re V, Ferro G, Baron T. Si-SiC core-shell nanowires. J Crystal Growth 2013;363(0):158–63.
[67] Ollivier M, Latu-Romain L, Salem B, Fradetal L, Brouzet V, Choi JH, Bano E. Integration of SiC-1D nanostructures into nano-field effect transistors. Mater Sci Semicond Process 2015;29:218–22.
[68] Rogdakis K, Lee SY, Bescond M, Lee SK, Bano E, Zekentes K. 3C-silicon carbide nanowire FET: an experimental and theoretical approach. IEEE Trans Electron Devices 2008;55(8):1970–6.
[69] Chen Y, Zhang X, Zhao Q, He L, Huang C, Xie Z. P-type 3C-SiC nanowires and their optical and electrical transport properties. Chem Commun 2011;47:6398–400.
[70] Rosaz G, Salem B, Pauc N, Gentile P, Potié A, Solanki A, Baron T. High-performance silicon nanowire field-effect transistor with silicided contacts. Semicond Sci Technol 2011;26(8):085020.
[71] Goldberger J, Hochbaum AI, Fan R, Yang P. Silicon vertically integrated nanowire field effect transistors. Nano Lett 2006;6(5):973–7.
[72] Wang D, Wang Q, Javey A, Tu R, Dai H, Kim H, McIntyre PC, Krishnamohan T, Saraswat KC. Germanium nanowire field-effect transistors with SiO_2 and high-k HfO_2 gate dielectrics. Appl Phys Lett 2003;83:2432–4.
[73] Wunnicke O. Gate capacitance of back-gated nanowire field-effect transistors. Appl Phys Lett 2006;89(8):083102.
[74] Choi JH, Bano E, Henry A, Attolini G, Zekentes K. Comparison of bottom-up and top-down 3C-SiC NWFETs, International Conference on Silicon Carbide and Related Materials; 2015.
[75] Roschke M, Schwierz F. Electron mobility models for 4H–, 6H–, and 3C–SiC. IEEE Trans Electron Devices 2001;48(7):1442–7.

[76] Fradetal L, Stambouli V, Bano E, Pelissier B, Wierzbowska K, Choi JH, Latu-Romain L. First experimental functionalization results of SiC nanopillars for biosensing applications, vol. 740–742. Stafa-Zurich, Switzerland: Trans Tech Publications Ltd; Materials Science Forum 2013 . p. 821.
[77] Fradetal L, Stambouli V, Bano E, Pelissier B, Choi JH, Ollivier M, Latu-Romain L, Boudou T, Pignot-Paintrand I. Bio-functionalization of silicon carbide nanostructures for SiC nanowire-based sensors realization. J Nanosci Nanotechnol 2014;14:3391.
[78] Fradetal L, Bano E, Montès L, Attolini G, Stambouli V. Silicon carbide nanowire devices for label-free electrical DNA detection. Mater Sci Forum (Trans Tech Publications) 2015;821–823:855–8.
[79] Stambouli V, Labeau M, Matko I, Chenevier B, Renault O, Guiducci C, Chaudouet P, Roussel H, Nibkin D, Dupuis E. Development and functionalisation of Sb doped SnO_2 thin films for DNA biochip applications. Sens Actuators B Chem 2006;113:1025.
[80] Le MH, Fradetal L, Delabouglise D, Mai AT, Stambouli V. Fluorescence and label free impedimetric DNA detection on SnO_2 nanopillars. Electroanalysis 2015;27:1210.
[81] Schoell SJ, Hoeb M, Sharp ID, Steins W, Eickhoff M, Stutzmann M, Brandt MS. Functionalization of 6H-SiC surfaces with organosilanes. Appl Phys Lett 2008;92:153301.
[82] Petrovykh DY, Kimura-Suda H, Whitman LJ, Tarlov MJ. Quantitative analysis and characterization of DNA immobilized on gold. J Am Chem Soc 2003;125:5219.
[83] Mateo-Martí E, Briones C, Pradier CM, Martín-Gago JA. A DNA biosensor based on peptide nucleic acids on gold surfaces. Biosens Bioelectron 2007;22:1926.
[84] Rogdakis K, Bano E, Montes L, Bechelany M, Cornu D, Zekentes K. Rectifying source and drain contacts for effective carrier transport modulation of extremely doped SiC nanowire FETs. IEEE Trans Nanotechnol 2011;10:980.
[85] Lenci S, Tedeschi L, Pieri F, Domenici C. UV lithography-based protein patterning on silicon: towards the integration of bioactive surfaces and CMOS electronics. Appl Surf Sci 2011;257:8413.
[86] Tanii T, Hosaka T, Miyake T, Zhang G-J, Zako T, Funatsu T, Ohdomari I. Preferential immobilization of biomolecules on silicon microstructure array by means of electron beam lithography on organosilane self-assembled monolayer resist. Appl Surf Sci 2004;234(1–4):102.
[87] Attolini G, Rossi F, Bosi M, Watts BE, Salviati G. The effect of substrate type on SiC nanowire orientation. J Nanosci Nanotechnol 2011;11:1.

CHAPTER 10

Silicon Carbide-Based Nanowires for Biomedical Applications

F. Rossi*, P. Lagonegro*, M. Negri*, F. Fabbri*, G. Salviati*, R. Alinovi, S. Pinelli**, F. Ravanetti†, A. Cacchioli†, E. Bedogni‡, F. Bigi‡, T. Rimoldi§, L. Cristofolini§, C. Galli¶, A. Smerieri¶ and G.M. Macaluso¶**

*IMEM-CNR, Institute of Materials for Electronics and Magnetism – Italian National Research Council, Parma University, Parma, Italy

**Department of Clinical and Experimental Medicine, Parma University, Parma, Italy

†Department ofVeterinary Science, Unit of Normal Veterinary Anatomy – Parma University, Parma, Italy

‡Chemistry Department – Parma University, Parma, Italy

§Physics and Earth Science Department – Parma University, Parma, Italy

¶Department of Biomedical, Biotechnological, and Translational Sciences – Parma University, Parma, Italy

Contents

10.1 INTRODUCTION

Nanotechnology promotes the research and development of innovative systems with unique properties and applications in several fields, from sensors to nanomedicine [1]. For healthcare applications, the ability to tailor the material properties allows for the design of new nanosystems with enhanced performance for diagnostics, imaging, and oncotherapy [2,3]. Nanowires (NWs) based on cubic silicon carbide (3C–SiC) have a strong potential to pave the way for the development of a novel generation of implantable nanodevices since they are chemically inert and compatible in the biological environment. Indeed, 3C–SiC is a proven bio- and hemocompatible material, and some prototypes for biomedical devices have been successfully realized by micromachining of thin films, as reviewed by Zorman et al. in Chapter 10 of the first edition of this book [4]. The combination of 3C–SiC with silicon dioxide (SiO_2) in core–shell NW structures (ie, core–shell 3C–SiC–SiO_2 NWs) opens more ways to engineer the surface via functionalization and decoration with macromolecules and nanoparticles (NPs) [5,6]. The potential use of core–shell NWs in nanomedicine is driven by the presence of the amorphous shell, since it can modify the material behavior in the biological environment. For instance, for blood-contact applications, silicon oxide could induce a stronger aggregation and activation of platelets, promoting clot formation and acute inflammatory processes [7,8]. Further, a peculiar feature of the core–shell NWs is their optical emission: the oxide shell enhances the core luminescence [9–11] when the nanosystem is excited by highly energetic sources, such as electron beams or X-rays. This property opens the possibility to exploit 3C–SiC–SiO_x NWs as radiation resistant scintillation nanostructures, which can be properly functionalized to play an active role for new oncotherapies, for example, X-ray-excited photodynamic therapy [12], in nanomedicine.

In this chapter we review our recent results on the growth and characterization of NWs (Section 10.2), the *in vitro* analysis of the cytocompatibility and cellular internalization mechanisms of 3C–SiC–SiO_2 NWs (Section 10.3), the exploitation of 3C–SiC–SiO_x NWs conjugated with an organic photosensitizer for *in vitro* X-ray-excited photodynamic therapy (Section 10.4), and, finally, the characterization of NW substrates in the biological environment, through *in vitro* cell cultures and interaction with platelet-rich plasma (PRP) (Section 10.5).

10.2 3C–SiC–SiO_2 CORE–SHELL NANOWIRES: GROWTH, STRUCTURE, AND LUMINESCENCE PROPERTIES

Several synthesis methods have been reported in the literature to realize SiC NWs. The most exploited are metal-assisted chemical vapor deposition (CVD) synthesis methods [13], direct reactions such as carbothermal reduction of silica [14], and catalyst-assisted polymeric precursor pyrolysis methods [15]. In addition, physical vapor transport [16], laser ablation [17], arc discharge [18], growth from solution [19] and carbon nanotube template-mediated growth (also called carbon nanotube-confined reaction methods) [20] have also been used. Complex processes and manipulation are involved in many of these synthesis approaches, while the advantages of the CVD technique are its low cost, the possibility of high growth rates on large areas, and the control of the shape and composition of the nanostructures by changing the growth parameters (ie, temperature, precursor flux, choice of the metal catalyst, etc.). Further, a selective growth of NWs can be achieved on patterned substrates or by patterning the catalyst deposition [21].

Core–shell 3C–SiC–SiO_2 NWs can be synthesized on silicon (Si) substrates in a CVD reactor using carbon monoxide (CO) as a single precursor. We defined a process with a gaseous CO precursor [22], different from that initially proposed by Park et al. [23], who employed a solid mixture of WO_3 and C as the CO source. Our process allows for the control of the precursor flow and concentration over time, and the growth temperature can be more precisely tuned, while when using the solid precursor, CO would be released at a fixed temperature [24]. The process is metal assisted, and the choice of the catalyst has to be made on the basis of the target application. Nickel-based compounds (eg, nickel nitrate ($Ni(NO_3)_2$)) perform as the most effective catalysts, but safety standards for applications in biological environments require one to minimize the employment of nickel, which can stimulate neoplastic transformations [25]. Therefore, we recently moved toward iron catalysts (eg, ferric nitrate ($Fe(NO_3)_3$)), which is far more suitable for biomedical applications.

Substrate preparation for growth is performed in three steps: (1) cleaning in an ultrasonic bath with organic solvents; (2) etching in hydrofluoric acid (HF:H_2O 1:20 for 120 s) to remove any native surface oxide. The sample is then rinsed in deionized water and dried in a nitrogen atmosphere; (3) dipping in catalyst solution and drying at 40°C in air with controlled humidity (less than 10%). To obtain uniform NW growth over an area of at least 10 cm^2, it is useful to dissolve the catalysts in ethanol and add a nonionic surfactant (eg, oleylamine) to enhance ethanol wetting of the silicon substrate. After the previously mentioned preparation protocol, the sample is loaded in the growth chamber of an open-tube CVD system, which is purged with N_2 to remove oxygen. The temperature is raised to about 1100°C and, upon reaching stabilization, CO is introduced with N_2 as the carrier gas at a concentration of about 4‰. The growth lasts for 15–30 min, and then the growth chamber is cooled down to room temperature in nitrogen atmosphere. During the heating phase, iron(III) nitrate undergoes a decomposition to iron(III) oxide [26] and a dewetting takes place, so that the catalyst forms droplets or islands on the substrate surface. The process proceeds then as conventional vapor liquid solid (VLS) growth [27].

The typical morphology of the NWs over the growth substrate is that of a dense network of long interwoven fibers (see Fig. 10.1a). Consistent with VLS growth, the NW tips are clearly distinguishable and composition-sensitive techniques, such as scanning electron microscope (SEM) imaging

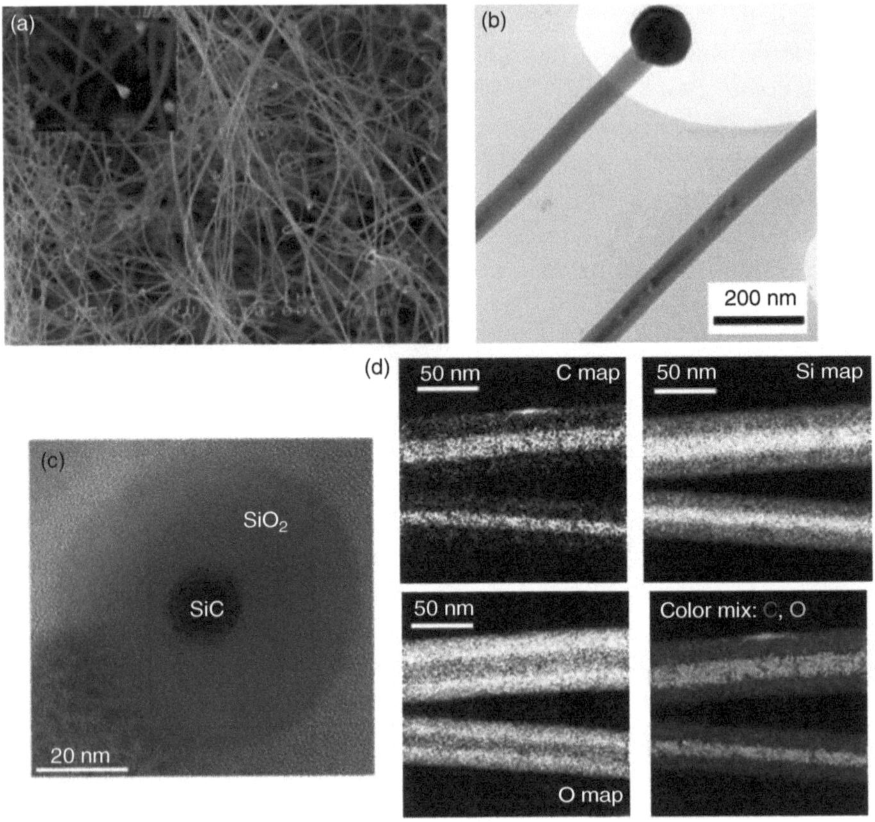

Figure 10.1 (a) SEM image of as-grown core–shell NWs on a Si substrate. Inset: detail of a NW tip. (b) TEM image of two neighboring NWs on a holey-carbon support film. (c) TEM image of a single NW in cross-section view (crystalline core). (d) Elemental maps acquired by TEM energy filtering (C K-edge at 284 eV, Si L-edge at 99 eV, O K-edge at 532 eV). *(Reprinted with permission from Ref. [51]. Copyright 2014 American Chemical Society.)*

with backscattered electrons and transmission electron microscopy (TEM) imaging in high-angle annular dark-field mode, show the presence of high-Z elements in the tip, which is mainly composed by crystalline iron silicides/silicates. From TEM studies (Fig. 10.1b–d), the NWs are shown to exhibit a core–shell structure, with a crystalline core, univocally identified as 3C–SiC by the crystal symmetry and the lattice spacing, wrapped by an amorphous SiO_x shell, with x slightly lower than or equal to 2. The NW axis defines the growth direction, which corresponds to the <111> direction of 3C–SiC. The occurrence of planar stacking faults in the core along (111) planes and of local stacking sequences typical of the 2H, 4H, and 6H polytypes is observed, as commonly found in 3C–SiC whiskers [28]. The average core diameter is around 20 nm, while the shell thickness can be tuned from about 20 nm (standard) to about 100 nm depending on the growth conditions [10]. Wet chemical etching of the SiC/SiO_2 NWs allows one to remove the amorphous shell in order to realize entirely bare crystalline 3C–SiC NWs. It has to be noted that the complete removal of the shell is not achievable by conventional HF etching [9], but it requires an ad hoc process through a standard RCA clean followed by chemical treatment with piranha solution (H_2SO_4:H_2O_2, 2:1) and a final etching (HCl:H_2O, 2:1, and HF:H_2O, 1:50) [11].

The core-to-shell ratio influences the luminescence of the nanosystem as proved, for instance, by cathodoluminescence (CL) spectroscopy. The standard luminescence of the core–shell NWs is a broad visible emission (see Fig. 10.2a). Gaussian deconvolution reveals that the broad emission is composed of three main features: the most intense at 2.69 eV, a shoulder at 2.38 eV, and a narrow emission at 2.00 eV. Two of these emissions are related to radiative centers of silicon oxide. The 2.69 eV emission is related to the silicon oxygen deficiency center (SiODCII) [29], which is the benchmark of the under-stoichiometry of the native silicon oxide shell. The 2.00 eV is related to the oxygen nonbridging center [29]. The 2.38 eV emission is instead related to the core and can be ascribed to 3C–SiC near-band edge (NBE) emission [30]. At a core-to-shell ratio of 1:1, the silicon oxide shell is beneficial to enhance the light emission yield of the 3C–SiC core [9]. Indeed, a type I band alignment [31] between 3C–SiC and SiO_2 can be hypothesized (Fig. 10.2b). The conduction and valence band offsets, experimentally determined in the case of bulk material, are equal to $\Delta E_C = 3.6$ eV and $\Delta E_V = 2.9$ eV, respectively [32]. Following electron beam excitation, an electron–hole pair is created across the SiO_x bandgap or oxide

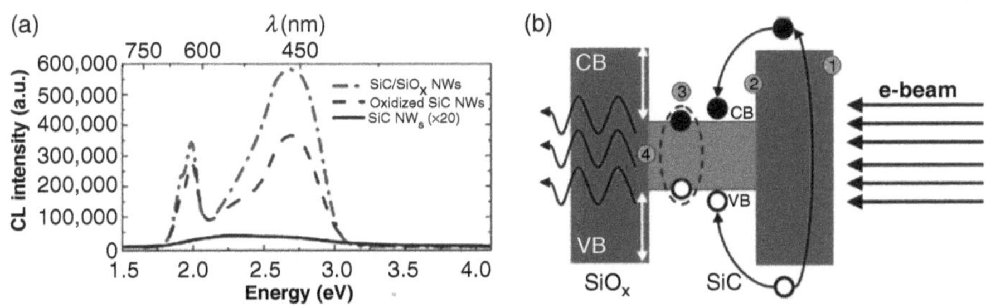

Figure 10.2 (a) Room temperature CL spectra of bare SiC NWs (solid line) and core–shell SiC–SiO_x NWs (dash-dotted line: as-grown, shell thickness 20 nm; dashed line: etched and reoxidized, shell thickness 5 nm). (b) Sketch of the enhancement mechanism of the 3C–SiC NBE emission. *(Adapted from Ref. [11].)*

intragap states (1) and the generated carriers diffuse to the 3C–SiC core (2) and there recombine (3), giving rise to the emission of a photon with the 3C–SiC NBE energy (4). In this way, the 3C–SiC luminescence at room temperature is enhanced in core–shell systems compared to bare SiC nanostructures. It is worth noting that this process occurs only with high-energy probes, such as an electron beam or X-ray photons, because the generation of carriers across the bandgap of the insulating oxide coating (E_{G-ox}, ~9 eV in the case of SiO_2) requires a probe energy at least three times higher ($3E_{G-ox}$) [33]. The diffusion of the carriers could be considered as an energy transfer from the shell to the core. Further, the shell acts not only as a carrier injector region but also as a passivation layer to reduce the nonradiative recombination related to surface states, as in entirely crystalline core–shell systems (eg, GaAs-based NWs) [34]. It has to be noted that the enhancement takes place mainly when the shell thickness is comparable to the core diameter. As the shell thickness increases, this beneficial effect is no longer observed and, instead, the shell becomes detrimental. At a shell thickness of about five times that of the core diameter (core-to-shell ratio 1:5), the SiO_x-related emissions play a dominant role [10].

10.3 *IN VITRO* CYTOCOMPATIBILITY OF 3C–SiC–SiO_2 NANOWIRES

Modern material research for nano/biomedical applications aims at developing multifunctional nanomaterials as vehicles for drug delivery [35], intracellular probes [36,37], electrical- and fluorescence-based platforms [38], etc. Any innovative nano/biomaterial in contact with animal and human cells and tissue must be cytocompatible, and to maximize the treatment efficacy it must also be internalized into the target cells [39]. Given these requirements, it is clear in the study of a nanosystem the significance of the ultrastructural study of the uptake process and the analysis of the interaction between nanomaterials and intracellular components.

Among 1D nanostructures (nanotubes and NWs), it has been assessed that some classes of materials, such as crystalline oxide NWs (eg, ZnO [40], SiO_2 [41], and TiO_2 [42]), induce cytostaticity or cell death, while others, such as metal NWs (eg, based on gold [43], iron [44], and silver [45]), are essentially cytocompatible. Contradictory results, highlighting the variability due to dimension, surface properties, biodurability, dose, and exposure time, have been reported for carbon nanotubes [46] (single walled and multiwalled) and semiconducting NWs (eg, Si [47], GaN [48], and other III–V

NWs [49,50]). Given the lack of previous *in vitro* and *in vivo* data about cytocompatibility or cytotoxicity of core–shell SiC–SiO_2 NWs, we recently started a study about the interaction of these NWs with different cell lines [51]. To perform the experiments the NWs were mechanically removed from the substrate through a mechanical process, so that the ultra-long NW fibers were broken up into short segments below 5 μm, and suspended in cell culture media for cell incubation. The NW behavior in water and in cell culture media was first studied by dynamic light scattering (DLS) and Z potential measurements on diluted suspensions. The size distribution estimated by DLS peaked around 900 nm, but showed tails down to 200 nm and up to 5 μm, in good agreement with the NW segment length measured by SEM on the dry product. The Z potential values were −(37 ± 3) mV in deionized water, −(12 ± 2) mV in Roswell Park Memorial Institute (RPMI) 1640 medium, and +(2 ± 3) mV in serum-supplemented medium. These values indicated that the material in water has a negative surface charge, likely related to O termination of the NW shell surface, and a good stability. The modification of the Z potential in cell culture media, where it becomes smaller and eventually changes sign, suggested that the medium components (eg, proteins) interact with the NWs and likely cover their surface causing a different charge compensation.

We analyzed how the NWs affect cell viability, morphology, and cell cycle progression, and specifically investigated oxidative stress, autophagy, and cellular uptake (internalization, compartmentalization, and cellular physical parameters). As representative cellular models, we selected a normal cell line (human dermal fibroblasts, HuDe, as noncancer control cells) and three human cancer cell lines: lung adenocarcinoma alveolar basal epithelial cells (A549), breast cancer cells (MCF-7), and the monocytic cell line derived from acute monocytic leukemia (THP-1). The choice of the A549 and MCF-7 models was based on the evaluation of high incidence tumors, which could be induced with standard protocols in nude mice for future *in vivo* studies. The additional THP-1 model was tested for its macrophage activity, which provides an insight into specific internalization pathways.

To inspect the midterm (up to 72 h) effects of the core–shell 3C–SiC–SiO_2 NWs on the selected cellular models, the cell viability was monitored as a function of NW concentration in the range 1−100 μg/mL. Fig. 10.3 shows the percentage of viable cells, as estimated by trypan blue exclusion and crystal violet staining assays, normalized to control untreated cells. The viability assays prove that only at the highest concentrations tested, that is,

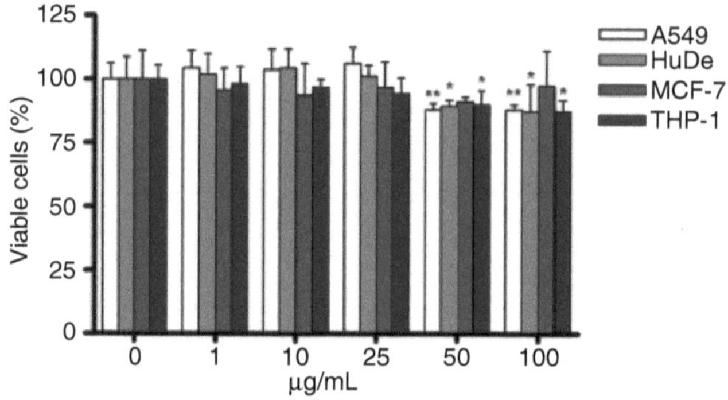

Figure 10.3 ***Cell viability assays for cells exposed to NWs from 1 to 100 μg/mL for 72 h.*** Values are normalized to controls. The figure shows the mean (±SD) of at least three separate experiments, each carried out in eight replicates. Significantly different from untreated control: *, $p < 0.05$; **, $p < 0.01$. *(Adapted from Ref. [51].)*

50 and 100 μg/mL, the cell proliferation shows a small downward trend. The effect is slightly cellular dependent: the decrease is being highly significant ($p < 0.01$) for A549 cells and significant ($p < 0.05$) for THP-1 and HuDe cells, whereas for MCF-7 the cell viability is almost not affected by exposure to the NWs.

The analysis of the cell cycle progression shows that the NWs do not induce any perturbation in any of the cellular models tested (Fig. 10.4). Further, no sub-G_0/G_1 peak, corresponding to apoptotic cells, was detected in the cytograms. The NWs do not induce apoptosis, as proved also by morphological analyses (absence of changes typical of cell apoptotic death, including nuclear condensation and apoptotic bodies) and by the lack of an increase of annexin-V FITC/propidium iodide-stained cells and of caspase-3 activation.

The oxidative stress was evaluated through quantification of reactive oxygen species (ROS), thiobarbituric acid reactive substances (TBARS), and carbonyl group levels (Fig. 10.5). While the NWs did not induce any ROS production in THP-1 and MCF-7 models, the intracellular ROS increased significantly ($p < 0.05$) and highly significantly ($p < 0.01$) in A549 and HuDe models after the first hour of treatment compared to control untreated cells. However, basal levels tended to be restored after a few hours and the recovery was complete after 24 h. As assessed by TBARS and carbonyl groups, the persistent intracellular presence of NWs caused, following 4-h incubation, oxidation of proteins and lipid peroxidation of membranes. The increase of carbonyl groups was significant ($p < 0.05$) only at the higher concentration tested (50 μg/mL) in A549, HuDe, and MCF-7 exposed cells. As for lipid peroxidation, the TBARS values were higher after

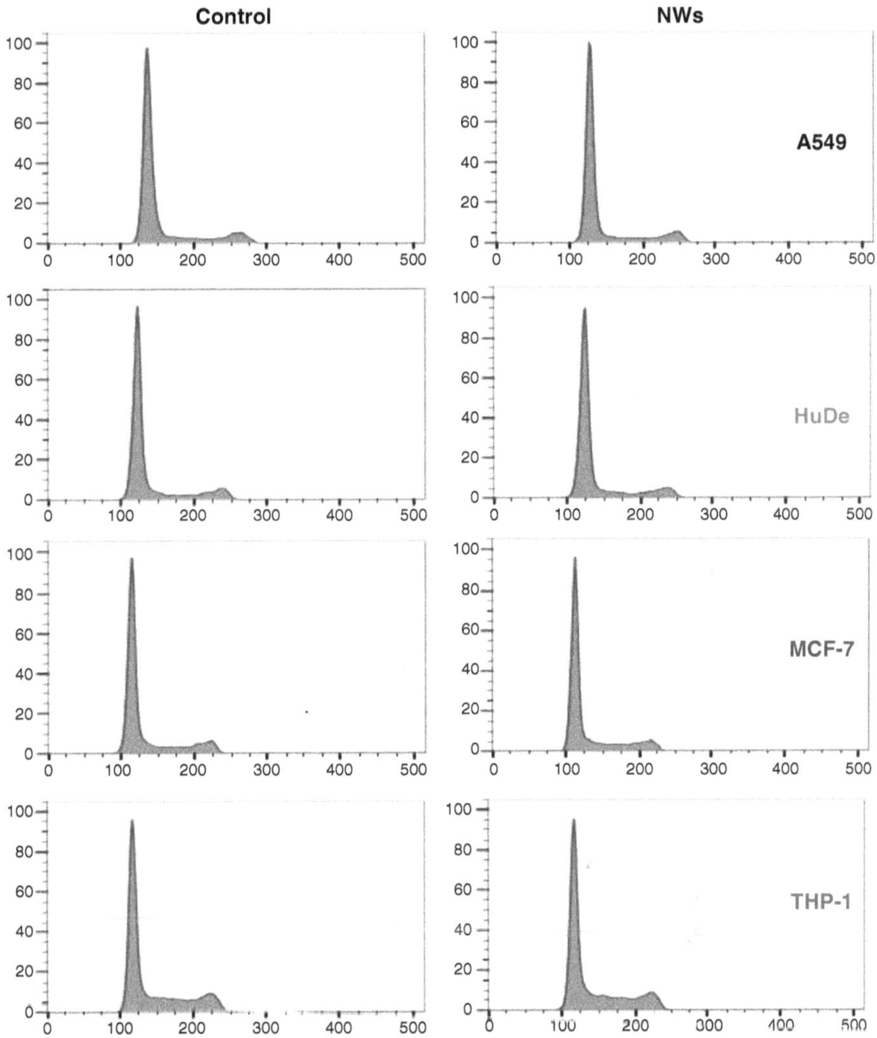

Figure 10.4 ***Monoparametric DNA analysis of the cellular cycle distribution (after 24 h, treatment with NWs at 50 μg/mL).*** The different peaks correspond to G_0/G_1, S, and G_2/M phases. *(Adapted from Ref. [51].)*

treatment with NWs in all the cellular models. A dose-dependent increase was observed in A549 and MCF-7, highly significant ($p < 0.001$) only for exposure to 50 μg/mL. In HuDe and THP-1 cultures, strong effects ($p < 0.001$) were detected also at the lowest concentration (10 μg/mL). These results indicate that the intracellular presence of NWs induced the same molecular events, that is, oxidation of proteins and peroxidation of membrane lipids, in all the cell lines, but to a different extent. These effects

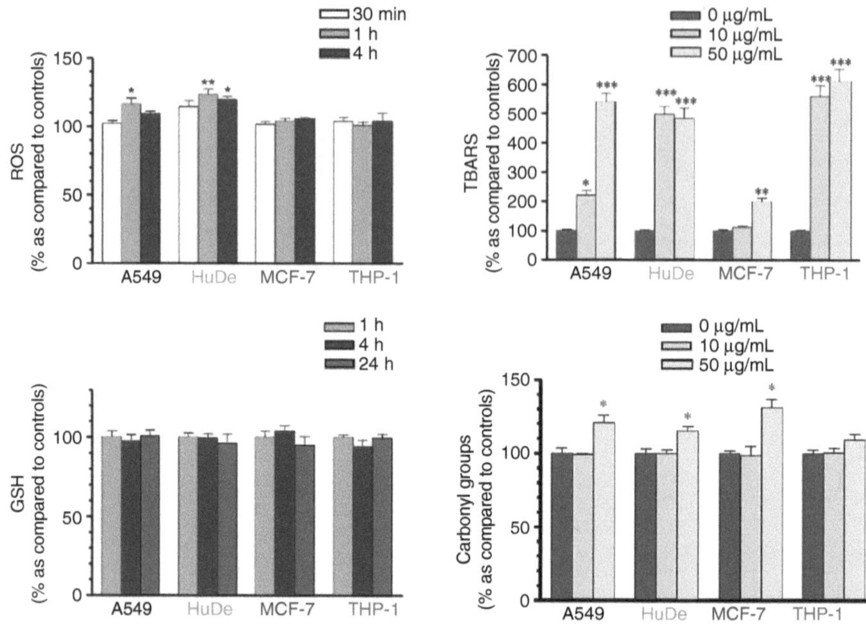

Figure 10.5 ***Effects of NW exposure on ROS, TBARS, carbonyl groups, and GSH.*** Values (percentage vs. control = [sample value/control value] × 100) are mean ± SD of three separate experiments, each carried out in triplicate. Significantly different from untreated control: *, $p < 0.05$; **, $p < 0.01$; ***, $p < 0.001$. *(Adapted from Ref. [51].)*

are late and may be limited by the activation of protection systems, so NWs do not elicit cytotoxic activity leading to irreversible cellular damage or death. The antioxidant status of the cells was examined over time by measuring the glutathione (GSH) contents, and no depletion of GSH was evidenced during the whole experimental period.

Since many biomedical applications (eg, biological sensing, drug delivery, electrophysiological experiments, etc.) require a biological interaction on a time scale of at least several days, we also tested the long-term (10 days) NWs' cytocompatibility on the three adherent cellular lines (A549, MCF-7, and HuDe). Following recent literature protocols [52], a clonogenic assay was performed. Fig. 10.6 shows the colony-forming ability of the cells exposed to the NWs, at the concentration of 10 and 50 μg/mL, for 10 days. No significant reduction was observed, confirming that even on a time scale of 10 days the core–shell 3C–SiC–SiO_2 NWs did not elicit any cytotoxic activity.

Flow cytometry revealed modifications of physical parameters correlated with NW uptake (Fig. 10.7). As proved in the literature [53,54], the side scatter (SSC) value is a reliable indicator of the presence of NPs and fibers

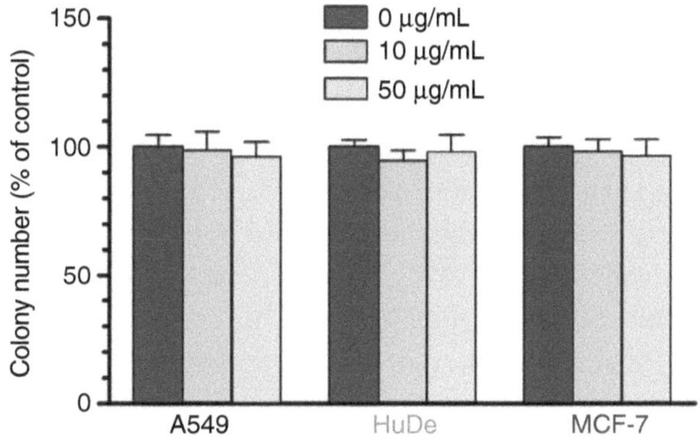

Figure 10.6 ***Effects of long-term exposure to NWs, evaluated by clonogenic assay.*** The histograms (means ± SD) report the colony-forming ability of A549, HuDe, and MCF-7 cells treated for 10 days with NWs at a concentration of 10 and 50 mg/mL, compared to control cells. *(Adapted from Ref. [51].)*

inside the cells. An increase of the SSC was detected during cell incubation with SiC–SiO_2 NWs (50 µg/mL), in particular for the A549 cellular model (mean SSC ratio [treated/control] statistically increased at 4 h [$p < 0.05$] and 24 h [$p < 0.001$]). When the NW concentration was varied, the mean SSC ratio exhibited a dose-dependent behavior, confirming that this parameter changes with the quantity of NWs internalized in cells.

No major alterations in cell morphology between cells treated with NWs and their controls were observed by SEM analysis. In particular, no morphological alteration related to cellular suffering and programmed cell death, such as membrane blebs and rounded apoptotic bodies, was detected.

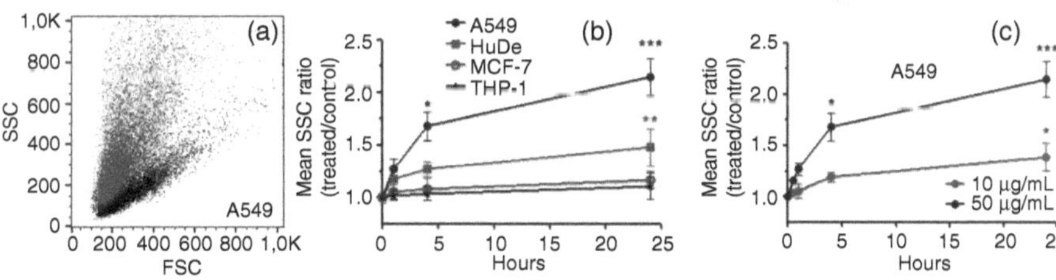

Figure 10.7 (a) Flow cytometry analysis of side scatter (SSC) versus forward scatter (FSC) in A549 cells. Unexposed cells were scanned as controls and compared to cells exposed to NWs at 50 µg/mL for 24 h. Representative cytograms are displayed. (b) Influence of the incubation time on SSC for the different cell lines. (c) Change of the SSC parameter over 24 h in A549 cells treated with NWs at 10 or 50 µg/mL. The curves indicate that uptake is fast and uptake rate is similar at the two doses. Data are expressed as mean SSC ratio (treated/control) + SD. Significantly different from untreated control: *, $p < 0.05$; **, $p < 0.01$; ***, $p < 0.001$. *(Adapted from Ref. [51].)*

However, the cellular spreading was higher in cells treated with the NWs, compared to controls. In addition cytoplasmic protrusions, in particular lamellipodia and filopodia, were detected as the structures of contact between the cell membrane and extracellular clusters of NWs. Single NWs, partly internalized inside the cells, were also detected [51].

TEM studies confirmed that the NWs did not induce shape/size changes or alterations of the nucleus/cytoplasm ratio. Clear images of NW internalization were obtained (Fig. 10.8). As discussed in the literature, the cellular uptake of a nanomaterial can occur by three main mechanisms: receptor-mediated endocytosis, pinocytosis, and phagocytosis [55]. In our

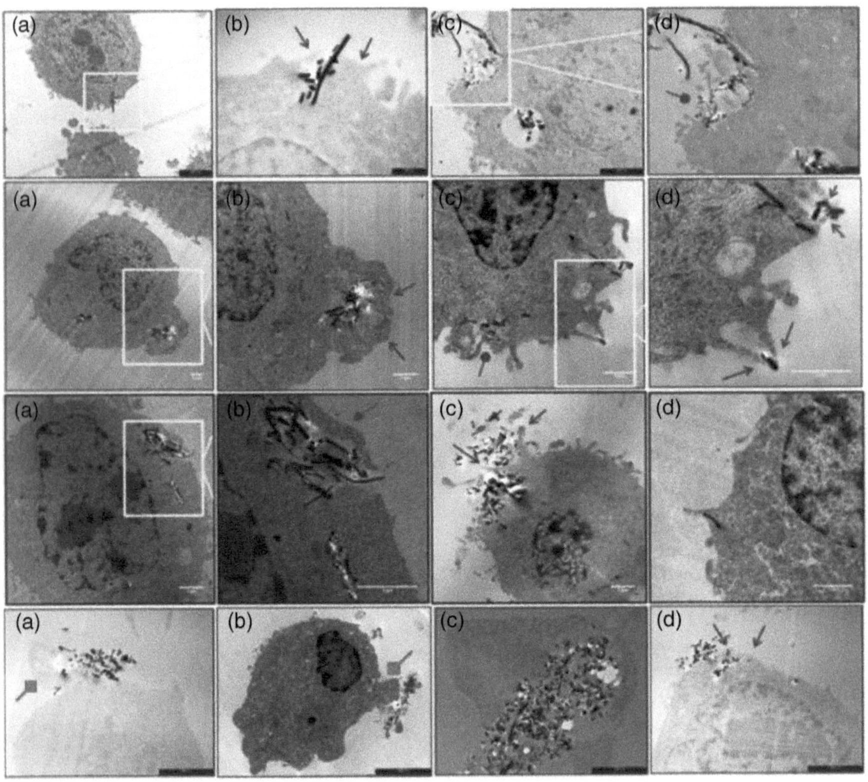

Figure 10.8 ***NW internalization analyzed by TEM in A549 cells (first row), MCF-7 cells (second row), HuDe cells (third row), and THP-1 cells (fourth row).*** The arrows indicate macropinocytosis with circular ruffles, the arrows with circular head indicate lamellipodia-like macropinocytosis, and the star indicates blebs of macropinocytosis. Pseudopodia engulfing NWs at an early (fourth row, a–b) and late (fourth row, c) stage of the phagocytosis process are indicated by arrows with square heads. A NW directly penetrating the cell membrane is visible in the micrograph (d) of the third row. *(Reprinted with permission from Ref. [51]. Copyright 2014 American Chemical Society.)*

samples, contacts between the plasma membrane and individual NWs, or small groups of NWs, were frequent for both concentrations of 10 and 50 μg/mL. In the proximity of the NWs, the plasma membrane presented cytoplasmic protrusions, characteristics of the process of endocytosis, which will expand to encompass the NWs in small vesicles. Analyzing the morphology of these protrusions, three types of macropinocytosis processes, such as lamellipodia-like, circular ruffles, and blebs, were identified. In detail, the lamellipodia-like morphology appears as a thin and long cytoplasmic evagination that surrounds the NW from one single side in the vicinity of the membrane. The structure of the circular ruffle is constituted by two symmetrical protrusions whose ends approach incorporating the NW while the structure in the blebs is recognizable by a bulging circular end.

The THP-1 cells showed, in addition to the micropinocytosis, a specific pathway for the internalization of NWs: phagocytosis. This particular mechanism, exclusive of macrophage-like cells, has been observed in different steps of advancement: in the initial phase the cells emit visible pseudopodia that subsequently surround agglomerates of several NWs in order to incorporate them into a single vacuole (Fig. 10.9). The analysis of cellular vesicular transport (Fig. 10.9) showed single or small groups of NWs in small vesicles, predominantly localized in the vicinity of the plasma membrane, whereas clusters of NWs are compartmentalized in big round vesicles with a perinuclear localization. The larger vesicles are presumably engaged in a compartmentalization function, whereas the smaller vesicles are used for vesicular transport [44–50,52–56]. Further, individual NWs are occasionally seen trapped in the cytoplasm and not included in vesicular structures. However, it must be stressed that the NWs were never observed inside the cell nucleus or in contact with the nuclear envelope.

As a final remark, the proved biocompatibility of the core–shell SiC–SiO_2 NWs, together with their good optical efficiency (Section 10.2) and ease of surface functionalization, pave the way to an application of these multifunctional nanosystems as future intracellular electrodes, biomolecule-delivery vectors, as well as drugs for photodynamic therapy (Section 10.4).

10.4 FUNCTIONALIZED 3C–SiC–SiO_x NANOWIRES FOR X-RAY-EXCITED PHOTODYNAMIC THERAPY *IN VITRO*

In the field of clinical oncotherapy, the standard protocols to control cancer growth and spread are still based on chemo- and radiotherapy combined with surgical resection. New strategies, such as the genetic approach, are

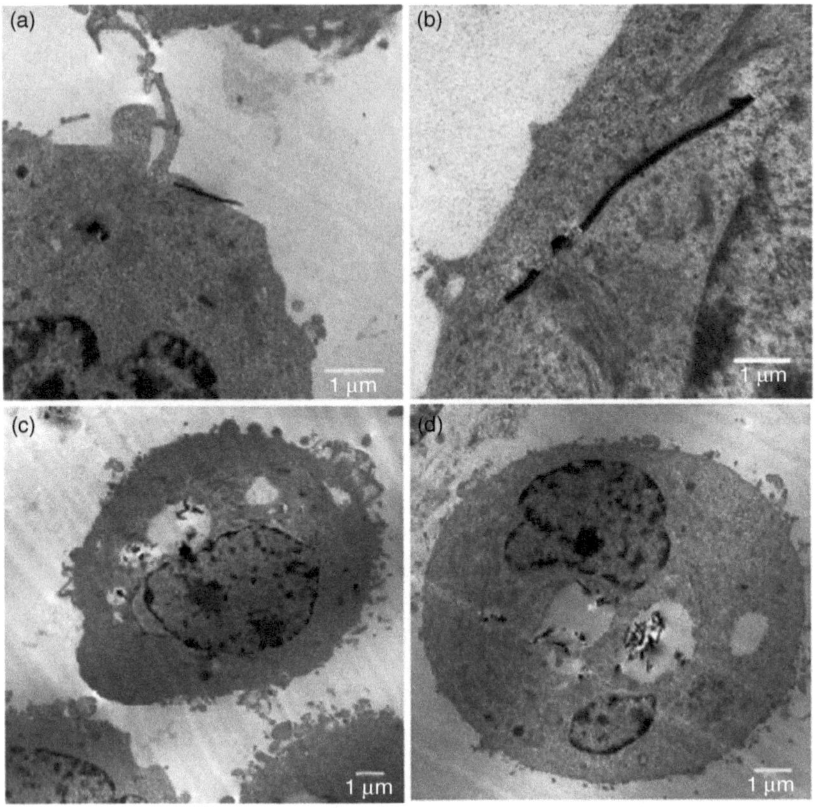

Figure 10.9 TEM micrographs showing cases of NWs that directly penetrate the cell membrane (a) and are trapped in the cytoplasm (b), and the vesicles' intracellular compartmentalization of NWs (c–d). *(Reprinted with permission from Ref. [51]. Copyright 2014 American Chemical Society.)*

nowadays in trial and are expected to deeply change cancer treatments in the near future. Alternative approaches aimed at killing tumors while minimizing side effects could be allowed by new classes of multifunctional nanomaterials and are presently under study. For instance, scintillation NPs were recently proposed in pilot studies [57,58] to activate self-lighted photodynamic therapy (SLPDT) to treat deep cancer. SLPDT is a variation of the well-assessed photodynamic therapy (PDT) usually used to "treat tumors on or just under the skin or on the lining of internal organs or cavities through the generation of an active form of oxygen (singlet oxygen, $^1O_2$) that destroys nearby cancer cells" (definition from the US-National Cancer Institute). In the previously mentioned pilot studies on SLPDT, photosensitizers are attached to the scintillation NPs and allow light generation under X-ray irradiation.

The photosensitizer (eg, organic dyes, aromatic hydrocarbons, porphyrins, phthalocyanines, and related tetrapyrroles) [59] promotes, through singlet oxygen generation, oxidative stress to kill cancer cells [60]. Since X-rays can penetrate through body tissue, deep tumors can be reached and treated with this method.

This approach requires the long-term biocompatibility of the scintillation NPs, which is not always verified in the literature, and the ability to properly conjugate the NPs with the photosensitizer, in order to enhance the photosensitizer absorption and reach an efficient singlet oxygen production. We recently proposed for SLPDT an innovative hybrid nanosystem based on core–shell 3C–SiC–SiO_x NWs, which are cytocompatible (see Section 10.2), functionalized with a tetracarboxyphenyl porphyrin derivative. Indeed, the NWs have an optical emission spectrum (see Section 10.2), which matches well the absorption bands (ie, for porphyrins the Q bands and the tail of the Soret band) of many organic photosensitizers. Among the different photosensitizers, the class of porphyrins has a well-established application in PDT, with initial approvals by the U.S. Food and Drug Administration. In our study we sought to link porphyrins to the SiC–SiO_x NWs by the formation of covalent bonds via 1,3-dipolar cycloaddition reaction. Therefore azide functional groups were introduced on the NW surface and carbon–carbon triple bonds in the phenyl rings of the porphyrin. In detail, to conjugate the porphyrin to the NWs a click-chemistry procedure was employed (Fig. 10.10). The as-grown NWs lying on Silicon substrates were first activated in aqueous HCl and then reacted with 3-azidopropyltrimethoxysilane by refluxing in toluene. This process leads to the binding of azide groups to the NWs, via condensation of alkoxysilyl groups with the hydroxyl groups present on the NW shell surface. The selected porphyrin was the tetra(4-carboxyphenyl) porphyrin (H_2TCPP), whose carboxyl groups were converted into amides containing a short chain with a terminal alkyne functional group. Indeed, to allow an efficient energy transfer from the NWs to the photosensitizers, they have to be linked together and the distance between the donor and the acceptor should be less than 10 nm [58]. The carboxyl groups of the porphyrin were first activated using EDCI/HOBt (EDCI, 1-(3-dimethylaminopropyl)-3-ethylcarbodiimide; HOBt, hydroxybenzotriazole), typical reagents employed for amino acid condensation. The hydroxybenzotriazole ester of H_2TCPP, formed as an intermediate, was in situ coupled with 2-propynylamine in the presence of dimethylaminopyridine in dimethylformamide at room temperature, affording

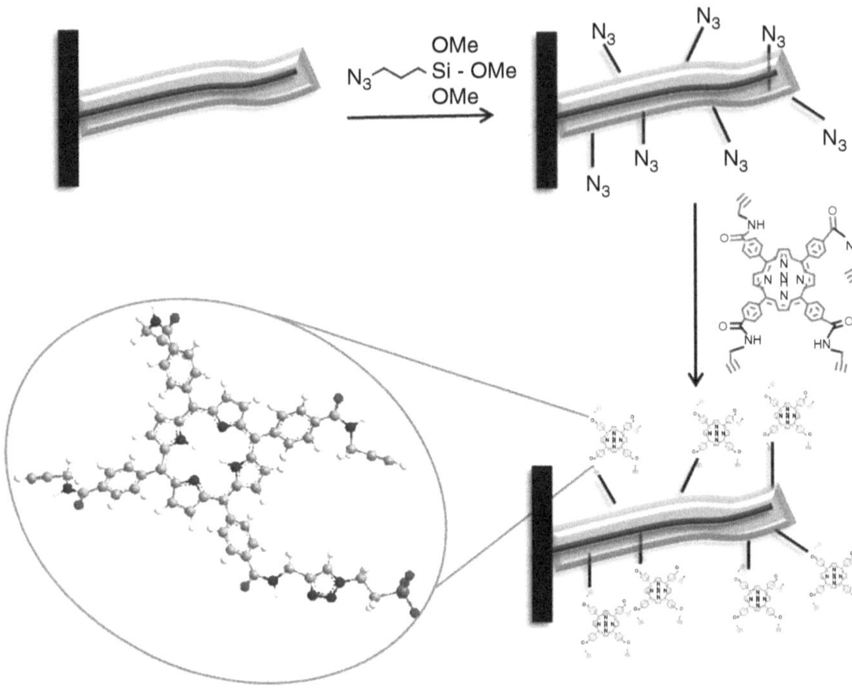

Figure 10.10 Design of the SiC–SiO_x–H_2TPACPP system: as-grown nanowires were functionalized with azide groups; the H_2TCPP porphyrin derivative containing C–C triple bonds (H_2TPACPP) was synthesized by converting the carboxyl groups into *N*-propynylamides; the nanohybrid was constructed by bonding H_2TPACPP to the NWs. *(From Ref. [12].)*

tetra(*N*-propynyl-4-aminocarbonylphenyl)-porphyrin (H_2TPACPP). The attachment of the photosensitizer to the NWs was accomplished on the basis of reactive group matching, that is, via the Huisgen 1,3 dipolar cycloaddition reaction giving the stable triazole (1,4- and 1,5-isomers) link. Thermal conditions, 130°C for 24 h, were preferred to Cu(I)-catalyzed click-reaction to avoid the metal insertion into the porphyrin core.

The hybrid NW–H_2TPACPP nanosystem has a room temperature fluorescence (see spectrum in Fig. 10.11a) characterized by the features of both the inorganic component, that is, the large and unstructured green band ascribed to the NBE 3C–SiC emission, and the organic photosensitizer, that is, the two orange and red emission peaks ascribed to the Qx(0,0) and Qx(0,1) porphyrin levels. The SiC NBE emission appears slightly red shifted in conjugated NWs compared to as-grown NWs. This red shift could be partly due to screening effects of the covering porphyrin due to Q-band absorption [61]. A small red shift and slight peak broadening occur also

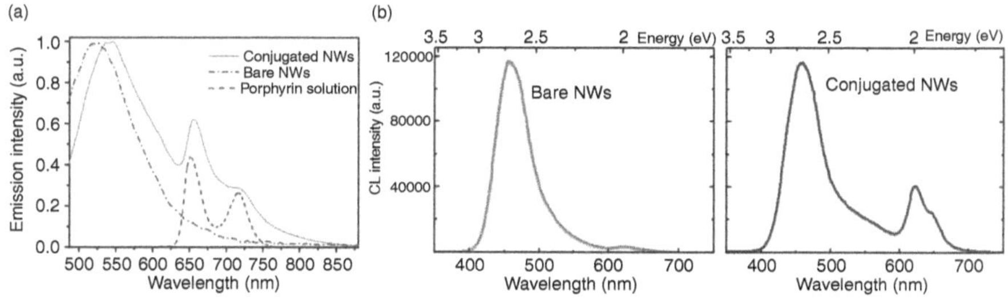

Figure 10.11 (a) Fluorescence spectra acquired at room temperature over as-grown SiC–SiO_x core–shell NWs (bare, green dashed-dotted line) and NWs functionalized with H2TPACPP (conjugated, orange solid line). The emission of H2TPACPP porphyrin in dichloromethane solution (red dashed line) is also shown for sake of clarity. (b) Room temperature CL spectra of the as-grown NWs (bare, left panel) and of the NW-H2TPACPP (conjugated, right panel). *(Adapted from Ref. [12].)*

for the porphyrin emission peaks in the nanosystem compared with the fluorescence of the free molecule in solution. This suggests important steric effects due to the bonding: the formation of covalent bonds and the adhesion to the locally flat NW surface (NW radius ~30 nm, much larger than the molecular size), can lead to a reduction of the tilting angle of the phenyl rings with respect to the macrocycle plane. According to this picture, the degree of conjugation of the molecule is increased for the bonded porphyrin compared to the free one. According to the literature [62], the electronic coupling and the steric molecule arrangement are expected to cause some red shift of the fluorescence spectrum. In a similar way, the slight broadening can be ascribed to intermolecular interactions in the solid state between H_2TPACPP molecules, besides a possible heterogeneous contribution from environments of different molecular packing.

The observation of porphyrin emission features in the fluorescence spectrum of the conjugated NWs is a key result. Indeed, the optical excitation for fluorescence measurements (laser wavelength 473 nm) is in the spectral region of NW absorption but out of the main absorption band of porphyrin. Therefore, the detected porphyrin emission can occur only if some form of electronic coupling between the NWs and the porphyrin takes place. This mechanism is confirmed by CL spectroscopy (Fig. 10.11b). The conjugated NWs show an orange CL band composed of a main peak at 620 nm and a shoulder at 652 nm. The 620 nm emission is ascribed to the nonbridging oxygen center (SiO) [63]. Then its increase in functionalized NWs, compared to as-grown NWs, is probably related to the residual SiOR group, used as a functionalization anchor. The 652-nm emission is due to the porphyrin luminescence, in particular the Qx(0,0) emission.

Considering the low energy release of the impinging electron beam to the porphyrin monolayer, the porphyrin emission is possible thanks to an energy transfer from the NWs to the molecule. Two possible energy transfer processes can take place: (1) a radiative energy transfer, in which the porphyrin absorbs the light emitted by the NWs, and then it reemits on its typical Q(0,0) emission [64], and (2) a Forster resonance energy transfer, in which the NWs, acting as the donor chromophore, transfer energy to the porphyrin, the acceptor, through nonradiative dipole–dipole coupling [65,66]. The hot electron transfer mechanism from the inorganic NW to the organic molecule can, instead, be ruled out, due to the presence of the wide bandgap oxide shell.

The energy transfer from the inorganic nanosystem component to the organic one is at the basis of its activity as a singlet oxygen source to induce oxidative stress in cancer cells (Fig. 10.12). The generation of $^1O_2$ by the hybrid NW–H_2TPACPP system was tested in water by exposing the

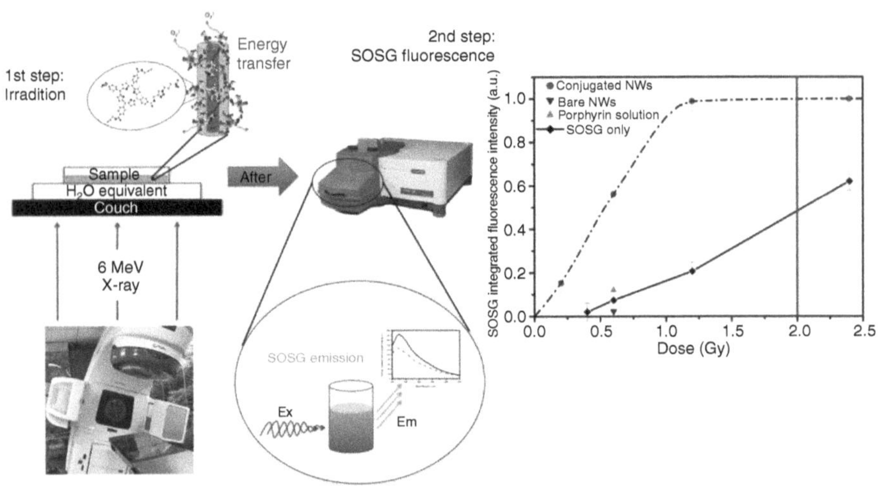

Figure 10.12 ***$^1O_2$ production excited by X-rays in a radiation therapy setup.*** Sketch of the experimental steps: a dish containing the sample solution is put on the couch and irradiated from the bottom, then it is transferred to a spectrophotometer to acquire the fluorescence spectrum of the SOSG marker. The plot reports the measured integrated fluorescence intensity, proportional to $^1O_2$ generated, as a function of the radiation dose. The experimental points are obtained from SOSG kit in water, with H2TPACPP-functionalized NWs (red circles) or without NWs (black diamonds). The up-triangle at 0.6 Gy is the experimental point obtained from SOSG kit in a water solution of mere porphyrin, while the down-triangle is obtained from SOSG kit in a water suspension of NWs as grown. The usual dose for clinical treatment (2 Gy) is indicated by the red vertical line. *(From Ref. [12].)*

NWs (detached from the substrate using an ultrasound microtip) to 6 MV X-rays in a clinical Linac Varian setup for radiation therapy. It has to be noted that no amorphous-to-crystal transition of the SiO_x layer, which is known to be cytotoxic in the crystalline phase, was induced by X-rays as verified by TEM studies. Singlet oxygen generation was revealed by a marker kit (Singlet Oxygen Sensor Green, SOSG [67]) based on fluorescein bound to a dimethylanthracene derivative, which shows a green fluorescence activated by interaction with $^1O_2$ and is highly selective to singlet versus other oxygen species (hydroxyl radicals, superoxide, etc.). In detail, a set of NWs (2 mg) was dispersed in a given aliquot (500 μL) of SOSG water solution (5 μM) and the solution was then exposed to radiation. After irradiation, the NWs were separated through ultracentrifugation and 200 μL of the remaining solution were sampled for fluorescence analysis. For each run, the fluorescence spectra were acquired on the as-prepared solution, the irradiated solution, and the irradiated solution treated with NWs. The spectrum of the not irradiated aliquot was acquired as a reference signal of the SOSG marker, while the spectrum of the irradiated aliquot as a background for the signal of the irradiated aliquot with NWs. An enhancement of the SOSG green fluorescence after treatment with the conjugated NWs exposed to radiation was observed, proving the production of $^1O_2$ by the NW–H2TPACPP system even in extremely low-dose irradiation conditions (0.4 Gy). On the contrary, no SOSG activation was observed in treatments with bare NWs, or with mere porphyrin solution, confirming that the $^1O_2$ production requires the conjugated NW–porphyrin system and occurs through the self-lighted photodynamic process. The $^1O_2$ concentration, as deduced from the integrated intensity of SOSG fluorescence, increases steeply as a function of the dose released up to 1 Gy. The dose–response relationship is shown in Fig. 10.12, from extremely low-dose values (fraction of Gy) to typical doses of radiation therapy in clinical cancer treatments (about 2 Gy per each session). The response of the nanosystem suspension (NW–H_2TPACPP in water, circles) is well above the control solution (SOSG kit alone in water, diamonds) and almost five times higher than the mere porphyrin response at the same dose (free porphyrin diluted in water, up-triangle) as well as the NWs as grown (suspension in water, down-triangle). The saturation in the SOSG fluorescence signal reached around 1 Gy, which can be reasonably attributed to the finite capacity of SOSG molecules to detect $^1O_2$ immediately after it has been produced. This is limited by the short lifetime of this species in water (4 μs) [68] and by the exchange—limited

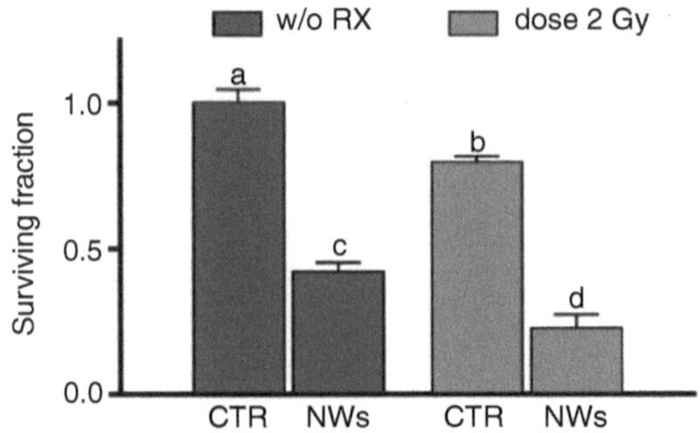

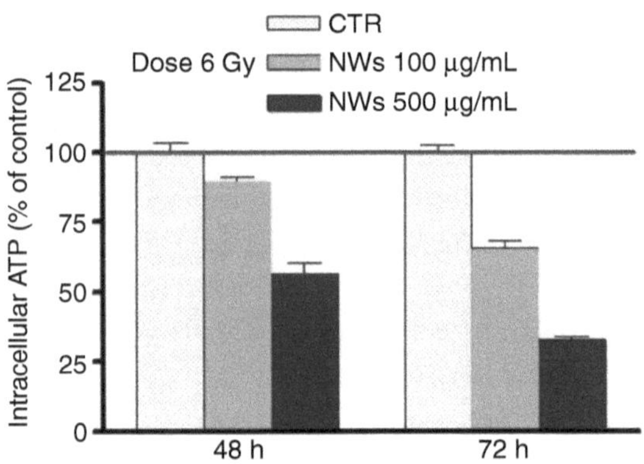

Figure 10.13 Top: Surviving fraction, normalized to control, of cells treated only with NW–H_2TPACPP (50 μg/mL), only with radiation (2 Gy, single session of 20 s) or a combination of NW–H_2TPACPP and radiation. Columns: mean of three determinations; bars: standard deviation. Each letter indicates a different level of significance ($p < 0.05$) after ANOVA analysis followed by a post-hoc test. Bottom: Intracellular ATP in A549 cells treated with NW–H_2TPACPP and radiation therapy (dose of 6 Gy, single session).

by diffusion—of SOSG molecules between the nanosystem surface and the bulk of the solution.

Finally, *in vitro* experiments (Fig. 10.13) were performed on A549 adenocarcinoma human alveolar basal epithelial cells to test the efficiency of the NW–H_2TPACPP nanosystem as an agent for SLPDT. The cells were incubated with H_2TPACPP-conjugated NWs and irradiated after 24 h with 6 MV X-rays at a dose of 2 Gy, chosen according to the standard conditions of clinical radiation therapy per single session. After treatment, a clonogenic survival assay (Fig. 10.13, top) was performed: cells were cultured for 12 days and then examined to verify whether the NW–H_2TPACPP nanosystem was able to enhance the effects of the ionizing radiation and to inhibit the cell proliferation. The histograms indicated that H_2TPACPP-conjugated NWs affect the survival of A549 cells, either with or without irradiation. The most striking result is that an additive antiproliferative effect was found compared to irradiation alone at the selected dose, and the cell-plating efficiency after the radiation treatment was significantly lower for cultures incubated with the NWs than for control cells (75%). To stress the effectiveness of our approach, an additional experiment was performed at increased NW concentration (100 and 500 μg/mL) and at increased dose (single exposure at 6 Gy), evidencing a remarkable reduction of the adenosine triphosphate (ATP) intracellular amount, which is directly correlated to the cellular viability (see Fig. 10.13, bottom). These results prove that the designed nanosystem enhances the effects of conventional radiation therapy and is therefore extremely promising in clinical oncotherapy to reduce the dose and/or the irradiation time, which means a reduction of target movements and of the psycho/physical discomfort of patients immobilized inside the isolated treatment room.

10.5 NANOWIRE PLATFORMS: *IN VITRO* CYTOCOMPATIBILITY AND PLATELET ACTIVATION

Recent studies underline the importance of micro- and nanoscale features on biomedical implant surfaces in determining cell viability and activity [69,70]. Complex surface patterns are usually defined by top-down processes to resemble the extracellular matrix structure; however, an innovative bottom-up approach to biomimicking can be offered by NWs [71,72]. In particular, NWs based on cubic silicon carbide, either bare (3C–SiC NWs) or surrounded by an amorphous shell (3C–SiC–SiO_2 core–shell NWs), may meet the chemical, mechanical, and electrical requirements for tissue engineering. They therefore have a strong potential to pave the way for the development of a novel generation of implantable nanodevices provided that they are compatible in the biological environment.

We recently proved SiC-based NW platforms as biomaterials [73]. As stated in the ISO 10993-5 guidelines [74] "an important quality issue of materials for medical devices having direct or indirect contact with the body tissue is the biocompatibility." Therefore any biomaterial with potential *in vivo* application must be tested through specific *in vitro* biocompatibility analyses, starting with cytotoxicity tests.

We initially performed indirect contact cytotoxicity tests, selecting as cellular models L929 mouse fibroblasts, which are one of the standard cell lines specified in ISO 10993-5, and the A549 human lung adenocarcinoma epithelial cells, which are widely documented for *in vitro* toxicity experiments [75]. Three thousand cells were seeded in 96 multiwell plates in DMEM and the plates were incubated at 37°C and 5% CO_2 in a humidified atmosphere for 24 h. The NW samples were cleaned with a piranha solution (H_2SO_4:H_2O_2 2:1), washed twice in deionized water, dipped in ethanol (99%), sterilized with UV light for 15 min, and finally soaked for either 24 h or 10 days in DMEM at 37°C in the incubator. To analyze the potential release of cytotoxic species from the sample, an aliquot was taken from the conditioned (ie, left in contact with the NW samples, for either 24 h or 10 days) medium and added to a pristine medium in different percentages (50, 70, and 100% of conditioned medium over total DMEM volume), following the ISO 10993-5 protocol for porous materials. The cells were cultured with the obtained medium for 24 h and then an MTT colorimetric assay, based on the 3-(4,5-dimethylthiazol-2-yl)-2,5-diphenyltetrazolium bromide dye, was performed to assess the cell viability. To confirm the results, a CellTiter-Glo luminescent cell viability assay, a homogeneous method of determining the number of viable cells based on quantitation of the ATP present, was performed. It was found (see Fig. 10.14) that the two viability assays were consistent with each other and the cell viability was

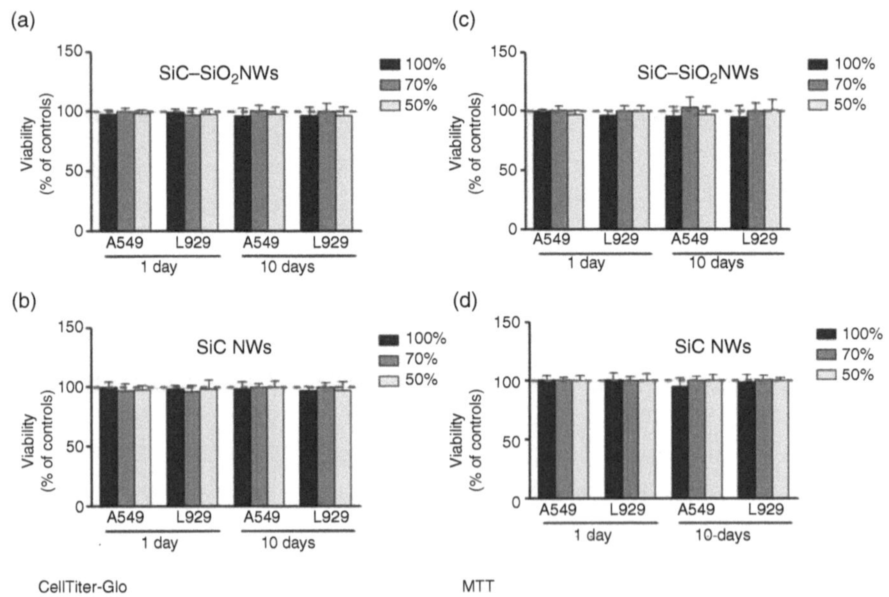

Figure 10.14 ***Cell viability (a, b: CellTiter-Glo assay; c, d: MTT assay) of cells cultured with conditioned medium (percentages of 50, 70, and 100 as indicated), for the indirect contact toxicity test.*** The bars are normalized to control cells cultured with nonconditioned medium.

always greater than 95%, indicating the absence of any toxic contamination. We also verified for both L929 and A549 cells that the conditioned medium did not affect normal cell-cycle progression. According to ISO 10993-5, a material is accepted as biocompatible if the viability is greater than 70%, and our findings demonstrate that this criterion was met by both the core–shell $SiC–SiO_2$ and bare SiC NWs.

A direct contact cytotoxicity test was then performed using the NW samples as platforms for culturing L929 and A549 cells. Twenty-thousand cells/mL were seeded on 1 × 1 cm NW samples (piranha-cleaned and sterilized as detailed earlier) and cultured in DMEM at 37°C and 5% CO_2 in a humidified atmosphere for 24 or 48 h. The cell viability was then analyzed via CellTiter-Glo assay, whereas MTT assays were not reliable due to direct chemical interactions of the reagent with the NW sample that alter the biological response. Fluorescence microscopy was employed to evaluate the adhesion and proliferation of live cells on the NW samples. For the analysis, the samples were thoroughly washed with DMEM and phosphate buffer solution (PBS), marked with a fluorescent dye (calcein AM, excitation/emission wavelengths of 495/515 nm), and observed after 30 min with an epifluorescence microscope. Fig. 10.15 shows the cell

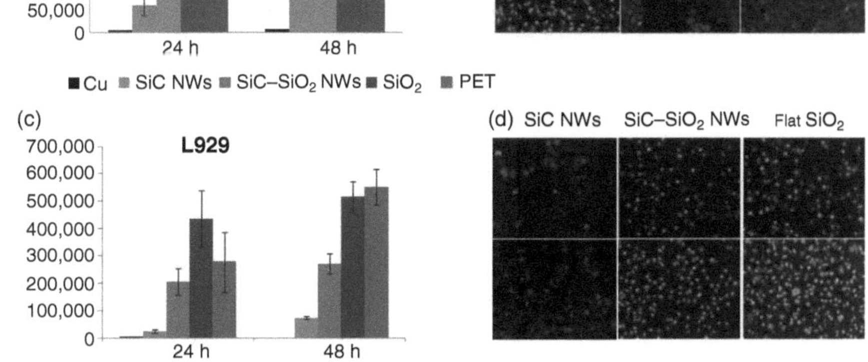

Figure 10.15 Cell viability histograms (a, c) and fluorescence images (b, d) acquired on A549 (a, b) and L929 (c, d) cells cultured for 24 and 48 h over different platforms: SiC NWs, $SiC–SiO_2$ core–shell NWs, flat SiO_2, and control PET and Cu surfaces.

viability (A549: panel (a), L929: panel (c)) at 24 and 48 h after cell seeding on the NW samples. In agreement with ISO 10993-5 guidelines, cell viability for direct contact tests on polyethylene terephthalate (PET), as the negative control, and copper, as the positive control, is reported for comparison. We also tested a flat SiO_2 (from natively oxidized Si) sample as a reference. Based on the observed viability (Fig. 10.15a and c) and fluorescence assays (Fig. 10.15b and d), cells appeared to be more numerous on the flat samples (PET and SiO_2) than on the NWs after 24 h of culture, likely indicating that cell adhesion was more difficult on the NWs.

Cell attachment and spreading on biomaterials are mediated by highly complex cellular structures called focal adhesions, which are multiprotein, microscale assemblies involving clustering of ligated transmembrane receptors called integrins [76]. It has been shown that integrin spacing is important to their clustering and focal adhesion formation. There is evidence that integrin clustering requires a 50–70-nm spacing between ligated integrins [77]. Nanotopographies with spacing larger than 100 nm have been shown to inhibit cell adhesion [78]. In our case, the NW diameter ranges between 20(for bare SiC NWs) and 60 nm (for core–shell NWs) and the distances between NWs vary from hundreds of nanometer to several micrometer. Hence, although integrin clustering may occur on a single NW, the microscale focal adhesion assembly can be impaired because of the long distances between NWs, which can explain why cell attachment was observed to be lower on the NWs than on the planar surfaces. However, fluorescence data (Fig. 10.15b and d) and SEM observation clearly show that both cell lines proliferate by 48 h. This suggests that the NW-coated substrates have a good biocompatibility and do not hinder cell growth. Additionally, it can be noted that A549 cells have almost the same viability, as assessed by chemiluminescence after 24 h, on SiC–SiO_2 and SiC NWs indicating that the adhesion and proliferation of this cell line was not affected by the NW type. On the contrary, L929 viability was significantly lower on SiC NWs compared to core–shell NWs, consistent with the idea that different cells may express different arrays of adhesion structures and be differently affected by the surface characteristics of the substrate.

Cellular adhesion and morphology on NW samples and the flat SiO_2 reference were analyzed by SEM, after proper fixation (2.5% glutaraldehyde in 0.1 M sodium cacodylate buffer for 30 min), dehydration in ethanol (EtOH), and gold coating. Figs. 10.16 and 10.17 report representative SEM micrographs of the A549 and L929 cells, respectively, imaged at 24 h (top line) and 48 h (bottom line) after seeding. For all samples, the cellular

Figure 10.16 SEM images of fixed A549 cells cultured 24 h (top line) and 48 h (bottom line) on different substrates: flat SiO_2 (a), SiC–SiO_2 core–shell NWs (b), and SiC NWs (c).

membrane displayed the typical morphology of healthy living cells. No morphological signs of cellular degradation or programmed cell death, such as membrane blebs and rounded apoptotic bodies, were detected. Compared to the flat SiO_2 samples, a larger number of rounded cells were present on the NW platforms (Figs. 10.16 and 10.17, top line). However, cells adhered to the NW platform after 48 h, with lamellipodia and filopodia, were observed to be entangled in the NW network.

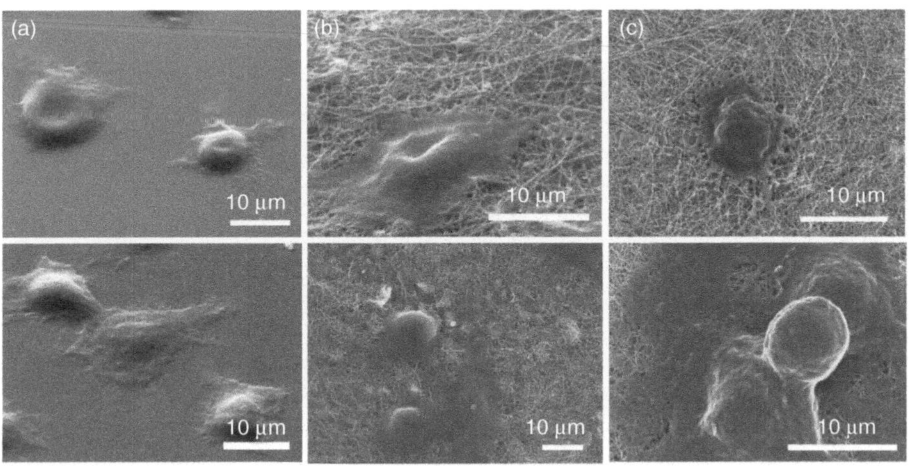

Figure 10.17 SEM images of fixed L929 cells cultured 24 h (top line) and 48 h (bottom line) on different substrates: flat SiO_2 (a), SiC–SiO_2 core–shell NWs (b), and SiC NWs (c).

Finally, we analyzed the blood compatibility of the NW platforms in terms of adhesion and activation of platelets on the NW bundles, in conformity with ISO 10993-4 [79] standards and as described in chapter: Study of the Hemocompatibility of 3C–SiC and *a*-SiC Films Using ISO 10993-4. Blood samples were collected by routine phlebotomy, using blood collection tubes containing 0.105 M (3.2%) buffered sodium citrate for anticoagulation, from young, healthy, female farm pigs weighing approximately 45 kg, free of hepatitis or other blood-borne disease. From these samples, PRP was obtained by centrifuging the whole blood at 1200 G for 10 min. The NW samples (piranha cleaned and sterilized as detailed earlier for cytotoxicity tests) were soaked in PRP for 15 min and rinsed in PBS. The adhesion of platelets on the sample surface and their shape, indicative of the activation status, were then examined by SEM after proper fixation (paraformaldehyde 4% or 2.5% glutaraldehyde in 0.1 M sodium cacodylate buffer for 30 min), dehydration (with EtOH at increasing concentrations), and gold coating. Typical SEM images are reported in Fig. 10.18 for bare and core–shell NWs, compared to a flat 3C–SiC sample (thin film grown by CVD [80]) and a flat SiO_2 sample. The inspection of the observed platelets indicates that the observed platelet activation state is material dependent and that nanostructured and bulk materials behave quite similarly. In particular, the majority of platelets on the 3C–SiC-only samples were resting discoid-shaped platelets (Fig. 10.18a) and b), whereas star-like shapes, characteristic of activated platelets, were observed for samples exposing an oxide surface (Fig.10.18c

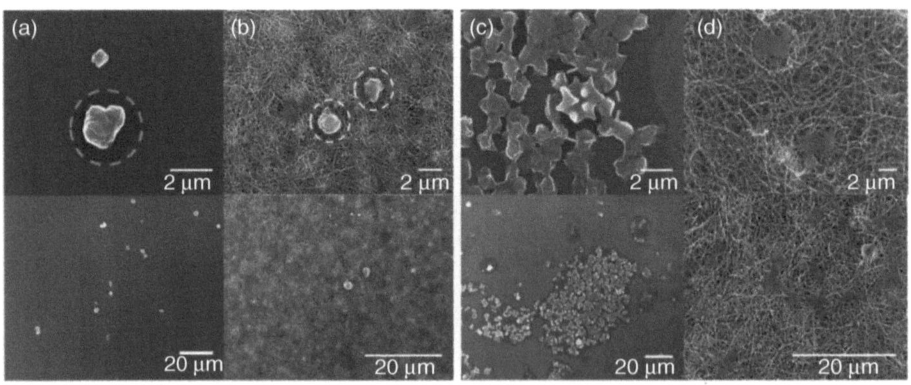

Figure 10.18 SEM micrographs of platelets (bottom line: overview, top line: detail) on different samples: flat 3C–SiC film (a); 3C–SiC NWs (b); flat SiO_2 (c); and core–shell NWs (d). The starry shape indicates an activation of the platelets.

and d). To confirm these results, an enzyme-linked immunosorbent assay was performed to evaluate P-selectin release. In detail, PRP (400 μL) was deposited with a micropipette on the surface of the NW samples (pre-warmed at 37°C) and they were then incubated at 37°C for 60 min. After incubation, the PRP was recovered and the plasma separated by centrifugation. Soluble P-selectin was quantified by specific immunosorbent assay (sP-Selectin ELISA kit) and concentrations were correlated to untreated PRP. The results (Fig. 10.19) attest that a higher platelet activation was induced by the nanostructured platform with respect to the corresponding planar material, and this activation was higher with core–shell NWs compared to bare 3C–SiC NWs. Indeed, the maximum release (approximately +80% with respect to the untreated PRP) occurs for the core–shell NWs, while 3C–SiC NWs induce, as in the bulk material (3C–SiC film), a significantly lower P-selectin level.

For implantable devices, the cytocompatibility and the limited platelet activation demonstrated for 3–SiC NWs makes them a biomaterial of interest for advanced application fields such as the brain–machine interface (see chapter: *In Vivo* Exploration of Robust Implantable Devices Constructed From Biocompatible 3C–SiC), in line with very recent results demonstrated on 3C–SiC needles [81]. Core–shell NWs are instead cyto, but not hemo, compatible. They provide a biomimetic cell growth substrate, likely because the NW network resembles the biological extracellular matrix, and promotes platelet activation, which makes it a suitable implantable biomaterial for tissue regeneration.

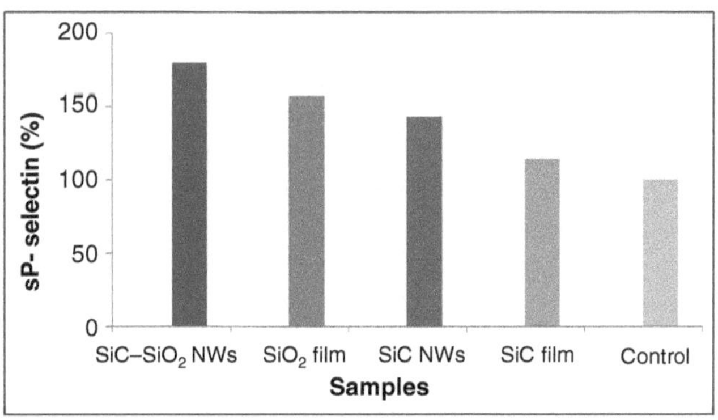

Figure 10.19 ***Soluble P-selectin released by platelets over different nanostructured samples (core–shell and bare NWs), compared with flat SiO_2 and 3C–SiC samples.*** Concentrations were related to untreated PRP (sample value/untreated PRP value) ×100 (control).

10.6 SUMMARY

This chapter presented an overview of the growth and optical properties (Section 10.2) and some *in vitro* applications of core–shell 3C–SiC–SiO_2 NWs and bare SiC NWs.

In Section 10.3 it was shown, by the analysis of cell proliferation, cell cycle progression, and oxidative stress, that the core–shell NWs are cytocompatible over a time of up to 10 days. They are effectively internalized by cells through the macropinocytosis mechanism (phagocytosis only in the THP-1 model) and sporadically by direct penetration. In all the cell lines, the intracellular presence of NWs induces the same molecular events: peroxidation of membrane lipids and oxidation of proteins. These effects are late and may be limited by the activation of protection systems, for instance the effect of ROS is not acute and effectively countered by the intracellular scavenger. Our results highlight that the core–shell 3C–SiC–SiO_2 NWs do not elicit either midterm (72 h) or long-term (10 day) cytotoxic activity leading to irreversible cellular damages or death.

In Section 10.4 it was shown that the hybrid nanosystem formed by core–shell NWs conjugated with an organic photosensitizer, a tetracarboxyphenyl porphyrin derivative, is an efficient source of singlet oxygen under 6 MV X-ray irradiation. Further, this nanosystem is effective for SLPDT tests *in vitro* on lung cancer cells cultured with the functionalized NWs and irradiated in a clinical linear accelerator for radiation therapy. The experimental results validate the nanosystem in view of nanomedicine applications aimed at the definition of more effective and synergic treatments of tumors deep inside the body.

In Section 10.5 it was shown that core–shell 3C–SiC–SiO_2 NWs and bare 3C–SiC NWs could be used as a biomimetic biomaterial for cell cultures, providing a biological interface that is biocompatible according to ISO 10993 standards. The blood compatibility of the NWs was also evaluated analyzing the interaction of the NW platforms with PRP, in agreement with ISO 10993-4 guidelines. The analysis of adhesion and activation of platelets on the NW bundles indicates that a higher platelet activation was induced by the core–shell structures compared to the bare ones. These data suggest that SiC–SiO_2 NWs may be suitable for tissue regeneration, while SiC NWs may be exploited for implantable prosthetic devices.

ACKNOWLEDGMENTS

This work has been supported by Fondazione Cariparma through the BioNiMed project (2012–14).

REFERENCES

[1] Linkov I, Bates ME, Canis LJ, Seager TP, Keisler JM. A decision directed approach for prioritizing research into the impact of nanomaterials on the environment and human health. Nat Nanotech 2011;6:784–7.

[2] Ferrari M. Cancer nanotechnology: opportunities and challenges. Nat Rev Cancer 2005;5:161–71.

[3] Doane TL, Burda C. The unique role of nanoparticles in nanomedicine: imaging, drug delivery and therapy. Chem Soc Rev 2012;41:2885–911.

[4] Saddow SE, editor. Silicon carbide biotechnology: a biocompatible semiconductor for advanced biomedical devices and applications. UK: Elsevier Ltd; 2011.

[5] Fabbri F, Rossi F, Melucci M, Manet I, Attolini G, Favaretto L, Zambianchi M, Salviati G. Optical properties of hybrid T3Pyr/SiO2/3C-SiC nanowires. Nanoscale Res Lett 2012;7(1):1–8.

[6] Rossi F, Fabbri F, Tallarida M, Schmeisser D, Modreanu M, Attolini G, Salviati G. Structural and luminescence properties of HfO_2 nanocrystals grown by atomic layer deposition on SiC/SiO_2 core/shell nanowires. Scripta Mater 2013;69:744.

[7] Corbalan JJ, Medina C, Jacoby A, Malinski T, Radomski MW. Amorphous silica nanoparticles aggregate human platelets: potential implications for vascular homeostasis. Int J Nanomed 2012;7:631.

[8] Steuer H, Krastev R, Lembert N. Metallic oxide nanoparticles stimulate blood coagulation independent of their surface charge. J Biomed Mater Res Part B 2014;102(5): 897–902.

[9] Fabbri F, Rossi F, Attolini G, Salviati G, Iannotta S, Aversa L, Verucchi R, Nardi MV, Fukata N, Dierre B, Sekiguchi T. Enhancement of the core near-band-edge emission induced by an amorphous shell in coaxial one-dimensional nanostructure: the case of SiC/SiO_2 core/shell self-organized nanowires. Nanotechnology 2010;21:345702.

[10] Fabbri F, Rossi F, Attolini G, Salviati G, Dierre B, Sekiguchi T, Fukata N. Luminescence properties of SiC/SiO_2 core–shell nanowires with different radial structure. Mater Lett 2012;71:137.

[11] Fabbri F, Rossi F, Lagonegro P, Negri M, Ponraj JS, Bosi M, Attolini G, Salviati G. 3C-SiC nanowires luminescence enhancement by coating with a conformal oxides layer. J Phys D Appl Phys 2014;47:394006.

[12] Rossi F, Bedogni E, Bigi F, Rimoldi T, Cristofolini L, Pinelli S, Alinovi R, Negri M, Dhanabalan S, Attolini G. Porphyrin conjugated SiC/SiOx nanowires for X-ray-excited photodynamic therapy. Sci Rep 2015;5:7606.

[13] Attolini G, Rossi F, Bosi M, Watts BE, Salviati G. Synthesis and characterization of 3C–SiC nanowires. J Non Cryst Solids 2008;354:5227–9.

[14] Seo W-S, Koumoto K, Aria S. Morphology and stacking faults of β-silicon carbide whiskers synthesized by carbothermal reduction. J Am Ceramic Soc 2004;83:2584–92.

[15] Yang W, Miao H, Xie Z, Zhang L, An L. Synthesis of silicon carbide nanorods by catalyst-assisted pyrolysis of polymeric precursor. Chem Phys Lett 2004;383:441–4.

[16] Sundaresan SG, Davydov AV, Vaudin MD, Levin I, Maslar JE, Tian YL, Rao MV. Growth of Silicon Carbide Nanowires by a Microwave Heating-Assisted Physical Vapor Transport Process Using Group VIII Metal Catalysts. Chem Mater 2007;19:5531–7.

[17] Shi W, Zheng Y, Peng H, Wang N, Lee CS, Lee ST. Laser Ablation Synthesis and Optical Characterization of Silicon Carbide Nanowires. J Am Ceramic Soc 2000;83:3228–30.

[18] Seeger T, Kohler-Redlich P, Rühle M. Synthesis of Nanometer-Sized SiC Whiskers in the Arc-Discharge. Adv Mater 2000;12:279–82.

[19] Yang G, Wu R, Chen J, Pan Y, Zhai R, Wu L, Lin D. Growth of SiC nanowires/nanorods using a Fe–Si solution method. J Nanotechnol 2007;18:155601.

[20] Dai H, Wong EW, Lu YZ, Fan S, Lieber CM. Synthesis and characterization of carbide nanorods. Nature 1995;375:769–72.

[21] Watts BE, Attolini G, Rossi F, Bosi M, Salviati G, Mancarella F, Ferri M, Roncaglia A, Poggi A. β-SiC NWs grown on patterned and MEMS silicon substrates. Mater Sci Forum 2011;508:679–80.
[22] Negri M, Dhanabalan SC, Attolini G, Lagonegro P, Campanini M, Bosi M, Fabbri F, Salviati G. Tuning the radial structure of core-shell silicon carbide nanowires. Crystal Eng Commun 2015;17:1258–63.
[23] Park B, Ryu Y, Yong K. Growth and characterization of silicon carbide nanowires. Surf Rev Lett 2004;11:373–8.
[24] Swift GA, Koc R. Tungsten powder from carbon coated WO3 precursors. J Mater Sci 2001;36:803–6.
[25] Denkhaus E, Salnikow K. Nickel essentiality, toxicity, and carcinogenicity. Crit Rev Oncol Hematol 2002;42:35–56.
[26] Elmasry MAA, Gaber A, Khater EMH. Thermal decomposition of Ni(II) and Fe(III) nitrates and their mixture. J Thermal Anal Calorim 1998;52:489–95.
[27] Wagner RS, Ellis WC. Vapor-liquid-solid mechanism of single crystal growth. Appl Phys Lett 1964;4:89.
[28] Seo WS, Koumoto K, Aria S. Morphology and stacking faults of β-silicon carbide whiskers synthesized by carbothermal reduction. J Am Ceramic Soc 2000;83:2584.
[29] Skuja L. Optically active oxygen-deficiency-related centers in amorphous silicon dioxide. J Non-Cryst Solids 1998;239:16.
[30] Ikeda M, Matsunami H. Free exciton luminescence in 3C, 4H, 6H, and 15R SiC. Phys Status Solidi A 1980;58:657.
[31] Pistol ME, Pryor CE. Band structure of core-shell semiconductor nanowires. Phys Rev B 2008;78:115319.
[32] Afanasev VV, Bassler M, Pensl G, Schulz MJ. Band offsets and electronic structure of SiC/SiO2 interfaces. J Appl Phys 1996;79:3108.
[33] Yacobi BG, Holt DB. Cathodoluminescence microscopy of inorganic solids. New York; Springer; 1990.
[34] Jabeen F, Rubini S, Grillo V, Felisari L, Martelli F. Room temperature luminescent InGaAs/GaAs core-shell nanowires. Appl Phys Lett 2008;93:083117.
[35] Smith DM, Simon JK, Baker JR. Applications of nanotechnology for immunology. Nat Rev Immunol 2013;13:592–605.
[36] Zhou W, Dai X, Fu TM, Xie C, Liu J, Lieber CM. Long term stability of nanowire nanoelectronics in physiological environments. Nano Lett 2014;14:1614–9.
[37] Alivisatos AP, Andrews AM, Boyden ES, Chun M, Church GM, Deisseroth K, Donoghue JP, Fraser SE, Lippincott-Schwartz J, Looger LL, Masmanidis S, McEuen PL, Nurmikko AV, Park H, Peterka DS, Reid C, Roukes ML, Scherer A, Schnitzer M, Sejnowski TJ, Shepard KL, Tsao D, Turrigiano G, Weiss PS, Xu C, Yuste R, Zhuang X. Nanotools for neuroscience and brain activity mapping. ACS Nano 2013;7:1850–66.
[38] Patolsky F, Zheng G, Lieber CM. Nanowire sensors for medicine and the life sciences. Nanomedicine (London, UK) 2006;1:51–65.
[39] Marches R, Mikoryak C, Wang RH, Pantano P, Draper RK, Vitetta ES. The importance of cellular internalization of antibody-targeted carbon nanotubes in the photothermal ablation of breast cancer cells. Nanotechnology 2011;22:095101.
[40] Muller KH, Kulkarni J, Motskin M, Goode A, Winship P, Skepper JN, Ryan MP, Porter AE. pH-dependent toxicity of high aspect ratio ZnO nanowires in macrophages due to intracellular dissolution. ACS Nano 2010;4:6767–79.
[41] Fu TM, Duan X, Jiang Z, Dai X, Xie P, Cheng Z, Lieber CM. Sub-10-nm intracellular bioelectronic probes from nanowire-nanotube heterostructures. Proc Natl Acad Sci USA 2014;111:1259–64.
[42] Park EJ, Shim HW, Lee GH, Kim JH, Kim DW. Comparison of toxicity between the different-type TiO_2 nanowires *in vivo* and *in vitro*. Arch Toxicol 2013;87:1219–30.

[43] Li N, Zhao P, Astruc D. Anisotropic gold nanoparticles: synthesis, properties, applications, and toxicity. Angew Chem Int Ed Engl 2014;53:1756–89.
[44] Song MM, Song WJ, Bi H, Wang J, Wu WL, Sun J, Yu M. Cytotoxicity and cellular uptake of iron nanowires. Biomaterials 2010;31:1509–17.
[45] El-Gamel NEA. Silver(I) complexes as precursors to produce silver nanowires: structure characterization, antimicrobial activity and cell viability. Dalton Trans 2013;42:9884–92.
[46] Lanone S, Andujar P, Kermanizadeh A, Boczkowski J. Determinants of carbon nanotube toxicity. Adv Drug Deliv Rev 2013;65:2063–9.
[47] Qing Q, Jiang Z, Xu L, Gao R, Mai L, Lieber CM. Freestanding kinked nanowire transistor probes for targeted intracellular recording in three dimensions. Nat Nanotechnol 2014;9:142–7.
[48] Li J, Han Q, Zhang Y, Zhang W, Dong M, Besenbacher F, Yang R, Wang C. Optical regulation of protein adsorption and cell adhesion by photoresponsive GaN nanowires. ACS Appl Mater Interfaces 2013;5:9816–22.
[49] Suyatin DB, Wallman L, Thelin J, Prinz CN, Jorntell H, Samuelson L, Montelius L, Schouenborg J. Nanowire-based electrode for acute *in vivo* neural recordings in the brain. PLoS One 2013;8:e56673.
[50] Persson H, Købler C, Mølhave K, Samuelson L, Tegenfeldt JO, Oredsson S, Prinz CN. Fibroblasts cultured on nanowires exhibit low motility, impaired cell division, and DNA damage. Small 2013;9:4006–16.
[51] Cacchioli A, Ravanetti F, Alinovi R, Pinelli S, Rossi F, Negri M, Bedogni E, Campanini M, Galetti M, Goldoni M, Lagonegro P, Alfieri R, Bigi F, Salviati G. Cytocompatibility and cellular internalization mechanisms of SiC/SiO_2 nanowires. Nano Lett 2014;14:4368–75.
[52] De Simone U, Manzo L, Profumo A, Coccini T. *In vitro* toxicity evaluation of engineered cadmium-coated silica nanoparticles on human pulmonary cells. J Toxicol 2013;2013:931785.
[53] Zucker RM, Daniel KM, Massaro EJ, Karafas SJ, Degn LL, Boyes WK. Detection of silver nanoparticles in cells by flow cytometry using light scatter and far-red fluorescence. Cytometry A 2013;83:962–72.
[54] Yamashita K, Nagai H, Kondo Y, Misawa N, Toyokuni S. Evaluation of two distinct methods to quantify the uptake of crocidolite fibers by mesothelial cells. J Clin Biochem Nutr 2013;53:27–35.
[55] Song MM, Song WJ, Bi H, Wang J, Wu WL, Sun J, Yu M. Cytotoxicity and cellular uptake of iron nanowires. Biomaterials 2010;31:1509–17.
[56] Wierzbicki R, Kobler C, Jensen MR, Lopacinska J, Schmidt MS, Skolimowski M, Abeille F, Qvortrup K, Molhave K. Mapping the complex morphology of cell interactions with nanowire substrates using FIB-SEM. PLoS One 2013;8:e53307.
[57] Chen W, Zhang J. Using nanoparticles to enable simultaneous radiation and photodynamic therapies for cancer treatment. J Nanosci Nanotech 2006;6:1159–66.
[58] Liu Y, Chen W, Wang S, Joly AL. Investigation of water-soluble x-ray luminescence nanoparticles for photodynamic activation. Appl Phys Lett 2008;92:043901.
[59] DeRosa MC, Crutchley RJ. Photosensitized singlet oxygen and its applications. Coordinat Chem Rev 2002;233-234:351–71.
[60] Bonnet R. Photosensitizers of the porphyrin and phthalocyanine series for photodynamic therapy. Chem Soc Rev 1995;24:19–33.
[61] Liu Y, Zhang Y, Wang S, Pope C, Chen W. Optical behaviors of ZnO porphyrin conjugates and their potential applications for cancer treatment. Appl Phys Lett 2008;92:143901.
[62] Akins DL, Zhu HR, Guo C. Fluorescence decay kinetics and structure of aggregated tetrakis(p-sulfonatophenyl)porphyrin. J Phys Chem 1996;100:14390–6.

[63] Fabbri F, Rossi F, Negri M, Tatti R, Aversa L, Dhanabalan SC, Verucchi R, Attolini G, Salviati G. Carbon-doped SiO_x nanowires with a large yield of white emission. Nanotechnology 2014;25:185704.
[64] Chandrasekhar S. Radiative transfer. Mineola, NY: Dover Publications; 1960.
[65] Helms V. Fluorescence resonance energy transfer. Principles of computational cell biology. Weinheim: Wiley-VCH; 2008.
[66] Shankar K, Feng X, Grimes CA. Enhanced harvesting of red photons in nanowire solar cells: evidence of resonance energy transfer. ACS Nano 2009;3:788–94.
[67] SOSG kit, produced by Molecular Probes and commercialized by Life Technologies, product information. Available from: http://tools.lifetechnologies.com/content/sfs/manuals/mp36002.pdf
[68] Egorov SYu, Kamalov VF, Koroteev NI, Krasnovsky AA, Toleutaev BN, Zinukov SV. Rise and decay kinetics of photosensitized singlet oxygen luminescence in water. Measurements with nanosecond time-correlated single photon counting technique. Chem Phys Lett 1989;163:421–4.
[69] Von der Mark K, Park J, Bauer S, Schmuki P. Nanoscale engineering of biomimetic surfaces: cues from the extracellular matrix. Cell Tissue Res 2010;339(1):131–53.
[70] Bruinink A, Bitar M, Pleskova M, Wick P, Krug HF, Maniura-Weber K. Addition of nanoscaled bioinspired surface features: a revolution for bone related implants and scaffolds? J Biomed Mater Res Part A 2014;102(1):275–94.
[71] Fischer KE, Alemán BJ, Tao SL, Daniels RH, Li EM, Bünger MD, Nagaraj G, Singh P, Zettl A, Desai TA. Biomimetic nanowire coatings for next generation adhesive drug delivery systems. Nano Lett 2009;9(2):716–20.
[72] Zhao N, Wang Z, Cai C, Shen H, Liang F, Wang D, Wang C, Zhu T, Guo J, Wang Y, Liu X, Duan C, Wang H, Mao Y, Jia X, Dong H, Zhang X, Xu J. Bioinspired materials: from low to high dimensional structure. Adv Mater 2014;26(41):6994–7017.
[73] Lagonegro P, Galli C, Smerieri A, Alinovi R, Pinelli S, Nezafati M, Rimoldi T, Negri M, Fabbri F, Bosi M, Attolini G, Cristofolini L, Macaluso G, Saddow SE, Rossi F, Salviati G. Nanowire platforms based on cubic silicon carbide for biomedical applications: cytocompatibility and platelet activation. Plos One, in press (2016)
[74] ISO, 10993-5. Biological evaluation of medical devices. Part 5: tests for *in vitro* cytotoxicity; 2009.
[75] Visalli G, Bertuccio MP, Iannazzo D, Piperno A, Pistone A, Di Pietro A. Toxicological assessment of multi-walled carbon nanotubes on A549 human lung epithelial cells. Toxicol *In Vitro* 2015;29(2):352–62.
[76] Bershadsky A, Kozlov M, Geiger B. Adhesion-mediated mechanosensitivity: a time to experiment, and a time to theorize. Curr Opin Cell Biol 2006;18(5):472–81.
[77] Cavalcanti-Adam EA, Micoulet A, Bluemmel J, Auernheimer J, Kessler H, Spatz JP. Lateral spacing of integrin ligands influences cell spreading and focal adhesion assembly. Eur J Cell Biol 2006;85(3):219–24.
[78] Yim EK, Darling EM, Kulangara K, Guilak F, Leong KW. Nanotopography-induced changes in focal adhesions, cytoskeletal organization, and mechanical properties of human mesenchymal stem cells. Biomaterials 2010;31(6):1299–306.
[79] ISO, 10993-4. Biological evaluation of medical devices. Part 4: selection of tests for interactions with blood; 2009.
[80] Bosi M, Attolini G, Negri M, Frigeri C, Buffagni E, Ferrari C, Rimoldi T, Cristofolini L, Aversa L, Tatti R. Optimization of a buffer layer for cubic silicon carbide growth on silicon substrates. J Cryst Growth 2013;383:84–94.
[81] Frewin CL, Locke C, Mariusso L, Weeber EJ, Saddow SE. Silicon carbide neural implants: *in vivo* neural tissue reaction. Sixth Annual international IEEE/EMBS conference on neural engineering (NER); 2013, p. 661–664.

INDEX

D

F

G

K

L

M

R

S

T

Kolibri 360 GmbH
Ferdinand-Jühlke-Straße 7
99095 Erfurt, Deutschland
produktsicherheit@kolibri360.de